Land Subsidence Analysis in Urban Areas

Springer Environmental Science and Engineering

For further volumes:
http://www.springer.com/series/10177

David G. Zeitoun • Eliyahu Wakshal

Land Subsidence Analysis in Urban Areas

The Bangkok Metropolitan Area Case Study

Springer

David G. Zeitoun
Jerusalem College of Technology
Jerusalem
Israel

Eliyahu Wakshal (Deceased)

ISSN 2194-3214 ISSN 2194-3222 (electronic)
ISBN 978-94-007-5505-5 ISBN 978-94-007-5506-2 (eBook)
DOI 10.1007/978-94-007-5506-2
Springer Dordrecht Heidelberg New York London

Library of Congress Control Number: 2012955030

Printed on acid-free paper

Springer is part of Springer Science+Business Media (www.springer.com)

To my father:
Mr. Josef Zeitoun who passed away in March 2000
To my mother,
To my wife Esther and my children

David Zeitoun

After writing much of the second chapter of this book and reviewing much of the modeling work done on subsidence, Elie Wakshal died in the hospital on September 27, 2012.

This is difficult news for all the people that know his devotion to hydrogeology and his love for others. Professor E. Wakshal was a nice man, an enthusiastic professor, a generous friend who always looked to help people.

"Give ear, O heavens, and I will speak; and hear, O earth, the words of my mouth. My doctrine shall drop as the rain; my speech shall distil as the dew, as the small rain upon the tender herb, and as the showers upon the grass" (Bible, Deuteronomy chapter 32, verse 1–3).

Professor E. Wakshal was a specialist of hydrochemistry and always combined mathematics with a deep knowledge in geology. His teaching and lectures were always connected with engineering projects and water system projects. He was able to understand the earth by the laws of physics and water flow through geology and chemistry.

He loved nature and humanities. He devoted his life to build this world in Israel, in Uruguay, and in Thailand and also to teach hydrogeology around the world.

Your work will stay for the next generation and your memory will stay forever.

David G. Zeitoun

Foreword

Soil subsidence is a worldwide engineering problem. In the Bangkok metropolitan area, land subsidence rates were 10 cm/year in the eastern suburbs, and 5–10 cm/year in central Bangkok, between 1978 and 1981.

After remedial measures were introduced in 1983 for controlling groundwater pumpage, a continuous recovery of the water level was observed in central Bangkok and its eastern suburbs. The annual subsidence rates between 1988 and 1989 were 3–5 cm in the eastern suburbs, and 2–3 cm in central Bangkok. Moreover between 1989 and 1990, the subsidence rates decreased to 2–3 cm in the eastern area, and 1–2 cm in the central area of Bangkok.

However, over the past 10 years, though the new policy directed by the DGR (Department of Groundwater Research) has succeeded in restoring the groundwater level, the subsidence in the upper layer continues to increase.

This is mainly due to the combined effects of groundwater pumping and loading of the buildings in the Bangkok vicinity. In most of the urban area, the man-induced subsidence is caused by several effects.

This situation raises various questions: What are the legal implications when a homeowner believes subsidence damaged his or her property? Who would be liable? What is the relative importance of each subsidence cause?

The book written by Dr. Wakshal and Dr. Zeitoun presents a new approach for the computation of the soil settlement. It permits to separate the different types of subsidence causes and answer the above questions.

Since 2008, the DGR is working with the authors on the implementation of the methodology presented in the book and the software built for the Bangkok Metropolitan Area.

This book provides the details of the engineering methods that are available and the new mathematical methodology developed by the authors. I congratulate the authors and encourage them to develop their ideas.

Also, I recommend this book to anyone working on subsidence projects. It is important to ensure that such valuable knowledge will be used in subsidence's studies.

Former Director General
Department of Groundwater Resources
Bangkok
Thailand

Ms. Somkid Buapeng

Preface

Subsidence may be defined as the sinking of the earth's surface in response to geologic or man-induced causes. When subsidence occurs, the features caused by the accumulation of sediments are termed geosynclines. Also subsurface hydrochemical activities, during cave formation, may lead to subsidence features at the ground surface, generally called, karst topography. Similar effects can be produced by mining or by the extraction of water, gas or petroleum by means of wells. Subsidence may also occur after irrigation of virgin areas of alluvial deposits.

Cities built on unconsolidated sediments consisting of clays, silt, peat, and sand are particularly susceptible to subsidence. Such areas are common in delta areas, where rivers empty into the oceans, along flood plains adjacent to rivers, and in coastal marsh lands. Building cities in such areas aggravates the problem for several reasons, as follows:

1. Construction of buildings and streets adds weight to the region causing additional soil deformations
2. Often the areas have to be drained in order to be occupied, which results in a lowering of the water table and leads to hydro-compaction
3. Often the groundwater is used as a source of water for both human consumption and industrial use
4. Levees and dams are often built to prevent or control flooding

Then, maximum permissible land subsidence (or consolidation) is a constraint in various management problems. Typical management problems are groundwater management (Ramnarong and Buapeng 1991, Brozovic et al. 2006) and planning of town and/or laws on building construction (Corwin et al. 1991). As emphasized by Barends (2010), "in order to develop a legal framework to claims and litigation, it is essential that direct and indirect causes of land subsidence effects can be quantified with sufficient accuracy from a technical and scientific point of view."

It is, therefore, necessary to quantify the contribution of each cause to soil subsidence of the ground surface in cities and urban areas.

Soil fracturing due to subsidence may provoke severe damages and affect the structural stability of buildings, and therefore legal control and assessment of damages is today required in large cities (Harter 2003, Ortiz et al. 2010).

Most existing methods and software applications treat the subsidence problem by analyzing one of the causes. For example, over-pumping creates large-scale subsidence, while building loading creates local subsidence/consolidation only.

In this book, an attempt is made to present a unified approach to the problem of subsidence. This unified approach is particularly important in urban areas where several factors are causing the soil subsidence.

Our main objective in this book is to provide groundwater hydrologists, soil mechanics engineers, and planners who deal with town development and management of groundwater resources with all the tools they need for monitoring and forecasting large-scale subsidence in urban areas. Also, we provide a theoretical background and numerical tools for computing the contribution of each cause to subsidence. This is particularly important for litigation problem relative to subsidence.

The difficulty of the presentation is due to the fact that the engineering practice and existing methods belong to two different engineering subjects and matters. On one hand, soil mechanics deals with soil consolidation, and on another hand, groundwater hydrology deals with groundwater movement and its effects on the ground.

In urban areas, the growth of population and consequently the town expansion is an important factor to account for man-induced subsidence.

The approach presented in this book belongs to three different engineering subjects:

(a) Groundwater hydrology
(b) Soil mechanics and consolidation
(c) Land use engineering

The general Biot equations for the consolidation of soil are the basic physical model used in this book.

Also, a large part of the book is dedicated to numerical solutions and computer implementation of the subsidence modeling.

We would like to express our thanks to Prof. G. F. Pinder for the valuable comments on this book.

The material of the book was presented in several seminars and courses. It was implemented in a research project with the DGR (Department of Groundwater Resources) and the Chulalongkorn University.

We would like to express our thanks for the cooperation with DG Somkid Buapeng, Dr. Vachi Ramnarong of the DGR and the Water Resources System Research Unit of the Chulalongkorn University directed by Prof. Sucharit Koontanakulvong. We thank S. Chokchai (of the CU-WRSRU) and all the academic and technical staffs of both institutes (the Chulalongkorn University and the DGR) for their active cooperation and help. Fruitful discussions were carried out with Prof. Ashim Das Gupta, Dr. Oranuj Lorphensri, senior hydro-geologist C. (Charoen) Chuamthaisong, and Jade Julawong.

References

Barends FBJ (2010) Review of subsidence management in the Netherlands. In: Proceedings of the EISOLS 2010, Queretaro, Mexico, October 2010, IAHS Publication 339

Brozovic N, Sunding DL, Zilberman D (2006) Optimal management of groundwater over space and time. In: Berga D, Goetz R (eds) Frontiers in water resource economics, natural resource management and policy series, vol 29. Springer-Verlag, New York

Corwin EJ, Alhadeff SC, Oggel SP, Shlemon RJ (1991) Earth fissures, urbanization and litigation. In: Proceedings of the fourth international symposium on land subsidence, May 1991. IAHS Publication – 200

Harter T (2003) Legal control of California's water resources, University of California, Division of Agricultural and Natural Resources, Oakland, CA

Ortiz JA, Alonso FA, Pacheco J, Zermeno ME, Araiza G, Mendoza E (2010) Assessment of the state and condition of damaged and structures affected by land subsidence. In: Proceedings of the EISOLS 2010, Queretaro, Mexico, October 2010, IAHS Publication 339

Ramnarong V, Buapeng S (1991) Mitigation of groundwater crisis and land subsidence in Bangkok. J Thai Geosci 2:125–137

References

Barends FBJ (2010) Review of subsidence management in the Netherlands. In: Proceedings of the EISOLS 2010, Querétaro, Mexico, IAHS Publ. 339, IAHS, Wallingford, [illegible]

[illegible]

[illegible]

[illegible]

[illegible]

Contents

Chapter 1
Introduction

Abstract In this section, the effect of land subsidence in urban area is presented on the example of the Bangkok area. In Section "The Effect of Land Subsidence in Urban Areas: The Bangkok Area", the different aspects of land subsidence are presented for the Bangkok area. Monitoring of soil settlements, determination of the causes, legal aspects are fundamental subjects of soil subsidence in urban areas. In Section "The Scope of the Book", a general presentation of the chapters is detailed.

Soil deformation can be either the sudden collapse of the ground and/or the slow subsidence and/or compaction of the sediments near the earth's surface. Sudden collapse events are rarely major disasters, certainly not anywhere near the scale of the earthquake, volcanic, tsunami, or landslide disasters, but the slow subsidence of areas can cause as much economic damage, although spread out over a longer period of time.

Cities built on unconsolidated sediments such as clays, silt, peat, and/or sand are particularly susceptible to subsidence. Such areas are common in delta areas, flood plains adjacent to rivers, and in coastal marsh lands. Building cities in such areas aggravates the subsidence problem for several reasons, as follows:

1. Construction of buildings and streets adds weight to the region causing additional soil deformation;
2. Often the areas have to be drained in order to be occupied, which results in a lowering of the water table and leads to hydro-compaction;
3. Often the groundwater is used as a source of water for both human consumption and industrial use; and
4. Levees and dams are often built to prevent or control flooding.

Most existing methods and software applications treat the subsidence problem by analyzing one of the causes. This is due to the fact that the causes appear at different

D.G. Zeitoun and E. Wakshal, *Land Subsidence Analysis in Urban Areas*,
Springer Environmental Science and Engineering, DOI 10.1007/978-94-007-5506-2_1,

scales. For example, over-pumping creates large-scale subsidence, while building loading creates local subsidence/consolidation only.

In this textbook, we are presenting an engineering approach to predicting the soil settlement due to subsidence, resulting from various causes.

A typical example of a city where several causes are causing subsidence is the city of Bangkok, Thailand.

1.1 The Effect of Land Subsidence in Urban Areas: The Bangkok Area

The Bangkok metropolitan area is situated on the flood plain of the Chao Phraya river which traverses the lower central plain of Thailand.

The geological layers involved in the transmission and confinement of groundwater in the Bangkok region are a basement complex and alluvial deposits which constitute the principal aquifers in Bangkok (Ramnarong and Buapeng 1992).

The topmost sediments, 15–30 m thick, consist of soft clay overlying stiff clay called the "Bangkok clay". The underlying sand, gravel, and clay constitute the principal aquifers of Bangkok. In the upper 600 m, the unconsolidated deposits are subdivided into eight principal artesian aquifers, separated from each other by thick confined clay or sandy clay layers. The current names of the aquifers are as follows:

1. Bangkok aquifer;
2. Phra Pradaeng aquifer;
3. Nakorn Luang aquifer;
4. Nonthaburi aquifer;
5. Sam Khok aquifer;
6. Phraya Thai aquifer;
7. Thon Buri aquifer; and
8. Pak Nam aquifer

Subsidence of land surface in the Bangkok area has been collected since 1978 using two different methods: (a) periodic leveling of a network of benchmarks; and (b) measuring compaction of sediments by observing the change in thickness of a particular sequence of deposits.

Results of the study during 1978 and 1981 showed that land subsidence rates were 10 cm/year in the eastern suburbs and 5–10 cm/year in central Bangkok.

After the remedial measures were implemented in 1983 for controlling groundwater pumpage, a continuous recovery of the water level was observed in central Bangkok and its eastern suburbs. The annual subsidence rates between 1988 and 1989 were 3–5 cm in the eastern suburbs and 2–3 cm in central Bangkok. Moreover, between 1989 and 1990, the subsidence rates decreased to 2–3 cm in the eastern area and 1–2 cm in the central area of Bangkok (Ramnarong and Buapeng 1992).

According to a study of the AIT (see Asian Institute of Technology 1982), soil compression during the period 1978–1981 of the top 50 m contributed 40% to the total subsidence, while 60% of the total subsidence results from a compression of the deeper zone 50–200 m.

Bangkok is underlain by soft marine clay that is in a stage of pre-consolidation. Therefore, the consolidation or compaction process is larger than for an over-consolidated soil. In a natural stage without human activities, this process is a long-time process. Soil scientists agree that a high rate of soil consolidation results in a soil matrix compression. Therefore, the analysis of subsidence is deeply connected to the human activities that cause the increase of soil consolidation.

Aquifer compaction, which causes land subsidence and its accompanying side effects, results from compression of grains and reduction of intergranular spaces. The compression is caused by a decrease in buoyant support of the grains and by a change in hydrostatic pressure between the pores in grains. A lowering of the water table or a decrease in the artesian head may causes rearrangement of the coarse grains, and/or compression and partial dewatering of compressible minerals, such as montmorillonite clay. The compaction stress at depth may be increased if a decrease in the artesian head at depth is accompanied by an increase in near-surface loading, as may be created by a near-surface water-level rise. The greater the clay content and void space between non-compressible grains, the greater is the potential for compaction of the material. Also, montmorillonite clay, which is the typical interstitial clay of the sediment in the basin, is more compressible than other clay minerals.

Two main causes for soil consolidation, soil compression, and consequently for land subsidence have been postulated by soil scientists, as follows:

1. The first possibility of land subsidence is due to the effect of deep well pumping. Extracting of water from aquifers is causing declines in the piezometric pressure in the normally consolidated Bangkok clay.

   Haley and Aldrich (1970) conducted the consolidation test and reported that the soft Bangkok clay is over-consolidated. By calculation, assuming a past pressure of 30–50 kPa greater than the existing effective stress, the increase in load of 30–50 kPa as a result of a decline of piezometric pressure would bring the effective normal stress to near preconsolidation pressure .This should result in the settlement from recompression of 0.05–0.07 m. Furthermore, from this critical stress, an additional 3–5 tms stress would have the magnitude of settlement about 10 times the re-compression settlement. Under the tested soil property, the rate of drawdown of 0.3 m per year in the clay would take 10–15 years to increase the average stress level by 3–5 t.m.s, respectively.

   In 1995, the JICA (Japan International Cooperation Agency) performed an extensive study of the groundwater management in the Bangkok area. The influence of over pumping on subsidence was considered using a one-dimensional model connected with the MODFLOW model (see Japan International Cooperation Agency (JICA) 1995).
2. The second possibility of land subsidence is due to the load of building and infrastructure on the ground surface and the pressure of deep foundations.

While the effect of the natural sedimentation is located near the Chao Phraya river, the load of the buildings is distributed throughout the whole Bangkok metropolitan area. Therefore, the computation of the load of the buildings is a difficult task.

- **A Brief History**
In 1969, land subsidence in Bangkok was given public attention for the first time when its evidence was pointed out in the newspapers. Many indications, such as protrusions of well casing and foundation of the buildings, were observed in many places.

The major cause of land subsidence was recognized and discussed in a series of seminars organized during the years 1980–2008 on this subject. Also, research projects on the management of groundwater and mitigation of land subsidence in Bangkok were initiated and undertaken by the DGR, the AIT, and other institutes. Also, a new ground surface leveling was conducted in mid-1978, in the Bangkok area.

Between 1978 and 1981, 51 new benchmarks were installed in addition to the 25 existing benchmarks belonging to the Bangkok Metropolitan Administration (BMA). Leveling was run every 3 months to measure the actual rate of subsidence by the Royal Thai Survey Department.

The center of land subsidence coincides with the center of depression of artesian pressure in the east to southeast of Bangkok at Lat Phrao, Bang Kapi, Hua Mak. Phra Khanong, and Bang Na. Since 1978, the rate of subsidence in these places was estimated at more than 10 cm/year, and as much as 1.14 m subsidence occurred between 1940 and 1980 (Ramnarong 1983). Between 1978 and 1992, the annual rates of subsidence at Dusit, Phaya Thai, Phra Nakhon, and Yan Nawa in central Bangkok ranged from 5 to 10 cm, and the subsidence rate at Thonburi in western Bangkok was estimated at less than 5 cm/year.

After the new groundwater management and the new policy were started in 1995, a reduction of the subsidence rate was observed in the Bangkok area. However, large rates are still measured in various districts.

- **Groundwater use**
Over the years, the pumpage steadily increased with additional private use as the public water supply facility could not cope with the demand. A historical record of pumpage for public water supply by the Metropolitan Waterworks Authority (MWA) was available, but for private pumpage, only estimates were provided by the Department of Mineral Resources (DMR), since there was no law regulating the use until 1978.

After the promulgation of the Groundwater Act in 1978, statistics of the total number of private wells and the total rate of withdrawal became available.

Also today, the DGR has an accurate estimation of the withdrawal from the public wells.

Groundwater utilization in Bangkok and adjacent areas are of the three following main types:

Domestic use

Since the MWA cannot cover its entire service area with surface water supply, groundwater is pumped from deep aquifers to supply houses, condominiums, hotels, restaurants, etc. A rapid urbanization accompanied by a recent economic boom hastened the growth of residential and shopping areas in the suburbs of the Bangkok metropolis. This situation has caused intensive pumping in the deeper aquifers.

Industrial use

Since the inception of industrial promotion in Thailand, the growth of industry has been remarkable in the last 30 years in Bangkok and its vicinity. Accordingly, the need for water in these industries grew rapidly, and most of the industries must invest for their own water supply by pumping groundwater.

Agricultural use

Although groundwater is used for agriculture in many places throughout the country, the quantity abstracted is not large, due to the fact that agricultural farms are using irrigation canals. Groundwater is used mostly as a supplementary source to irrigate cash crops after harvesting the rice.

However, in the future, this use may increase seriously.

- **Land subsidence**

  From the data collected and analyzed, it can be summarized that:

  1. Subsidence rates of less than 2 cm/year existed in most parts of Bangkok;
  2. In Central Bangkok, a subsidence rate of more than 2 cm/year was observed in the years 1978–1996;
  3. Subsidence rates of 2 to more than 5 cm/year were observed in Samut prakarnon in the south, Bang Pli in the southeast, and Samut Sakhon in the southwest.;
  4. Recently, subsiding areas almost coincided with the new groundwater depression zones shown; and
  5. Although the groundwater water level rose in Bangkok, land subsidence is still occurring.

  Groundwater levels in Bangkok area were affected by a regional decline of groundwater levels caused by heavy withdrawals in the surrounding areas, which resulted in the continuous occurrence of land subsidence in the metropolis. Serious land subsidence may now take place in the surrounding areas unless groundwater withdrawals are strictly regulated.

- **The legal aspect: the Groundwater Act**

  The uncontrolled and excessive pumping of groundwater has caused the widespread decline of the water level and land subsidence in the Bangkok Metropolis and its vicinity since the 1970s. As a legal measure against land subsidence, the Government enforced the Groundwater Act 1977. The underlying principle of this Act is the government control of groundwater activities. It is implemented since 1978.

Also, in recent years, the government is now enlarging the groundwater act to other causes such as excess of building and/or urban planning and town development acts.

Development of DSS tools for the detection of aquifers sensitivity to subsidence are under development by the DGR (Propovitch 1991; Zeitoun and Melloul 2008).

1.2 The Scope of the Book

The main purpose of this book is to describe a methodology for the engineering calculation of the subsidence. Also, the methodology presented in this book permits the estimation of the relative contribution of the different causes of the total subsidence.

Compaction of a soil layer may be due to several causes. When applying an external or internal pressure on the soil matrix, the resulting soil displacement may be of two types, as follows:

1. A reversible soil displacement where the soil matrix is not changed, and the soil will come back to the initial position when no more pressure is applied; or
2. A residual compaction (or consolidation) where the soil matrix is compact and the displacement results in a reduction of the void ratio. This type of displacement is not reversible (see Sneed and Galloway 2000).

Both displacements are referred to as soil subsidence, but their analysis is different.

The reversible soil displacement is generally caused by non-constant or periodic loading, such as traffic loading on roads, or a wetting and drying process in the unsaturated zones. This type of cyclic displacements at the soil surface is not in the scope of our work.

This book concentrates on the estimations of residual compaction resulting from the following three main causes:

1. The natural consolidation (river sedimentation);
2. The relative contribution of the decline in piezometric pressure; and
3. The overloading due to urbanization of the total land subsidence.

In Chap. 2, we review the different types of subsidence appearing in the world.

In Chap. 3, we present the various mathematical models currently used for the analysis of soil deformation for each type of subsidence. The different models of subsidence are presented in terms of:

(a) A geometrical model of the porous media is needed and a definition of representative element volume (REV);

(b) The variables of the model and the governing equations for the fluid and/or the gas flow as well as the mechanical behavior of the soil skeleton;
(c) The boundary conditions associated with the partial differential equations appearing in the model; and
(d) The type of loadings and the way it acts on the soil matrix.

Then models are described for four types of subsidence commonly encountered in practice, as follows:

- Subsidence due to water withdrawal in saturated soil;
- Subsidence due to oil and gas withdrawal in non saturated soils;
- Subsidence due to sea water encroachment; and
- Subsidence appearing in fracture rocks.

In Chap. 4, we describe the Terzaghi theory of consolidation as a way of decoupling the general Biot equation using the assumption of constant total stress. We detailed the computation of the water pressure and the soil displacement. The classical one-dimensional theory is presented with solved exercises. Also the three-dimensional theory is described. Then we discussed the limitation of the theory in the case of internal loadings.

We also discuss the use of the theory of consolidation for large-scale problems such as subsidence due to large-scale groundwater pumping. We review the different approaches that were used in the literature for the numerical modeling of the subsidence due to groundwater pumping. We concentrate on the two following main points:

1. The mechanical behavior of the clayey soil; and
2. The connection between groundwater movement and stress pressure in the soil.

Then, a critical review of the consolidation theory for large-scale problems is presented in terms of the principle of superposition of loadings.

In Chap. 5, we present the constitutive equations of the Biot model. The main difficulties of this model are the coupling of the equations and the application of the superposition principle. We present the general solution of the coupled problem resulting from the Biot model. This solution is based on a new method for decoupling the equations. This decoupling approach permits to derive a superposition principle of the loadings. Several types of loadings and boundary conditions for the water pressure and the displacement field are discussed. Analytical solutions are also derived and discussed.

In Chap. 6, we first review iterative methods proposed for the numerical solution of the coupled Biot equations. Then, we describe the numerical scheme based on the decoupling method of the solution of the Biot model. The numerical method is called the compartmental model and is similar to the finite volume method. We emphasize the treatment of the different types of boundary conditions.

In Chap. 7, the interface of the software is described in general terms and we have detailed the computation of the different types of loadings and the input data.

In Chap. 8, the implementation of the methodology is described on a modeling study of the Bangkok area. The use of historical satellite maps of the urbanization of the Bangkok area is connected with the GIS system. The building of the model data is described with details. We analyze separately the different types of effects. The computational of the total subsidence and each component is presented. Also, the effect of pumping from a deep layer on subsidence in the upper layer is presented.

References

Asian Institute of Technology (1982) Investigation of land subsidence caused by deep well pumping in the Bangkok area: comprehensive report 1978–1981. NEB. Pub. 1982–002

Haley & Aldrich Inc. (1970) Effect of deep well pumping on land subsidence in Bangkok. In: Master plan, water supply and distribution, metropolitan Bangkok, Thailand, vol 4, Report by Camp, Dresser & McKee Inc., Metropolitan Water Works Association, Bangkok

Japan International Cooperation Agency (JICA) (1995) The study on management of groundwater and land subsidence in the Bangkok metropolitan area and its vicinity. Final report submitted to the Department of Mineral Resources, Bangkok, Thailand

Propovitch NP (1991) Detection of aquifers susceptibility to land subsidence. In: Proceedings of the fourth international symposium on land subsidence, May 1991. IAHS Publication – 200

Ramnarong V (1983) Environmental impacts of heavy groundwater development in Bangkok, Thailand. International conference on groundwater and man, Australian Government Publishing Service, Canberra 2, pp 345–350

Ramnarong V, Buapeng S (1992) Groundwater resources of Bangkok and its vicinity: impact and management. In: Piancharoen C (ed) Proceedings of a national conference on geologic resources of Thailand: potential for future development, Bangkok, Thailand

Sneed M, Galloway DL (2000) Aquifer-system compaction and land subsidence: measurements, analysis, and simulations – the Holly site, Edwards Air Force Base, Antelope Valley, CA, US Geological Survey Water-Resources Investigations report 00–4015

Zeitoun DG, Melloul A (2008) Decision support system (DSS) based on automatic water balance computation for groundwater management planning: the case of Israel's coastal aquifer. In: Joshi PK (ed) Geoinformatics for natural resource management. Nova Science, Hauppauge

Chapter 2
The Subsidence Phenomenon Throughout the World

Abstract In this section, the different types of land subsidence are discussed. They are separated into natural causes and man-induced causes. The natural causes are mainly geological causes, such as formations of caves, sinkholes, and karst topography. The man-induced causes are mainly mining or fluid withdrawal (groundwater and/or oil or gas). Also, loads of buildings are an important cause of consolidation in urban areas. These causes are explained in general term, and examples of subsidence in the United States are briefly referred to.

2.1 Type of Subsidence

Subsidence may be defined as the gradual and/or sudden sinking of landforms to a lower level as a result of a subsurface movement of earth materials.

Subsidence is a worldwide global problem. For example, in the United States, more than 17,000 square miles in 45 states, an area roughly the size of New Hampshire and Vermont combined, have been directly affected by subsidence. The principal causes are aquifer-system compaction, drainage of organic soils, underground mining, hydro- compaction, natural compaction, sinkholes, and thawing permafrost (National Research Council 1991). More than 80% of the identified subsidence in the US is a consequence of the exploitation of underground water. Also the increasing development of land and water resources threatens to increase land subsidence problems. In many areas in the US underlain by soluble rocks such as limestone, gypsum, or salt, land subsidence is a common environmental consequence of our land- and water-use practices.

The most common type of a sudden collapse is due to erosion of underground soil and/or rock caused by leaking human-made sewer pipes or water mains. This rarely destroys large areas but commonly swallows up vehicles.

The second most common type of a sudden collapse involves dissolution of carbonate rocks (limestone) beneath the surface.

D.G. Zeitoun and E. Wakshal, *Land Subsidence Analysis in Urban Areas*,
Springer Environmental Science and Engineering, DOI 10.1007/978-94-007-5506-2_2,

Such a catastrophic subsidence is most commonly triggered by groundwater-level declines caused by pumping, or by diversion of surface runoff or groundwater flow through susceptible rocks. Although the collapse features tend to be highly localized, they can introduce contaminants into the aquifer system and, thereby, have lasting regional impacts. Collapse features tend to be associated with specific rock types having hydro-geologic properties that render them susceptible to the formation of cavities. Human activities can facilitate the formation of subsurface cavities in these susceptible materials and trigger their collapse, as well as the collapse of preexisting subsurface cavities.

The natural causes may be separated into two main categories:

(a) the geological causes and
(b) man-induced causes.

2.2 The Geological Causes of Land Subsidence

In terms of land area affected, underground mining accounts for about 20% of the total land subsidence in the United States. Most of this fraction is associated with underground mining for coal. Subsidence over underground coal workings develops as a gradual down warping of the overburden into mine voids. It is generally unrelated to subsurface water conditions. Salt and gypsum are, respectively, almost 7,500 and 150 times more soluble than limestone. Underground salt and gypsum mines are also subject to down warping of the overburden, but these evaporite minerals are susceptible to rapid dissolution by water. Also, limestone is often associated with sinkhole formation.

2.2.1 Carbonate Dissolution and Karst Topography

Carbonate rocks such as limestone, composed mostly of the mineral calcite ($CaCO_3$) are very susceptible to dissolution by groundwater during the process of chemical weathering. Such dissolution can result in systems of caves, sinkholes, and eventually to karst topography.

2.2.2 Dissolution

Water in the atmosphere can dissolve small amounts of carbon dioxide (CO_2). When rain falls on the earth, the rain water has an amount of carbonic acid (H_2CO_3). As the water infiltrates into the groundwater system and encounters carbonate rocks

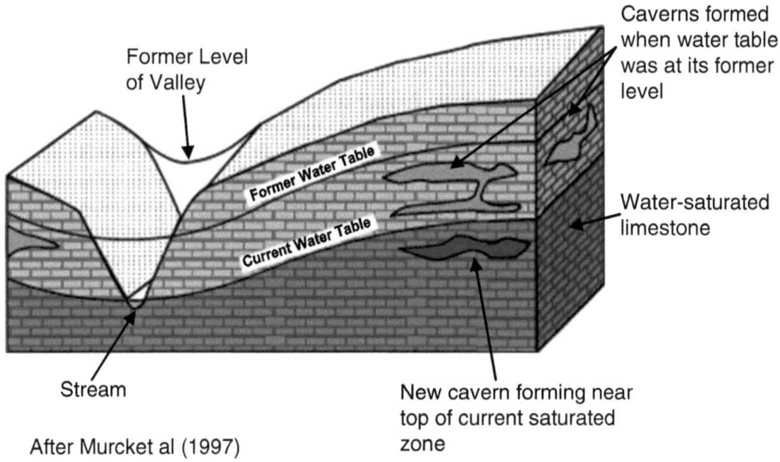

Fig. 2.1 An example of cave formation (After Murck et al. 1997)

like limestone, the calcite is dissolved in the limestone by the following chemical reaction:

$$CaCO_3 + H_2CO_3 = Ca^{+2} + 2HCO_3^{-2}$$

This relation states that calcite reacts with carbonic acid to produce dissolved calcium ion plus dissolved bicarbonate ion.

This reaction takes place as the water flows through any types of opening in the rock such as fractures. Over a long period of time, a large part of the limestone is dissolved resulting in the formation of cavities.

2.2.3 Caves and Cave Formation

Caves are large underground open spaces. A cave system of many interconnected chambers is called a ***cavern***. Most caves are formed by the chemical dissolution process described above, as a result of groundwater flow. The dissolution begins along fracture systems in the rock, widening the fractures and connecting them to other fractures, until a cave is formed (Fig. 2.1).

2.2.3.1 Sinkholes

A ***sinkhole*** is a large dissolution cavity that is open to the earth's surface. Some sinkholes form when the roofs of caves collapse; others can form at the surface

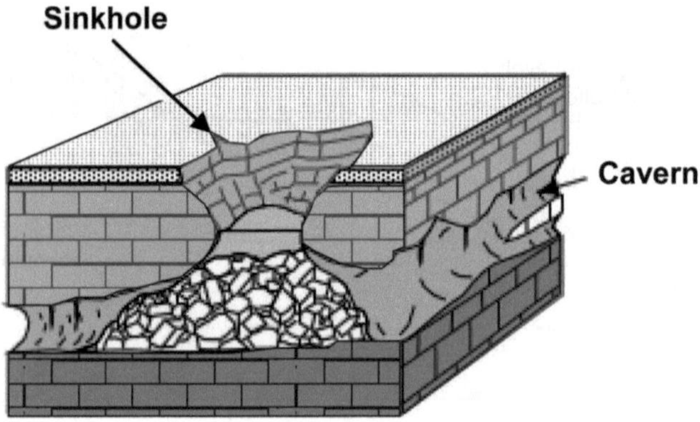

Fig. 2.2 An example of a sinkhole (After Nelson 2000)

by dissolving the rock downward. Because we are here concerned with subsidence disasters and hazards, we will concentrate on the formation of sinkholes by collapse.

Sinkholes are common in areas underlain by limestone. Central Florida is such an area, and in one small area of about 25 km^2, over 1,000 sinkholes have formed by collapse in recent years.

Sinkholes may form as a result of lowering the water table by excessive pumping for human use of the water (see the Florida case in National Research Council 1991). Caverns that were forming just below the water table were filled with water. The water table was lowered over the years resulting in the level of groundwater in the caverns to become lower. While the water table was high, the water in the cavern helped to support the roof of the cavern. This support is removed when the water table is lowered, and thus, the unsupported roof eventually becomes unstable and collapses to form a sinkhole (Fig. 2.2).

Sinkholes may also form by the slow enlargement of caverns by continued dissolution of the limestone. This may occur no matter what the level of the water table is. When sinkholes collapse to expose the water table at the surface, the sinkhole will be filled with water forming small circular lakes.

Although common in areas underlain by limestone, sinkholes can form in any area where highly water-soluble rocks occur close to the surface. Such rocks include rock salt made of the mineral halite and gypsum deposits, both of which easily dissolve in groundwater.

2.2.3.2 Karst Topography

In areas where highly water soluble materials lie close to the surface, dissolution below the surface can eventually lead to the formation of caverns and sinkholes. As the sinkholes begin to coalesce, the surface topography will become chaotic, with many enclosed basins and streams that disappear into sinkholes, run underground, and reappear at springs. Such a chaotic topography is known as ***karst topography***.

Karst topography starts out as an area with many sinkholes, but eventually, as weathering and dissolution of the underlying rock continue, the ground surface may be lowered, and areas that have not undergone extensive dissolution stand up as towering pillars above the surrounded terrain. The latter type of karst is called ***tower karst***.

2.3 Man-Induced Causes

2.3.1 Mining

2.3.1.1 Removal of Solids and Mine-Related Collapse

Humans can play a large role causing a collapse of the surface. Mining activities that remove material from below the surface can result in a collapse if precautions are not taken to ensure that there is adequate support for the overlying rocks.

2.3.1.2 Removal of Salt

- Salt occurs beneath the surface in areas that were once below sea level in restricted basins where extensive evaporation caused the concentration of salt in seawater to become so high that the salt was precipitated on the bottom. This occurred, for example during the Jurassic Period (about 150 million years ago) in the area of the Gulf of Mexico. Later deposition of other sediments on top of this salt resulted in low-density salt underlying higher-density sediments. Since salt is rather ductile, it began to flow upward toward the surface, and in many cases, became detached from the original layer of salt at a depth to form what is called a salt dome. Since the salt now occurs close to the surface, it can dissolve and collapse to form sinkholes.
- The salt can also be mined to produce salt for human usage. One mining technique involves injecting fluids into the salt to dissolve it. The fluids are then recovered and the salt re-precipitated from the solutions. Such mining, because it dissolves large cavities in the salt, can lead to instability and collapse. Such solution mining resulted in a sinkhole 300 m in diameter in Hutchinson, Kansas.

2.3.1.3 Coal Mining

Since mining often removes material from below the surface without dissolution, mining can create voids that may become unstable and collapse.

Coal occurs beneath the surface as extensive layers called coal seams. These seams were once swampy areas on the surface where much vegetation flourished, died, and became buried before it could decay. Processes acting over long periods of geologic time have turned dead vegetation into coal. Other useful substances are mined by digging tunnels in a rock, but in most mining techniques, the useful substance occurs along narrow zones, and only these enriched zones need to be removed. In mining coal, however, all of the material is useful, so large masses of material are removed. The technique used in coal mining is referred to as "room-and-pillar" mining. The rooms are where the coal has been removed, and the pillars are left to support the overlying rock. Sometimes, too few pillars are left, and the overlying rock collapses into the mine. This is not only dangerous to the miners but can also cause hazards to areas on the surface where the collapse occurs

- Underground fires in coal mines can also lead to collapse hazards. Fires can start by spontaneous ignition of coal dust or methane gas released from the coal. Such fires are difficult to extinguish, and often are left to burn for years. In Pennsylvania, for example, coal mine fires have burned for more than 25 years. Burning of the coal results in removal of the coal, and thus, may undermine support for the roof of the mine resulting in a collapse.

Formation of subsurface cavities by dissolution requires:

1. bedrock composed in large part of soluble minerals;
2. a water source that is unsaturated with respect to these minerals and, therefore, can dissolve them;
3. an energy source in the form of a hydraulic gradient to move the water through the rock; and
4. an outlet for the escaping, mineralized water.

Once a through-flowing passage develops in the soluble rock, erosion and further dissolution enlarges the passage, further enhancing the throughflow. Once established, subsurface cavities may provide habitat for populations of species specially adapted to cave environments—a cave ecosystem.

2.3.1.4 Evaporite Rocks Can Form Cavities Within Days

Evaporites are sediments deposited from natural waters that have been concentrated as a result of evaporation. Evaporite rocks such as salt and gypsum underlie about 35–40% of the contiguous United States. Natural solution-related subsidence has occurred in each of the major salt basins (Ege 1984), perhaps most notably in the Permian basin of Texas, New Mexico, Oklahoma, and Kansas, and the smaller Holbrook basin of northeast Arizona. Although evaporites underlie most of the Michigan-Appalachian and Gulf Coast basins, naturally forming collapse features are much less common in these areas.

Man-induced collapse cavities are relatively uncommon in gypsum deposits and are more likely to develop above salt deposits, where they are associated with both purposeful and accidental dissolution of salt.

2.3.1.5 Carbonate Rocks Form Cavities Over Centuries

Natural cavities in carbonates (limestone and dolomite) develop by the same processes that form cavities in evaporite rocks, albeit much more slowly. The high solubilities of salt and gypsum permit cavities to form in days to years, whereas cavity formation in carbonate bedrock is a very slow process that generally occurs over centuries to millennia. The slow dissolution of carbonate rocks favors the stability and persistence of the distinctively weathered landforms known as karst.

Both dissolution and erosional processes play roles in the maturation of karst in carbonates; if not for a balance between mechanical erosion and dissolution, the distinctive landscapes could not persist.

The high strength of carbonate rocks confers resistance to mechanical failure despite progressive weakening by chemical dissolution. The potential for dissolution is controlled by the amount of water available and also by the level of saturation of that water with respect to calcium carbonate. Where the potential for dissolution is low, mechanical erosion dominates the morphology of carbonates.

For example, in the arid southwest, limestone exposures tend to erode as cliffs rather than form karst. Carbonate karst landscapes comprise about 40% of the contiguous United States east of the longitude of Tulsa, Oklahoma (White et al. 1995). In these more humid landscapes, surface and subsurface drainage pathways converge in discrete conduits formed in the carbonate bedrock. Sinkholes, swallows (where streams disappear into the subsurface), and springs are linked to form an interconnected surface and subsurface drainage network.

Thus, karst aquifer systems are directly affected by variabilities in the timing and magnitude of surface runoff. Surface runoff carries all the components of streamflow into the conduit flow system, including suspended sediment, dissolved contaminants, immiscible fluids, and micro- and macrobiological agents. The slower infiltration of surface water through porous soil and rock to the water table, which helps to protect ground water from contamination in most areas, is short-circuited in karst landscapes.

2.3.2 *Fluid Withdrawal*

We have seen how fluids (particularly water) in the subsurface can dissolve rock to undermine support and cause collapse of the surface. Here, we look at another role that fluids may play in causing subsidence. Any fluid that exists in the pore spaces or fractures of rock is under pressure due to the weight of the overlying rock. So long as the pressure of the fluid is enough to support the overlying rock, no subsidence at the surface will occur. But, if fluids are withdrawn from below the surface, a decrease in fluid pressure may occur resulting in the removal of support and possible a collapse. The two most important fluids that occur beneath the surface are water (in the form of groundwater) and petroleum (in the form of oil and natural gas). Both of these fluids are often withdrawn for human use, and thus, humans are often responsible for fluid withdrawal-related subsidence. But, such withdrawal can also occur by natural processes.

2.3.2.1 Swelling

Many types of clay particles are capable of absorbing large volumes of water, causing them to increase substantially in size compared with their dry state. Repeated cycles of dry and wet conditions cause such soils to form networks of fissures, resulting in the familiar cracked, polygonal landscape typical of non-sandy desert areas. Individual clay particles may shrink and swell up to 1,000%. While overall soil swelling is not nearly this extreme, volumetric changes of even 5% can damage human structures.

2.3.2.2 Compaction

Compaction is the process by which the porosity of a given form of sediment is decreased as a result of its mineral grains being squeezed together by the weight of overlying sediment or by mechanical means.

2.3.2.3 Hydro-Compaction

Hydro-compaction is essentially the opposite of swelling. In dry areas with a mix of sand, silt and clay, dry clay or silt particles act as a sort of cement, holding the sand particles in an open matrix with air spaces in between. This open soil structure is vulnerable to water infiltration in the form of soaking rain, irrigation, or broken pipes. The sudden influx of water dissolves the silt and clay, causing the air spaces between the sand articles to collapse and the affected area to sink, often dramatically.

Most subsidence occurs as a result of hydro-compaction, discussed previously under mass-wasting processes. Hydro-compaction occurs when the sediments lose water. Since lowering of the water table involves a loss of water, hydro-compaction often occurs.

Hydro-compaction means that water absorbed on and within clay minerals are removed by withdrawal or drying, and the clays shrink. Shrinkage of clays results in less volume, so the surface will subside as the clays become more tightly compacted (Fig. 2.3).

2.3.2.4 Groundwater

Groundwater occurs nearly everywhere below the surface of the earth, where, as we have said above, it fills the pore spaces and fractures in rock at levels below

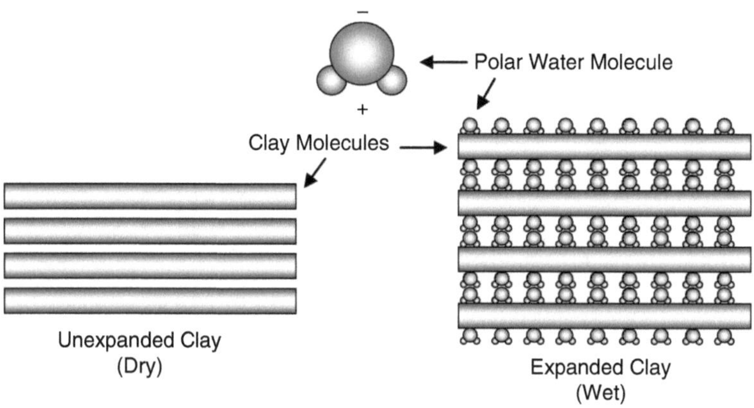

Fig. 2.3 The swelling process in the clay expansion (After Coch 1995)

the water table. The zone beneath the water table is called the saturated zone. Groundwater flows into the saturated zone by percolation downward from rainfall on the surface. Surface bodies of water, like streams, lakes, and swamps, are areas where the water table is exposed at the surface. Springs are also areas where the water table is exposed at the surface. If one digs or drills a well to intersect the water table, water will flow into the well and fill it to the level of the water table. The level of the water table can change as a result of changing amounts of input in the groundwater system (called ***recharge***) and output from the groundwater system (called ***discharge***). Recharge takes place by water infiltrating down from the surface. Discharge occurs as a result of outflow through surface bodies of water, springs, and wells. During the wet season, the water table is generally higher because recharge exceeds discharge. During dry seasons, the water table is depressed because discharge exceeds recharge. Likewise, during periods of drought, the water will be lower (Fig. 2.4).

Groundwater moves through the saturated zone both downward and upward. The downward flow occurs due to gravity, and the upward flow occurs because fluids tend to flow towards areas of lower pressure.

Subsidence can be caused by any process that results in lowering of the water table. Thus, drought, dry seasons, and excessive withdrawal of groundwater by humans can cause the water table to move to deeper levels and result in subsidence (Figs. 2.5 and 2.6).

Most hydro-compaction is an elastic deformation process. Recall that elastic deformation is reversible, so that when the clays or peat dry out, they compact, but when they become wet once again, they expand. Compaction, however, can become inelastic, which is not reversible. In such a case as the pores are closed by compaction, they cannot be restored when new fluids are pumped in.

The rate of subsidence relative to rate of fluid withdrawal can sometimes show when material passes from elastic compaction to inelastic compaction. If the rate of fluid withdrawal is large but the rate of subsidence is small, this is

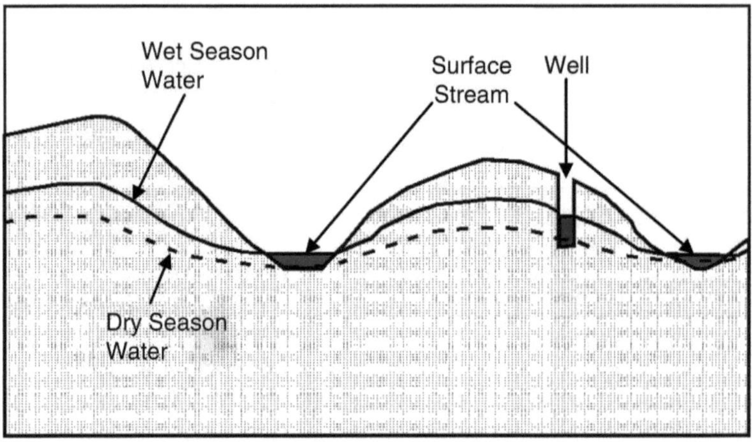

Fig. 2.4 The water level in groundwater

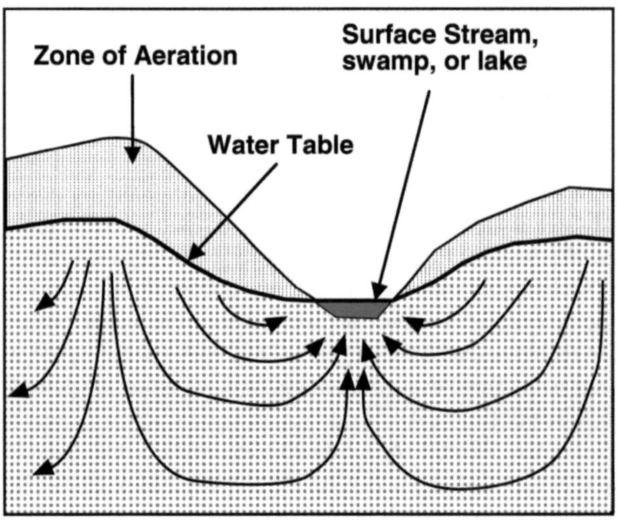

Fig. 2.5 A flow stream in groundwater

usually an indication of elastic compaction. If, however, there is a large amount of subsidence with only small amounts of fluid withdrawal, inelastic compaction is likely occurring.

2.3.2.5 Oil and Gas Withdrawal

Oil and natural gas are both fluids that can exist in the pore spaces and fractures of rock, just like water. When oil and natural gas are withdrawn from regions in the

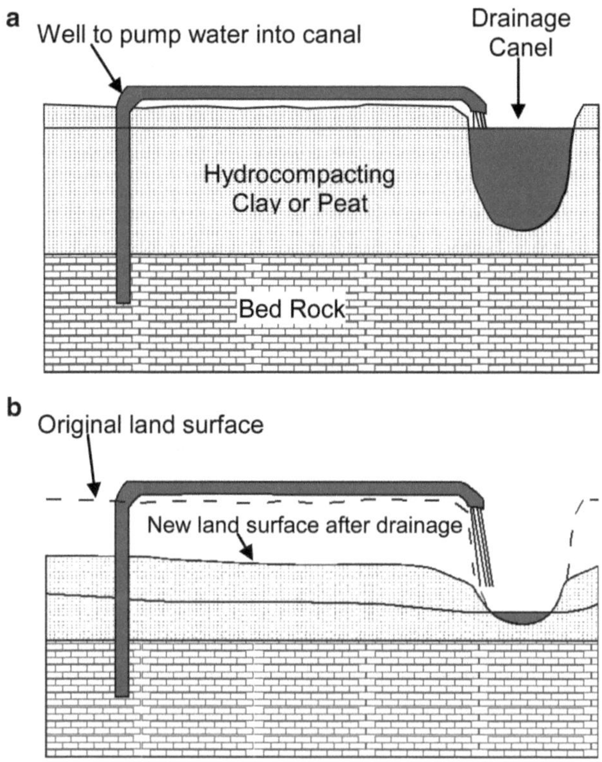

Fig. 2.6 (**a**) First stage of subsidence due to hydro-compaction. (**b**) Second stage of subsidence due to hydro-compaction

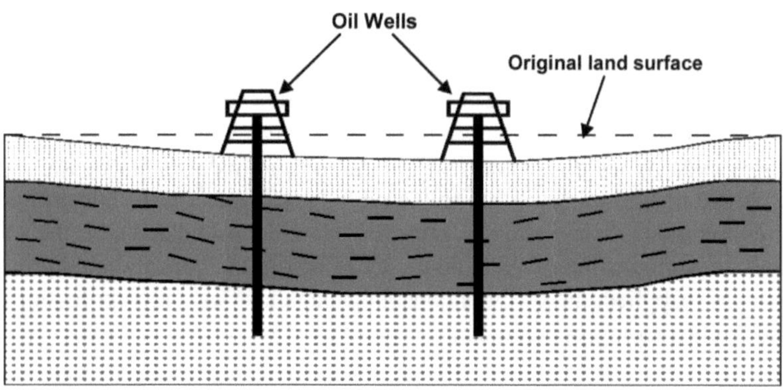

Fig. 2.7 Subsidence in an oil field

earth near the surface, the fluid pressure provided by these fluids is reduced. With a reduction in fluid pressure, the pore spaces begin to close, and the sediment may start to compact, resulting in subsidence of the surface (Fig. 2.7).

This has occurred recently in the oil fields of southern California. For example, in the Wilmington oil field of Long Beach, California, subsidence was first recognized in 1940 due to the withdrawal of oil from the subsurface. The area affected was about 50 km^2. Near the center of this area, the surface subsided by up to 9 m. In 1958, repressurization of the area was attempted by pumping fluids back into the rocks below. By 1962, further subsidence had been greatly reduced, and the area continuing to subside had been reduced to 8 km^2. Still, up to this point, very little uplift had occurred to restore the area to its original elevation. This subsidence event has cost over $100 million.

2.3.2.6 Coastal and Inland Flooding

In a coastal aquifer, sea water encroachment and sea movement may cause soil subsidence. The increase in water density causes an increase of pressure on the soil skeleton resulting in compaction of the soil.

2.3.3 The Soil's Loading

In urban areas, the building of houses and towers are causing a land settlement, often called the consolidation process.

2.3.3.1 Consolidation

According to Terzaghi (1943), "consolidation is any process which involves a decrease in water content of a saturated soil without replacement of water by air". In general, it is the hydro-compaction process in which reduction in volume takes place by expulsion of water under long term static loads. It occurs when stress is applied to a soil that causes the soil particles to pack together more tightly, therefore reducing its bulk volume. When this occurs in a soil that is saturated with water, water will be squeezed out of the soil. Soils are tested with an oedometer test to determine their compression index. When stress is removed from a consolidated soil, the soil will rebound, regaining some of the volume it had lost in the consolidation process. If the stress is reapplied, the soil will consolidate again along a recompression curve, defined by the recompression index.

2.4 Subsidence in Cities

Cities built on unconsolidated sediments consisting of clays, silt, peat, and sand are particularly sensitive to subsidence. Such areas are common in delta areas, where rivers empty into the oceans, along floodplains adjacent to rivers, and in coastal

Table 2.1 Known places of subsidence

City	Maximum subsidence (m)	Area (km^2)
Long Beach/Los Angeles CA	9.00	50
San Joaquin Valley, CA	8.80	13,500
Mexico City, Mexico	8.50	225
Tokyo, Japan	4.50	3,000
San Jose, CA	3.90	800
Osaka, Japan	3.00	500
Houston, TX	2.70	12,100
Shanghai, China	2.63	121
Niigata, Japan	2.50	8,300
Nagoya, Japan	2.37	1,300
New Orleans, LA	2.00	175
Taipei, China	1.90	130
Bangkok	1.00	800
Venice, Italy	0.22	150
London, England	0.30	295

Source: Nelson (2000)

marsh lands. In such settings, subsidence is a natural process Sediments deposited by the rivers and oceans get buried, and the weight of the overlying, newly-deposited sediment, compacts the sediment and the material subsides. Building cities in such areas aggravates the problem for several reasons.

1. Construction of buildings and streets adds weight to the region and further adds stress on the soil.
2. Often the areas have to be drained in order to be occupied. This results in lowering of the water table and leads to hydro-compaction and consolidation.
3. Often the groundwater is used as a source of water for both human consumption and industrial use. This also results in lowering the water table and further hydro-compaction.
4. Levees and dams are often built to prevent or control flooding. This shuts off the natural supply of new sediment to the area. In a natural setting, sedimentation resulting from floods helps replenish the sediment that subsides and thus, builds new material over the subsiding sediment, decreasing the overall rate of subsidence. When the sediment supply is cut off, the replenishment does not occur, and the rate of subsidence in enhanced.

The table below shows a list of cities throughout the world that have been experiencing subsidence problems. Note that most of these cities are coastal cities like London, Houston, and Venice, or are built on river flood plains and deltas, like New Orleans, Baton Rouge, and the San Joaquin Valley of central California. Mexico City is somewhat different in that it was built in a former lake (Table 2.1).

2.5 Predicting and Mitigating Subsidence Hazards

In an ideal world, all areas susceptible to subsidence would be mapped. Then human actions could be taken to either avoid the subsidence to occur, or avoid the inhabitance of such areas if they are prone to natural subsidence.

The exact place and time of a disaster related to subsidence cannot usually be predicted with any degree of certainty. This is true of both slow subsidence related to fluid withdrawal and a sudden subsidence related to sinkhole formation or mine collapse. Therefore, the goal of the actions should be reduce the destruction potential of the subsidence. In other words, the goals of the actions should be to "mitigate" the effect of the subsidence. Mitigation is the best approach to these subsidence hazards.

For subsidence caused by a sudden collapse of the ground to form sinkholes, several measures can be taken. First, geologists can make maps of areas known to be underlain by rocks like limestone, gypsum, or salt, that are susceptible to dissolution by fluids. Based on knowledge of the areas, whether active dissolution is occurring or has occurred in the recent past, and knowing something about the depth below the surface where these features occur, hazard maps can be constructed.

Once these areas have been identified, detailed studies using drill holes, or ground-penetrating radar can be used to locate open cavities beneath the surface. These areas can then be avoided when the time comes for making decisions about land use.

In areas where there is a possibility of sudden collapse, one should be aware of any cracks that form in the ground. Especially, when the cracks start to form a circular or elliptical pattern, a collapse event is imminent.

In areas located above known mining operations or former mining operations, maps can be constructed based on knowledge of the actual locations of open cavities beneath the surface. Such maps can then be used as a guide for land use planning. Currently laws are in place to prevent active mining beneath urban areas, but these laws did not always exist, and older mines could still cause problems.

Where fluid withdrawal is the main cause of subsidence, information on the rate of fluid withdrawal should be determined and combined with studies of the material in the subsurface based on sampling with drill core methods. If subsidence is suspected or observed, human activities can be modified to prevent further subsidence. For example, new sources of water can often be found, or wastewater can be treated and pumped back into the ground to help maintain the level of the water table, maintain fluid pressure, or re-hydrate hydro-compacting clays and peat.

Fluid withdrawal problems are complicated in the United States where laws are in conflict. Rights to withdrawal of an underground resource like water or oil usually take precedence over the rights to sue for damages that might result from subsidence.

In the next chapter, a description of the various physical and mathematical modeling of the different types of subsidence is presented.

References

Coch NK (1995) Geohazards: natural and human. Prentice-Hall, Englewood Cliffs
Ege JR (1984) Mechanisms of surface subsidence resulting from solution extraction of salt. In: Holzer TL (ed) Man-induced land subsidence. Geol Soc Am Rev Eng Geol 6:203–221
Murck BW, Skinner BJ, Porter SC (1997) Dangerous earth: an introduction to geologic hazards. Wiley, New York
National Research Council (1991) Mitigating losses from land subsidence in the United State. National Academy Press, Washington, DC
Nelson SA (2000) Subsidence: dissolution and human-related causes. University of Tulane, http://www.tulane.edu/~sanelson/geol204/subsidence.htm
Terzaghi K (1943) Theoretical soil mechanics. Wiley, New York
White WB, Culver DC, Herman JS, Kane TC, Mylroie JE (1995) Karst lands. Am Sci 83:450–459

Chapter 3
Mechanical Modeling of Porous Media

Abstract In this chapter, the various mathematical models currently used for the analysis of soil deformation for each type of subsidence are presented. The various models of subsidence are presented in terms of

- a geometrical model of the porous media is needed and a definition of representative element volume (REV);
- the variables of the model and the governing equations for the fluid and/or the gas flow as well as the mechanical behavior of the soil skeleton;
- the boundary conditions associated with the partial differential equations appearing in the model; and
- the type of loadings and the way it acts on the soil matrix.

Then, the models are described for different types of subsidence commonly encountered in practice:

- subsidence due to water withdrawal in saturated soil;
- subsidence due to oil and gas withdrawal in non-saturated soils;
- subsidence due to sea water encroachment;
- subsidence appearing in fracture rocks.

3.1 Description of Porous Media

- A ***soil*** consists of an aggregation of variously sized solid particles (SP) and the "pore spaces" (PS) between the particles. (see Fig. 3.1: SP – dark gray, PS – grey)
- When the "pore space" is completely full of water, then the soil is *saturated.*
- When the "pore space" contains both water and air, then the soil is *unsaturated.*

A ***porous media*** is an array of a large number of various sized fixed solid particles. Also, the volume concentration of solids is not small (soil).

It is characterized by its ***porosity*** n (i.e., the pore volume fraction) and its ***grain size*** d. The latter characterizes the "coarseness" of the medium.

D.G. Zeitoun and E. Wakshal, *Land Subsidence Analysis in Urban Areas*,
Springer Environmental Science and Engineering, DOI 10.1007/978-94-007-5506-2_3,

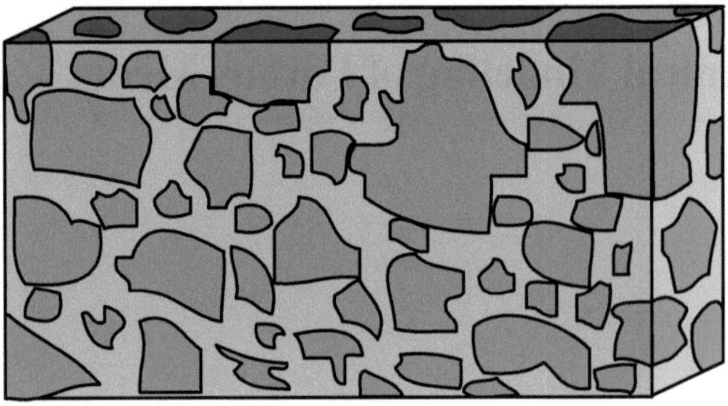

Fig. 3.1 Typical porous media

In the literature, porous media with simple characteristics have been proposed in order to analyze the flow and compression of general porous media.

Most of the models of subsidence are based on various mechanical modeling of porous media. A model of porous media includes various steps, as follows:

(a) A geometric model of porous medium. In practice, the pore networks that exist in actual porous media are too complex to measure and are difficult to represent mathematically. In the next section, we will review some classical models used in porous media.

(b) For the fluid and gas flowing in the pores, fluid properties must be defined. In fluid dynamics, there are six properties of the flow that an engineer is usually interested in – pressure p, density ρ, the three velocity components (u,v,w), and the temperature T. For gases and mixtures of gases (e.g., air) the equation of state links p, ρ, and T by:

An equation of state:

$$P = \rho RT \tag{3.1}$$

We need to come up with five additional equations linking the six properties. These five equations are PDEs and turn out to be:

- conservation of mass or continuity;
- conservation of u- momentum;
- conservation of v-momentum;
- conservation of w-momentum; and
- conservation of energy.

These elements of gas dynamics are presented in Liepmann and Roshko (1957).

(c) In classical hydrology, the soil matrix is considered as a non-deformable solid, and no variables associated with soil movement are needed. When subsidence occurs, the solid matrix is deformable; then, the variables associated with the soil deformation must be defined. In a displacement formulation of the soil matrix behavior, soil displacement is chosen. This displacement is represented as a three-dimensional displacement vector.

(d) The boundary conditions associated with each equation.

1. A model of a porous medium as a **bundle of capillary tubes**:

   The simplest model just considers the pores to be equivalent to a collection of straight capillary tubes. Then, the radii of the tubes can be varied. Tortuosity can be introduced and the pore radius can vary along each tube. Finally, tubes can be connected to one another (see Bear 1972, p. 162).

   Beginning with the simplest useful model, consider a collection of capillary tubes of varying diameters placed into an open body of water. The water will rise to different heights depending on the tube radii. At any elevation, the percentage of the cross- sectional area of all of the tubes that are water-filled defines the saturation.

   Limitations of this model for representing porous media:

   - The capillaries do not vary in diameter along their length.
   - The pores are all perfectly connected vertically and are not connected horizontally.
   - The model cannot represent an appropriate residual saturation.

2. Periodic porous media

   A ***periodic porous media*** is a porous media having the property that the fixed solid particles are identical, and the whole media is a periodic system of cells which are replicas of a standard (representative) cell (experimental column).

   Several models for flow through porous media are based on a periodic array of spheres.

   - Hasimoto (1959) obtained the periodic fundamental solution to the Stokes problem by Fourier series expansion and applied the results analytically to a dilute array of uniform spheres.
   - Zick and Homsy (1982) use Hasimoto's fundamental solution to formulate an integral equation for the force distribution on an array of spheres for arbitrary concentration. By numerical solution of the integral equation, solutions for packed spheres and for several porosity values were obtained (see also Sorensen and Stewart 1974; Tartar 1980).
   - Continuous variation of porosity was examined only when the particles are in suspension.
   - A general model for the flow through periodic porous media has been advanced by Brenner (1980) (see also Balint and Balint 2007). In fact, Brenner showed how Darcy's experimental law and the permeability tensor can be computed from a canonical boundary value problem in a standard (representative) cell.
   - Lee et al. (1996) derived averaged permeability of a periodic porous media.

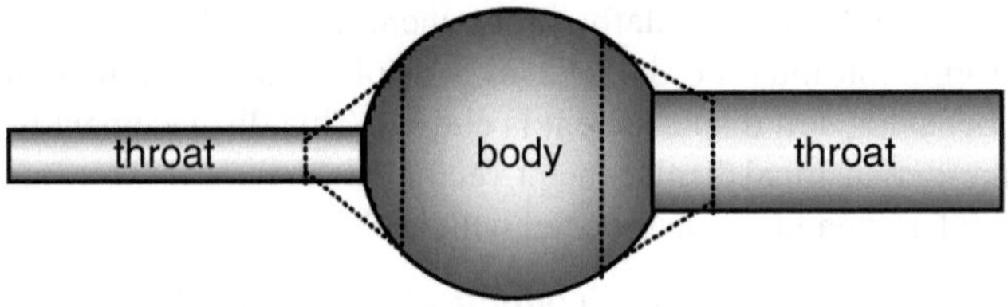

Fig. 3.2 A typical pore network

Table 3.1 Classifications of a particle size

Classification	Grain diameter, mm
Gravel	> 2.0
Sand	0.5–2.0
Silt	0.002–0.5
Clay	<0.002

3. Pore networks

   A pore network can be represented as a series of pore bodies connected by pore throats of various diameters (Fig. 3.2).

   This model shows the influence of pore connectivity, including the possibility of trapping of the wetting phase. However, this trapping does not represent the observed residual saturation in soils. This model can be used to investigate the influence of interfacial area on topics such as inter-phase mass transfer (see Ferre 2005; Snyder and Stewart 1966). Also it may be used to understand the roles of microstructures during shrinkage and/or swelling processes of soils (Audiguier et al. 2010).

   The limitations to this model are the inability to model transient processes and questions as to the applicability of the results to larger scales (larger REVs).

4. General porous media

   Generally, a porous medium can be defined as anything that is comprised of a solid matrix and voids. For the porous media of greatest interest in hydrogeology, the matrix solids include native rock, deposited sediments (gravel, sand, silt or clay), and organic materials. These solids may be cemented, aggregated, or unconsolidated.

While there is no single standard definition for grain size classification, the size classes shown in Table 3.1 are a good guideline (see Ferre 2005).

A porous medium is characterized by the mixture of particle sizes that comprise the matrix. This can be described as simple fractions of gravel, sand, silt, and clay, or more precisely as a particle size distribution (PSD), which defines the cumulative fraction of the solids that is smaller than a given grain diameter.

The **porosity**, n $[L^3L^{-3}]$, of a porous medium is defined as the volumetric fraction of the medium that is occupied by the voids:

$$n = \frac{V_v}{V_T} = \frac{V_v}{V_s + V_v}, \tag{3.2a}$$

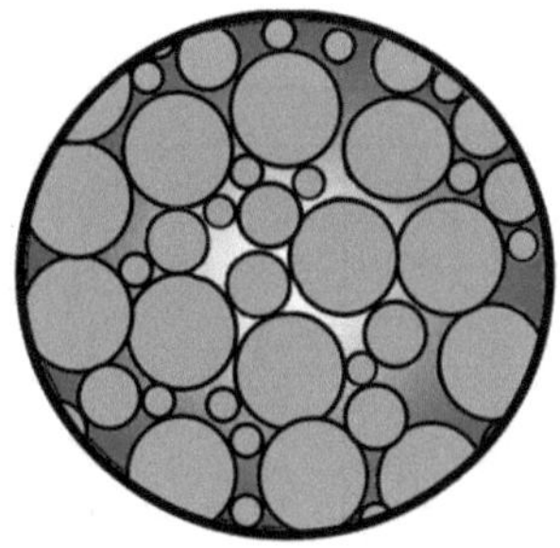

Fig. 3.3 A cross-section of porous medium

where V_v, V_s and V_T are the volumes of the voids, matrix solids, and total medium, respectively. The porosity also defines the average fraction of a cross section through a medium that is occupied by voids (Fig. 3.3). By definition, the porosity can only range from 0 to 1. (Note that ϕ is also commonly used to denote porosity, but n will be used in these notes.) Typical porosity values [m^3m^{-3}] for unfractured, unconsolidated geologic media are: gravel (0.25–0.40); sand (0.25–0.50); silt (0.35–0.50); and clay (0.40–0.70) (see Freeze and Cherry 1979).

The fraction of the area occupied by voids is equal to the porosity.

The **primary porosity** of a medium occurs between the solid particles. The **secondary porosity** is caused by fracturing, root holes, animal or insect burrows, and dissolution. The primary porosity of a medium depends on the structure of the matrix, including the shapes of the solid particles, the distribution of the particle sizes, the degree of compaction of the medium, and, to a lesser degree, the average size of the solid particles (Fig. 3.4). **Sorting** describes the range of sizes of the matrix solids. In a perfectly sorted medium, all of the matrix particles have the same diameter. The more poorly sorted a medium, the more efficiently the particles can be packed, leading to a decrease in the porosity of the medium.

Some of the pore space (such as intra-granular porosity) is not connected. The **effective porosity**, n_e [L^3 L^{-3}], describes only that porosity that is connected, allowing for fluid flow through it and is defined as:

$$n_e = \frac{V_{vc}}{V_T}, \tag{3.2b}$$

V_{vc} is the volume of the connected voids.

The **void ratio** of a medium, e [L^3 L^{-3}], can also be described as:

$$e = \frac{V_v}{V_s}. \tag{3.3}$$

Unlike the porosity, the void ratio can be greater than one. This parameter is used more often in soil engineering.

Water, gases, and dissolved compounds travel predominantly through soil voids. Note from Fig. 3.4 that the shapes and distributions of the pore spaces change with

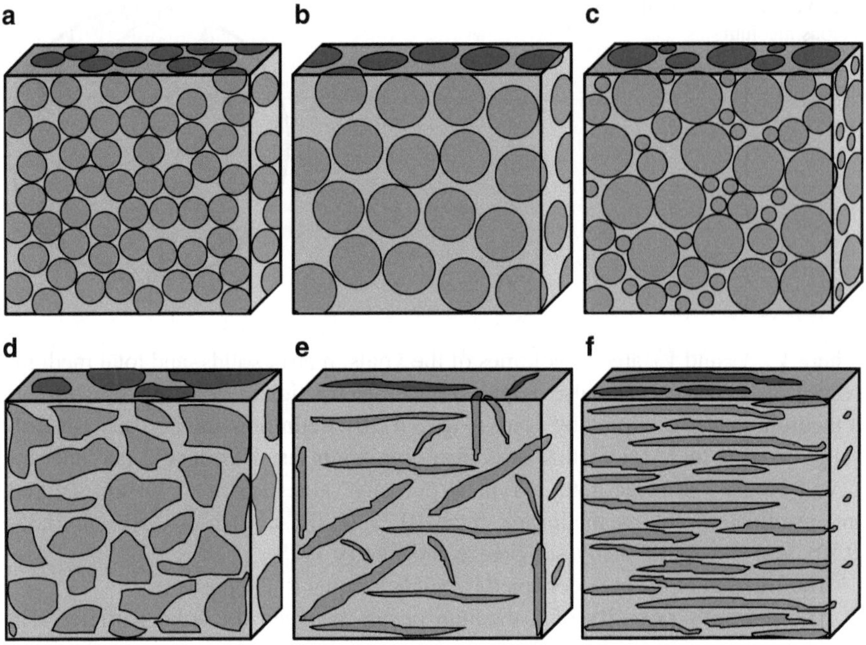

Fig. 3.4 Typical grain packings showing: (**a** and **b**) well sorted, well rounded grains; (**c**) poorly sorted, well rounded grains; (**d**) moderately well sorted, angular grains; (**e**) platy uncompressed grains; and (**f**) platy compressed grains

the packing, shape, and degree of compaction of the soil particles. This clearly impacts the average length of the path that a water, gas, or solute particle must take to travel between any two points that are separated by a fixed distance. The ratio of the average travel path to the distance of separation is known as the **tortuosity**, τ [LL^{-1}]. A medium with a high tortuosity has a more convoluted, less well-connected pore space.

Many important physical and chemical processes take place at the surface of the solid particles. The **specific surface area** [$L^2\ L^{-3}$] describes the area of the interface between the matrix solids and the void space per unit volume of porous medium. The value of this property increases with the roughness of the surface of the particles and with a decrease in the size of the particles.

The void spaces in a porous medium can be filled with a variety of fluids. Commonly, water and air fill most of the pore space. But other fluids, including gases and fluids that do not dissolve in water, may also be present (e.g., petroleum). The **volumetric water content**, θ [$L^3\ L^{-3}$], of a medium is defined as the ratio of the volume of water, V_w, to the total volume of a sample, V_T:

$$\theta = \frac{V_w}{V_T}. \tag{3.4}$$

The maximum volumetric water content of a medium is equal to the porosity. The fraction of pore spaces that are filled with water is known as the **water saturation**, S_w [−], and is defined as the ratio of the volumetric water content to the porosity:

$$S_w = \frac{\theta}{n}. \tag{3.5}$$

Fluid contents and fluid saturations can be calculated in a similar manner for any fluids within the soil pores.

Fluid contents can also be defined on a mass basis. For instance, the **gravimetric water content**, θ_g [M M^{-1}], of a medium is equal to the ratio of the mass of water in a sample, M_w, to the mass of the oven-dried soil, M_s:

$$\theta_g = \frac{M_w}{M_s}. \tag{3.6}$$

The **bulk density**, ρ_b [M L^{-3}], of a porous medium is equal to the ratio of the total mass of a sample to its total volume. Considering a medium with only air and water in the pores, the bulk density is then equal to the following composite of the densities of the medium components:

$$\rho_b = \frac{M_T}{V_T} = (1-n)\,\rho_s + \theta\rho_w + (n-\theta)\,\rho_a. \tag{3.7}$$

The **dry bulk density**, ρ_{bd} [M L^{-3}], of a medium is defined as the mass of the soil solids in a sample divided by the sample volume:

$$\rho_{bd} = \frac{M_s}{V_T}. \tag{3.8}$$

Combining Eqs. (3.4), (3.6), and (3.8) shows that the volumetric water content can be related to the gravimetric water content as:

$$\theta_g = \theta\frac{\rho_w}{\rho_{bd}} \tag{3.9}$$

For most soils, the dry bulk density is greater than one, so the gravimetric water content is lower than the volumetric water content. Exceptions include media with very high porosities and very low particle densities, such as peat.

The **compressibility**, α [L^{2}M^{-1}L^{-1}T^2], of a medium is defined as the change in volume of a unit volume of medium under a unit applied pressure. Physically, as a medium compresses, it experiences a reduction in the volume (and consequent increase in density) of the solids and/or a reduction in porosity. Both changes lead to an increase in the bulk density of the medium.

In addition to the properties of the medium solids, three fluid properties are of primary importance in the study of subsurface hydrogeology. The **fluid**

Table 3.2 Typical permeability values for common porous media

Classification	Permeability, Darcy (9.87×10^{-9} cm^2)
Gravel	10–1,000
Sand	1–100
Silt	0.001–0.1
Clay	0.000001–0.001

density ρ_f [ML^{-3}] (ρ_w for water density) is the ratio between the mass of a fluid sample to the volume of the sample. The **dynamic viscosity**, μ [ML^{-1}T^{-1}], describes the resistance to flow presented by a flowing fluid:

$$\mu = \frac{\sigma}{dv/dy}, \tag{3.10}$$

where σ is the shear stress applied to the fluid, y is the distance perpendicular to a solid surface, and v is the velocity parallel to the surface. The third fluid property of interest, **fluid compressibility**, β [L^{2}M^{-1}L^{-1}T^2], describes the resistance of the fluid to changing its volume in response to a change in the applied pressure per unit volume of fluid:

$$\beta = -\frac{1}{V_0}\frac{dV}{dP}, \tag{3.11}$$

where V_0 is the sample volume before compression. For an **incompressible** fluid, $\beta = 0$. Values of β and α are compared in Table 2.5 of Freeze and Cherry (1979).

The two mechanisms contributing to the storage capacity of a medium are combined as the **specific storage**:

$$S_s = \rho g\,(\alpha + n\beta)\,. \tag{3.12}$$

The specific storage represents the volume of water released per unit volume of porous medium per unit decrease in the hydraulic head and has the units [L^{-1}]. Comparing the value of β (3.84×10^{-10} m^2/N) to typical values for α (gravel, 10^{-10} to 10^{-8}; sand, 10^{-9} to 10^{-7}; clay, 10^{-8} to 10^{-6} m^2/N) shows that the majority of the water released from storage in a confined aquifer comes from compaction of the aquifer material, except in clean gravel aquifers. The reported values of specific storage show that 1 m^3 of water pumped from a well requires that the hydraulic head must be decreased by 1 m throughout 100 m^3 of clay or throughout 10,000 m^3 of sand.

One of the main focuses of the field of hydrogeology is the study of the movement of water through porous media. Therefore, a key characteristic of a medium is its **permeability**, k [L^2], which describes the ability of the medium to transmit a fluid under an applied potential gradient. See Freeze and Cherry (1979) and Tables 3.1 and 3.2 for further typical permeability values.

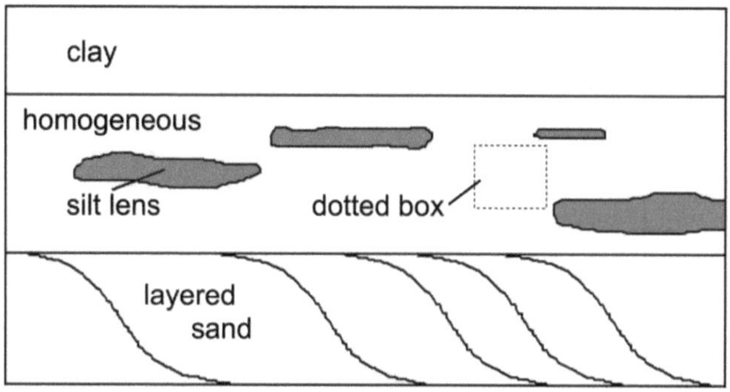

Fig. 3.5 A structured porous medium

Fluid flowing through a porous medium can be visualized most simply as a fluid flowing through a collection of interconnected tubes. At the walls of each tube, the fluid velocity is zero. The velocity increases with distance away from the tube walls as the "drag" from the walls exerts less influence on the fluid. In a smaller tube, a larger fraction of the fluid is adjacent to the walls and the maximum distance from the walls is smaller; therefore, the average velocity in the tube is lower. In a porous medium, the average diameter of the "tubes" through the void spaces is directly proportional to the average pore diameter, which is directly related to the average particle diameter. Given that the ability to conduct a fluid is directly proportional to the average pore diameter, the permeability [L^2] should be directly related to the average particle diameter [L]. Dimensional analysis suggests a relationship of the form:

$$k = Cd^2, \tag{3.13}$$

where d is the average grain diameter and C is a unitless constant. The Kozeny-Carmen equation (see Bear 1972 for a further development) replaces the constant, C, with an expression that depends on the porosity:

$$k = \frac{n^3}{(1-n)^2} \frac{d_m^2}{180}, \tag{3.14}$$

where the value of d_m can be determined from a particle size distribution (PSD) plot.

In a natural environment, the properties of the porous medium will vary in space. In a **homogeneous** medium, the porous medium properties described above are the same regardless of the location from which a sample is collected. In a **heterogeneous** medium, the medium properties vary with the location of sampling. For example, the dotted box on Fig. 3.5 only includes homogeneous sand. If the box were located farther to the right or to the left, it would include a portion of a silt lens and the average medium properties would be different. Because the

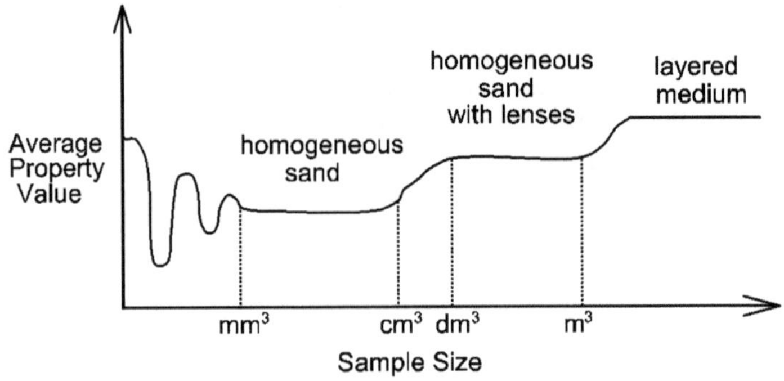

Fig. 3.6 Variability of any porous medium property within a sample as a function of the sample size

porous medium properties (e.g. the porosity or permeability) of a sample taken from the box depend on the location of the sample, the medium is considered to be heterogeneous. Similarly, in the underlying layered sand the average properties of a sample may depend on the location of the sample within each layer and on which layer is sampled. A closer examination of the medium within the dashed box would show small-scale variations in the exact distribution of particles and voids. As the box becomes very small, on the order of the grain size, it may only contain a void or a soil particle, depending upon its location. Consider a series of samples represented by a box that continually increases in size. The heterogeneity within the sample as a function of the size of the box (scale of measurement) can be shown schematically in Fig. 3.6.

In general, both the value and the variability of a porous medium property depend upon the scale of measurement. The term **representative elementary volume** (**REV**) has been used to describe a scale at which the measured porous medium properties are representative of the porous medium. An example of this may be a measurement made in one of the "flat" regions in Fig. 3.6. However, it is evident from Fig. 3.6 that there may not be a single REV for a given medium. Rather it is better to consider a medium at a macroscopic scale and, ideally, to make measurements at a scale that is appropriate for a specific application. Despite this limitation of the concept of an REV, this term will be used in these notes as a convenient way to describe some volume within which medium properties are assumed to be constant.

Some properties can vary depending upon the direction of measurement. For instance, consider the porous medium shown in the lower right frame of Fig. 3.4. The tortuosity of this medium is clearly higher in the vertical direction than in the horizontal. Therefore, it is likely that fluid could flow more easily in the horizontal direction than in the vertical because the fluid does not have to flow "around" the flat plates arranged in the horizontal plane. A medium that has directionally dependent properties is said to be **anisotropic**. A medium that has properties that are directionally independent is **isotropic**.

3.2 Classification of the Different Types of Subsidence Based on Mechanics

3.2.1 Generality on Fluid Mechanics

3.2.1.1 General Fluid Mechanics

These equations may be derived using a Lagrangian approach or an Eulerian approach (see Fowler 1998).

In the Lagrangian approach, the governing equations for the motion of fluid particles are written following a fixed set of fluid particles (e.g., a cloud, a tornado, tip vortices from an aircraft). This is somewhat like tracking satellites and missiles in space, using equations to describe their position in space and the forces acting on them.

In the Eulerian approach, the motion equations are developed for a (usually) fixed or (sometimes) moving volume (called "the control volume") in space surrounded by permeable boundaries. The equations describe the fluid motion inside the control volume as new fluid enters, and old fluid particles leave. The Eulerian approach is the preferred approach in most fluid dynamics applications and is chosen in the following.

Conservation of Mass (Continuity)

The conservation of mass stems from the principle that mass cannot be created or destroyed inside the control volume. Obviously, in some situations (e.g., nuclear reactions), the mass may be converted into energy.

Let V be a control volume, and let us assume that it and its surface S remain fixed in space. The surface is permeable so that fluid can freely enter in and leave. The continuity equation claims that "the time rate of change of mass within the control volume V equals the rate at which mass enters V through the boundary S".

Let us assume that the control volume V is made of several infinitely small (infinitesimal) volume elements dV. The mass of the fluid inside each of these elements is ρdV, where ρ is the fluid density. The density is free to change from point to point, from one sub-element dV to another within V. Thus,

$$\text{Total mass within V} = \iiint_{V} \rho dV \tag{3.15a}$$

In the above equation, the three integral signs indicate a three-dimensional integration, or volume integration. The subscript V says that this integration takes place inside V. Then,

$$\textit{Time rate of change of mass within the control volume } V = \frac{d}{dt}\iiint\limits_{V} \rho \mathrm{dV} \tag{3.15b}$$

In calculus, the order of integration or differentiation may be interchanged so that these operations do not interfere with each other. Since our control volume is fixed in space, the limits of the volume integral are not functions of time, and there is no interaction between the two operations. Thus,

$$\textit{Time rate of change of mass within the control volume } V = \iiint\limits_{V} \frac{\partial \rho}{\partial t} \mathrm{dV} \tag{3.16}$$

Notice that we are using the partial derivative inside the integral since ρ is a function of (x, y, z, t) and we are only interested in its variation inside each sub-element dV with respect to time, while the spatial location (x, y, z) of the sub-element dV remains fixed.

We next turn our attention to the amount of fluid that enters V through the surface S. For this purpose, we assume that the surface S is made of many quilt-like infinitesimal patches dS. At the center of each element is a unit normal vector (i.e., a vector of length unity, normal to the surface) $\vec{n}$. By common convention, this normal is pointing away from the surface dS. The normal component of fluid velocity pointing towards the control volume (entering the control volume) is given by $-\vec{V} \bullet \vec{n}$. Notice the negative sign. We are interested in the component of velocity pointing towards the control volume, not away from it.

Let us compute the rate at which mass enters the control volume through dS as the product of density times normal velocity times area. Thus, the rate at which mass enters the control volume is given by:

$$\mathrm{dS} = -\rho\vec{\mathrm{V}} \bullet \vec{\mathrm{n}}\,\mathrm{dS} \tag{3.17a}$$

Then, the rate at which mass enters the control volume through the entire surface is given by:

$$S = -\oiint\limits_{S} \rho\vec{\mathrm{V}} \bullet \vec{n} dS \tag{3.17b}$$

In (3.17b), a two-dimensional integration is performed over a surface; thus, the two integrals. The little circle over the integral signs tells us that our surface is a closed surface. The subscript S just reminds us that the integration is over the surface S.

Equating the relation (3.16), which is the left-hand side of continuity with the relation (3.17b), which represents the right-hand side of the continuity equation, we get:

$$\iiint\limits_{V} \frac{\partial \rho}{\partial t} dV = -\oiint\limits_{S} \rho \vec{V} \bullet \vec{n} dS \tag{3.18a}$$

Bringing the expression on the right hand side to the left, we get the conservation of mass equation in integral form:

$$\iiint\limits_{V} \frac{\partial \rho}{\partial t} dV + \oiint\limits_{S} \rho \vec{V} \bullet \vec{n} dS = 0 \tag{3.18b}$$

Note that Eq. (3.18b) holds for a steady or unsteady, viscous or in viscid, compressible or incompressible, two-dimensional or three-dimensional flow. It may be used in all speed regimes from flow around a submarine in water to hypersonic flow over the space shuttle during reentry.

The Divergence Theorem

From calculus, one can derive the following equation (which has nothing directly to do with fluid mechanics) called the divergence theorem. Let $\vec{F}$ be any three-dimensional vector, which is a general function of (x, y, z, and t). Then, the divergence theorem applied to a control volume V surrounded by a surface S states:

$$\oiint\limits_{S} \vec{F} \bullet \vec{n} dS = \iiint\limits_{V} \vec{\nabla} \bullet \vec{F} dV \tag{3.19}$$

The inverted triangle $\vec{\nabla}$ is called a "del" operator. It is a collection of derivatives in the x-, y- and z- directions arranged to form a vector: In a Cartesian coordinate system, this operator is given by:

$$\vec{\nabla} = \vec{i}\frac{\partial}{\partial x} + \vec{j}\frac{\partial}{\partial y} + \vec{k}\frac{\partial}{\partial z} \tag{3.20}$$

In a cylindrical coordinate system, this operator is given by

$$\vec{\nabla} = \vec{e}_z\frac{\partial}{\partial z} + \vec{e}_r\frac{\partial}{\partial r} + \vec{e}_\theta\frac{1}{r}\frac{\partial}{\partial \theta} \tag{3.21}$$

When this operator operates on a function g(x, y, z, and t), the resulting vector is called the gradient of g, or simply "del g". The gradient of a scalar in a Cartesian coordinate system is:

$$\vec{\nabla} g = \vec{i}\frac{\partial g}{\partial x} + \vec{j}\frac{\partial g}{\partial y} + \vec{k}\frac{\partial g}{\partial z} \tag{3.22}$$

Since the "del" operator is a vector, it can act on other vectors giving a dot product or cross product. The dot product between $\vec{\nabla}$ and $\vec{F}$ is called a divergence of $\vec{F}$, and the cross product between $\vec{\nabla}$ and F is called a curl of $\vec{F}$. These operations are performed as follows:

$$Let \quad \vec{F} = F_1\vec{i} + F_2\vec{j} + F_3\vec{k} \tag{3.23a}$$

Then,

$$\textit{Divergence of } \vec{F} = \vec{\nabla} \bullet \vec{F} = \frac{\partial F_1}{\partial x} + \frac{\partial F_2}{\partial y} + \frac{\partial F_3}{\partial z} \tag{3.23b}$$

$$\textit{Curl of } \vec{F} = \vec{\nabla} \times \vec{F} = \begin{vmatrix} \vec{i} & \vec{j} & \vec{k} \\ \frac{\partial}{\partial x} & \frac{\partial}{\partial y} & \frac{\partial}{\partial z} \\ F_1 & F_2 & F_3 \end{vmatrix} \tag{3.24}$$

We have shown these operations in a Cartesian coordinate system. Similar derivation may be derived in the cylindrical coordinate system or radial coordinate system.

Applying Eq. (3.19) to the surface integral in the continuity Eq. (3.18) one can get:

$$\iiint_V \frac{\partial \rho}{\partial t} dV + \iiint_V \vec{\nabla} \bullet \left(\rho\vec{V}\right) dV = 0 \tag{3.25}$$

or,

$$\iiint_V \left[\frac{\partial \rho}{\partial t} + \vec{\nabla} \bullet \left(\rho\vec{V}\right)\right] dV = 0 \tag{3.26}$$

Equation (3.25) holds for any arbitrarily shaped control volume V, at any instance in time for all flows. Then, necessarily the integrand is zero:

$$\frac{\partial \rho}{\partial t} + \vec{\nabla} \bullet \left(\rho\vec{V}\right) = 0 \tag{3.27}$$

Equation (3.27) is called the differential form (PDE form) of a continuity equation.

For steady flows, the time derivative vanishes. For incompressible flows, r is a constant. Thus, the continuity equation for incompressible flows becomes:

$$\vec{\nabla} \bullet \left(\vec{V}\right) = 0 \tag{3.28}$$

3.2.1.2 Conservation of Momentum

The u-momentum equation is an extension of Newton's law which states that "the rate of change of momentum of a particle (i.e., a system with fixed mass) is equal to the force acting on the particle." For open systems, i.e., control volumes, the mass may change with time as a result of fluid entering or leaving the control volume. We therefore say

> The rate of change of the u-momentum of particles within a control volume = forces along the x-direction acting on the control volume + net rate at which the u-momentum enters the control volume through the boundaries.

A similar volume integral derivation leads to the following equation (see White 1991):

The u-momentum equation in PDE form thus becomes:

u-momentum equation:

$$\rho\left(\frac{\partial u}{\partial t} + u\frac{\partial u}{\partial x} + v\frac{\partial u}{\partial y} + w\frac{\partial u}{\partial z}\right) + \frac{\partial p}{\partial x} = 0 \tag{3.29a}$$

By a similar process, the v- and w-momentum equations become:

v-Momentum Equation:

$$\rho\left(\frac{\partial v}{\partial t} + u\frac{\partial v}{\partial x} + v\frac{\partial v}{\partial y} + w\frac{\partial v}{\partial z}\right) + \frac{\partial p}{\partial y} = 0 \tag{3.29b}$$

w-Momentum equation:

$$\rho\left(\frac{\partial w}{\partial t} + u\frac{\partial w}{\partial x} + v\frac{\partial w}{\partial y} + w\frac{\partial w}{\partial z}\right) + \frac{\partial p}{\partial z} = 0 \tag{3.29c}$$

These equations may be summarized in a vectorial form as:

$$\rho\frac{\partial \vec{u}}{\partial t} + \rho\left(\vec{u}\,\vec{\nabla}\right)\cdot\vec{u} = -\nabla p \tag{3.30}$$

When the fluid is viscous and external forces are present, this last equation is known as the Navier stokes equation (White 1991):

$$\frac{\partial \vec{u}}{\partial t} + \left(\vec{u}\,\vec{\nabla}\right) \cdot \vec{u} = -\frac{\vec{\nabla} p}{\rho} + \mu \cdot \Delta \vec{u} + \vec{f} \tag{3.31}$$

μ is the viscosity coefficient and $\vec{f}$ the external forces.

3.2.2 *Effective Stress and Water Pressure in Saturated and Unsaturated Soils*

3.2.2.1 The Principle of Effective Stress in Saturated Soils

Consolidation refers to the process of increasing the density of a saturated soil by applying external stress to "squeeze out" the water. There are two major issues:

- for a particular stress increment, *how much compression* will result; and
- for a particular stress increment, *how long will it take* for this compression to occur.

The ***amount*** of compression depends on the ***compressibility*** of the soil, and the ***time*** required depends on the ***consolidation*** properties.

We shall describe the process using the spring/piston/water *analog* (a mechanical model). The soil structure or *soil skeleton* is represented by a spring and all the forces involved are applied in the vertical direction $\vec{k}$ (See Fig. 3.7a).

At all stages, the applied load in the vertical direction is shared between the spring and the water. The equilibrium of the piston requires that:

$$\vec{P} \cdot \vec{k} = \vec{F} \cdot \vec{k} + pA \tag{3.32a}$$

Where $\vec{P}$ is the applied force; $\vec{F}$ is the force taken by the spring; p is the pressure in the water; and A is the cross-sectional area of the piston. In soil mechanics, the water pressure is usually denoted by the letter u instead of P. Equation (3.32a) can be re-written as:

$$\vec{F} \cdot \vec{k} = \vec{P} \cdot \vec{k} - uA \tag{3.32b}$$

Equation (3.32b) states that the force in the spring (soil) is the applied force less the force due to the pore water.

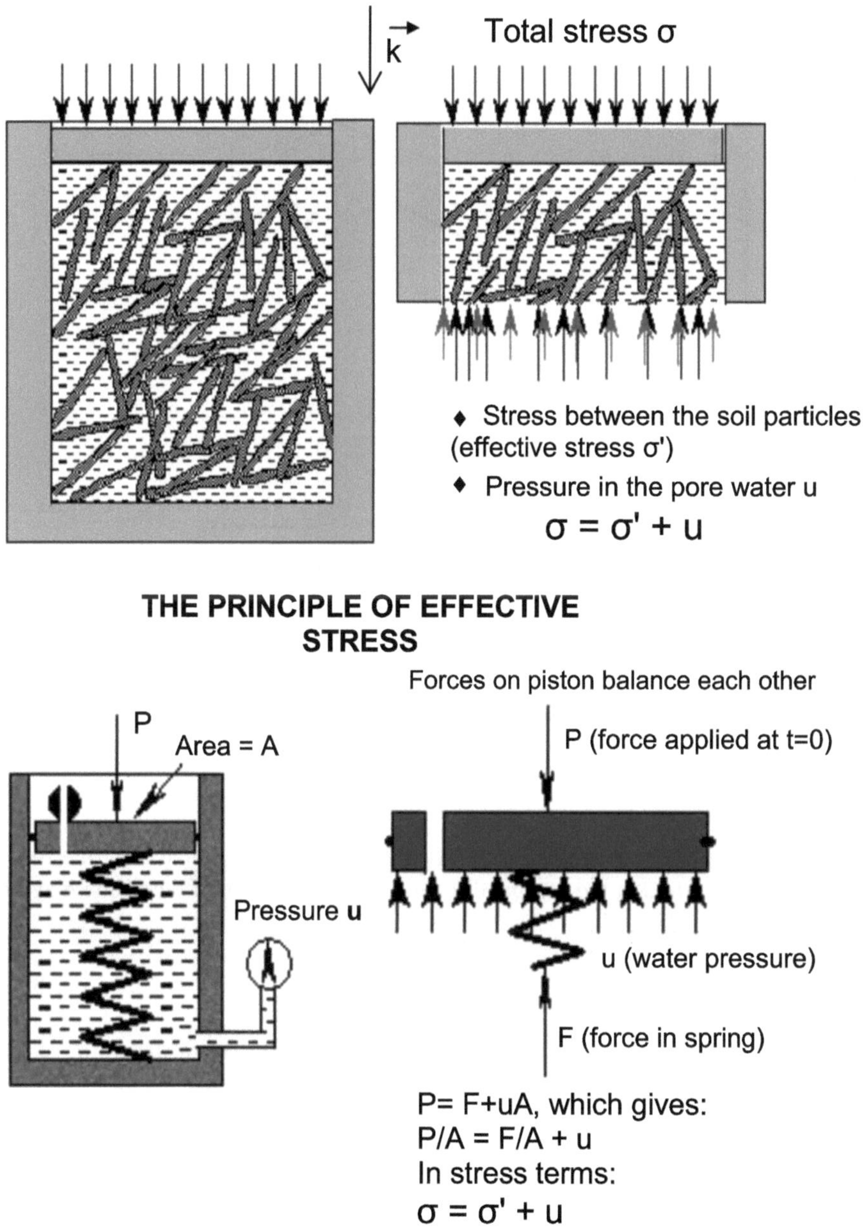

Fig. 3.7 (**a**) The principle of effective stress. (**b**) Stresses components on faces of "stress cube" aligned with reference frame axes. (**c**) Component stress reversals on going from positive to negative faces in the ox, oy plane

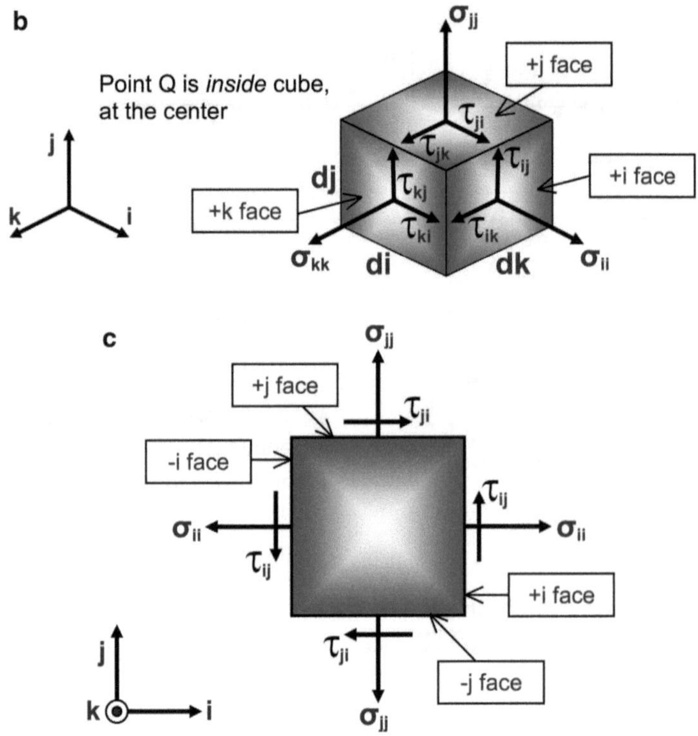

Fig. 3.7 (continued)

If we treat all forces as being "spread over" the cross-sectional area, the force per unit area is *stress* in the vertical direction, the equilibrium of the piston may be expressed as:

$$\sigma'_v = \sigma_v - u \tag{3.32c}$$

Where σ'_v is the stress carried by the soil skeleton (the ***effective stress***), σ_v is the ***total stress*** applied to the soil, and **u** is the ***pore water pressure*** (or just the *pore pressure*).

Equation (3.32a) is called the ***Principle of Effective Stress***. In Hueckel (1992), this principle is discussed for clays soils.

3.2.2.2 The State of Stress

Equation (3.32c) is derived for forces acting in the vertical direction. The generalization of the equation for forces acting in any directions requires the definition of the state of stress in the soil.

Consider a unit surface in the X-Y plane of area $A = dx \cdot dy$, the normal to this surface is in the positive Z direction $\vec{k}$. A given force at point Q acting on this surface has three components $\vec{F} = F_x\vec{i} + F_y\vec{j} + F_z\vec{k}$. The vectors $(\vec{i}; \vec{j}; \vec{k})$ corresponds to the reference frame (Ox,Oy,Oz). These force components per unit area A corresponds to the state of stresses acting on the surface A. By definition, the stresses components

$$\sigma_{zz} = \lim_{A \to 0} \frac{F_z}{A} \tag{3.33a}$$

$$\tau_{zy} = \lim_{A \to 0} \frac{F_y}{A} \tag{3.33b}$$

$$\tau_{zx} = \lim_{A \to 0} \frac{F_x}{A} \tag{3.33c}$$

The stress component σ_{zz} is acting in the direction normal to the surface. However, the two others components τ_{zy} and τ_{zx}, called shear stress, are acting inside the surface. In Fig. 3.7b, all the different components of the stress tensor are described using the convention of positive stress in the principles axes directions.

Note that stresses are forces per unit area. On Fig. 3.7c, the principal stress components and the shear stresses are shown on the three positive faces (ox direction, oy direction) and the three opposite negative faces.

All the nine components of stresses may be written as a tensor called the stress tensor $\overline{\sigma}$:

$$\overline{\sigma} = \begin{bmatrix} \sigma_{xx} & \tau_{xy} & \tau_{xz} \\ \tau_{yx} & \sigma_{yy} & \tau_{yz} \\ \tau_{zx} & \tau_{zy} & \sigma_{zz} \end{bmatrix} \tag{3.34}$$

From moment equilibrium conditions on the infinitesimal stress cube it may be shown that (See Germain and Muller 1995):

$$\tau_{xy} = \tau_{yx}; \tau_{xz} = \tau_{zx}; \tau_{yz} = \tau_{zy}; \tag{3.35}$$

Property (3.35) is known as shear stress reciprocity. It follows that the stress matrix (3.34) is symmetric and only six independent components exists:

$$\overline{\sigma} = \begin{bmatrix} \sigma_{xx} & \tau_{xy} & \tau_{xz} \\ \tau_{xy} & \sigma_{yy} & \tau_{yz} \\ \tau_{xz} & \tau_{yz} & \sigma_{zz} \end{bmatrix} = \begin{bmatrix} \sigma_{xx} \\ \sigma_{yy} \\ \sigma_{zz} \\ \tau_{yz} \\ \tau_{zx} \\ \tau_{xy} \end{bmatrix} \tag{3.36a}$$

In solid mechanics the shear stress is denoted with the letter τ, like in Eq. (3.36a). However, fluid mechanics, only the letter σ is used.

$$\overline{\sigma} = \begin{bmatrix} \sigma_{xx} & \sigma_{xy} & \sigma_{xz} \\ \sigma_{xy} & \sigma_{yy} & \sigma_{yz} \\ \sigma_{xz} & \sigma_{yz} & \sigma_{zz} \end{bmatrix} = [\sigma_{ij}]_{\{i=1,2,3;j=1,2,3\}} = \begin{bmatrix} \sigma_{11} & \sigma_{12} & \sigma_{13} \\ \sigma_{21} & \sigma_{22} & \sigma_{23} \\ \sigma_{31} & \sigma_{32} & \sigma_{33} \end{bmatrix};$$

$$\sigma_{ij} = \sigma_{ji} \tag{3.36b}$$

3.2.2.3 The Principle of Effective Stress in Matrix Form

The deformation of porous medium solids containing water within its voids is defined by the principle of effective stress (Terzaghi 1943), which states that the total stress, $\overline{\sigma}$, consists of the sum of the effective stress (inter-granular stress that is exerted within the skeleton grains), $\overline{\sigma'}$, and the pore pressure, P, as

$$\overline{\sigma} = \overline{\sigma'} + p\overline{I}$$

$$\overline{I} = \begin{bmatrix} 1 & 0 & 0 \\ 0 & 1 & 0 \\ 0 & 0 & 1 \end{bmatrix} \tag{3.37}$$

For a dual-porosity medium, the fluid pressure within the system is comprised of the pressure within the porous phase, p_P, and the pressure within the fractured phase, p_F (subscripts P and F referring to porous phase and the fractured phase, respectively). Thus, the three-dimensional matrix equation for the state of stress in the porous phase becomes:

$$\overline{\sigma_P}' = \overline{\sigma_P} - p_P\overline{I} \tag{3.38a}$$

and the fractured phase is

$$\overline{\sigma_F}' = \overline{\sigma_F} - p_F\overline{I} \tag{3.38b}$$

At equilibrium, the stresses in the fracture and the porous phases must equal to the total stress within the system, therefore;

$$\overline{\sigma} = \overline{\sigma_P} = \overline{\sigma_F} \tag{3.39}$$

More sophisticated models for stress distribution between soil and fluid in porous media are presented in Shanz (2009).

3.2.2.4 Calculating the Initial Vertical Effective Stress in the Ground

The process of soil compression is highly non-linear (the “spring” load-compression relationship is not linear) – i.e., to work out how much compression we get when we apply a load (resulting in a change in effective stress), we must know the starting value of the vertical effective stress.

Total vertical stress σ_v**:** the weight per unit plan area of everything above a point in the ground, including soil and water.

Initial pore water pressure $\mathbf{u_o}$**:** for “hydrostatic” conditions, this is found by knowing how far below the water table the point is, and knowing that the unit weight of water is γ_w.

The **effective vertical stress** σ'_v is then:

$$\sigma'_v = \sigma_v - \mathbf{u_o}$$

In the example shown in Fig. 3.8, the soil above the water table is assumed to have a unit weight $\gamma = \mathbf{17\ kN/m^3}$, and the saturated soil below the water table to have a unit weight of $\gamma_{sat} = \mathbf{21\ kN/m^3}$. The calculation of the effective vertical stress is shown in Fig. 3.8.

3.2.2.5 The Principle of Effective Stress in Non-saturated Soils

Bishop (1959) suggested that the behavior of an unsaturated soil can be described in terms of a single effective stress which may be written in a tensor form as,

$$\overline{\sigma'} = \overline{\sigma} - (p_a - \chi(p_a - p_w))\overline{I} \tag{3.40}$$

where, $\overline{\sigma'}$ is Bishop’s stress tensor; $\overline{\sigma}$ is the total stress tensor; p_a is pore air pressure; p_w is pore water pressure; χ is a variable often associated with the degree of saturation.

The analysis by Houlsby (1997), for an unsaturated granular material subject to simultaneous deformation and fluid flow, resulted in a form of Bishop’s equation with $\chi = S_r$ (where S_r is the degree of saturation). The analysis relied on the following equilibrium equation for the total stress components in an unsaturated soil:

$$\sigma_{ij} = n[S_r\, p_w + (1 - S_r)\, p_a]\,\delta_{ij} + (1 - n)s_{ij} + T_{ij} \tag{3.41}$$

n is the porosity; s_{ij} is the average stress in the soil grains, T_{ij} represents the forces in the contractile skin, δ_{ij} is the Kronecker delta

Discussion on the validity of such relation has been discussed by Murray (2002).

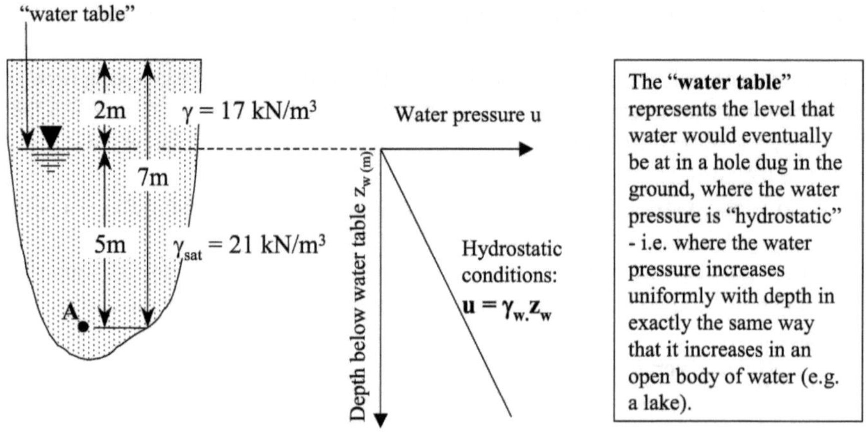

"**Hydrostatic**" means exactly that - "static" - i.e. no flow. That means that the total head is equal at all depths (the pressure head increases with depth, but the elevation head reduces with depth). So, it follows that for vertical flow to occur, the water pressure distribution must be different from hydrostatic. This difference is what we will call later ***excess pore pressure.***

For Point A:

Total vertical stress	$\sigma_v = 2 \times 17 + 5 \times 21$	$= 139$ kPa
Pore (water) pressure	$u = 5 \times 10$	$= 50$ kPa
	$(\gamma_w = 9.81\ \text{kN/m}^3 \approx 10\ \text{kN/m}^3)$	
Therefore, effective vertical stress	σ'_v	$= 89$ kPa

Fig. 3.8 Computation of effective stress

3.2.3 Mass Conservation Equation

As explained before, the flow of any fluid is governed by mass and momentum conservation. In the case of groundwater flow, it may be shown using averaging methods on periodic porous media and/or pore networks, that the Darcy's law expresses the momentum conservation (Eq. 3.30-See Whitaker 1986).

3.2.3.1 Derivation of the Flow Equations

(see Barenblatt et al. 1960; Plotkowiak 2005).
Assume that:

1. No mass transfer between phases,

$$A(y + \Delta y)$$

- No sources or sinks,
- Darcy flow range.

Consider an REV:

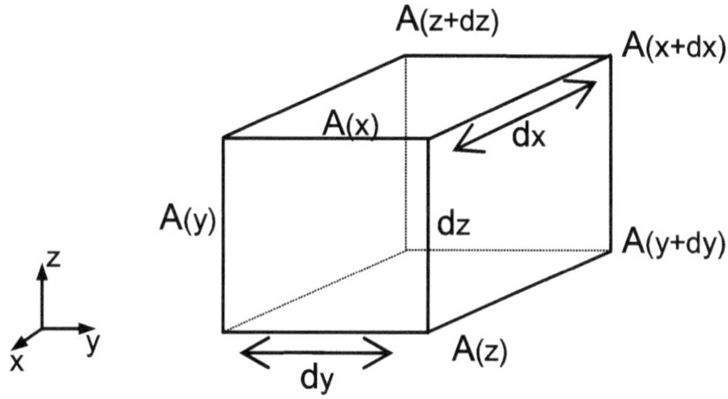

Conservation of mass for each phase requires:

mass rate in − mass rate out = rate of mass accumulation.

The rate of mass accumulation is:

$$\rho qA \text{ with the units, } \left[M/L^3\right]\left[L^3/TL^2\right]\left[L^2\right] = [M/T].$$

Taking the flow to be positive in the positive x, y, and z directions, and summing the net mass accumulation over all six faces gives:

$$\begin{aligned} &\rho(x)q(x)A(x) - \rho(x+\Delta x)q(x+\Delta x)A(x+\Delta x) \\ &\quad + \rho(y)q(y)A(y) - \rho(y+\Delta y)q(y+\Delta y)A(y+\Delta y) \\ &\quad + \rho(z)q(z)A(z) - \rho(z+\Delta z)q(z+\Delta z)A(z+\Delta z) \\ &\quad = \frac{\partial\left(\overline{n\rho}\right)}{\partial t}\Delta x\Delta y\Delta z \end{aligned} \tag{3.42}$$

where $\overline{\phi}$ = average porosity, $[L^3/L^3]$,
$\overline{\rho}$ = average density, $[M/L^3]$,
$\overline{q}$ = volume flux, $[L^3/TL^2]$,
A = cross − sectional area, $[L^2]$.

Dividing the equation by the unit volume of the REV, $\Delta x \Delta y \Delta z$ gives:

$$\frac{\rho(x)q(x)A(x)}{\Delta x \Delta y \Delta z} - \frac{\rho(x+\Delta x)q(x+\Delta x)A(x+\Delta x)}{\Delta x \Delta y \Delta z} + \frac{\rho(y)q(y)A(y)}{\Delta x \Delta y \Delta z} - \frac{\rho(y+\Delta y)q(y+\Delta y)A(y+\Delta y)}{\Delta x \Delta y \Delta z} + \frac{\rho(z)q(z)A(z)}{\Delta x \Delta y \Delta z} - \frac{\rho(z+\Delta z)q(z+\Delta z)A(z+\Delta z)}{\Delta x \Delta y \Delta z} = \frac{\partial\left(\overline{n\rho}\right)}{\partial t} \tag{3.43}$$

Taking the limit as Δx, Δy and Δz approach zero and recognizing that

$$A(x+\Delta x) = A(x) = \Delta y \Delta z, \text{ and}$$

$\lim\limits_{x \to 0} \frac{F(x+\Delta x)-F(x)}{\Delta x} = \frac{\partial F(x)}{\partial x}$, gives:

$$-\frac{\partial(\rho q_x)}{\partial x} - \frac{\partial(\rho q_y)}{\partial y} - \frac{\partial(\rho q_z)}{\partial z} = \frac{\partial(n\rho)}{\partial t} \tag{3.44a}$$

Expanding the derivative terms by the chain rule gives:

$$-\rho\frac{\partial(q_x)}{\partial x} - q_x\frac{\partial(\rho)}{\partial x} - \rho\frac{\partial(q_y)}{\partial y} - q_y\frac{\partial(\rho)}{\partial y} - \rho\frac{\partial(q_z)}{\partial z} - q_z\frac{\partial(\rho)}{\partial z} = n\frac{\partial(\rho)}{\partial t} + \rho\frac{\partial(n)}{\partial t} \tag{3.45}$$

Assume incompressible fluids:

$$\frac{\partial\rho}{\partial x} = \frac{\partial\rho}{\partial y} = \frac{\partial\rho}{\partial z} = \frac{\partial\rho}{\partial t} = 0,$$

The same development can be followed to define a relationship for each fluid in the system, when sink/source are present. In this case, Eq. (3.45) may be written as (Bear 1979):

$$\frac{\partial(\rho n)}{\partial t} = -\rho\nabla\cdot\vec{v}_w + \rho Q_h \tag{3.46}$$

Where ρ the fluid density (kg/m^3) is, n is the porosity (volume of fluid per volume of soil), $\vec{v}_w$ is the fluid flux (m^3/s), and Q_h is the sink/source mass flux (m^3/s).

As the state variable for the flow equation, both the pressure p (Pa) and the equivalent freshwater head h_f (m) can be used.

$$\mathrm{h_f} = \frac{\mathrm{p}}{\rho_0 \mathrm{g}} + \mathrm{z} \tag{3.47a}$$

Assuming that Darcy's law governs the pore water flow, we have

$$\begin{aligned} \mathrm{n}\left(\vec{\boldsymbol{v}}_{\mathrm{w}} - \vec{\boldsymbol{v}}_{\mathrm{s}}\right) &= -\frac{\mathrm{k}}{\gamma_{\mathrm{w}}}\left(\frac{\partial \mathrm{p}}{\partial x}\boldsymbol{i}_{\mathrm{x}} + \frac{\partial \mathrm{p}}{\partial y}\boldsymbol{i}_{y} + \frac{\partial \mathrm{p}}{\partial z}\boldsymbol{i}_{z}\right), \\ &= -\frac{\mathrm{k}}{\gamma_{\mathrm{w}}}\vec{\nabla}\mathrm{p} = -\frac{\rho_0 \mathrm{gk}}{\gamma_{\mathrm{w}}}\vec{\nabla}\mathrm{h_f} \end{aligned} \tag{3.47b}$$

n is the porosity of the porous aquifer; $\vec{\boldsymbol{v}}_w$ and $\vec{\boldsymbol{v}}_s$ are the velocities of pore water and solid matrix, respectively. Notice that when the velocity of pore matrix is nearly zero $\vec{\boldsymbol{v}}_s \approx \vec{0}$ then Darcy' law may be expressed as:

$$\begin{aligned} \vec{\boldsymbol{v}}_{\mathrm{w}} &= -\frac{\mathrm{k}}{\mathrm{n}\gamma_{\mathrm{w}}}\left(\frac{\partial \mathrm{p}}{\partial x}\boldsymbol{i}_{\mathrm{x}} + \frac{\partial \mathrm{p}}{\partial y}\boldsymbol{i}_{y} + \frac{\partial \mathrm{p}}{\partial z}\boldsymbol{i}_{z}\right) \\ &= -\frac{k}{n\gamma_w}\vec{\nabla}\mathrm{p} = -\frac{\rho_0 \mathrm{gk}}{\mathrm{n}\gamma_{\mathrm{w}}}\vec{\nabla}\mathrm{h_f} \end{aligned} \tag{3.47c}$$

k denotes the permeability of the saturated aquifer and γ_w is the unit weight of pore water; the symbols $\vec{\boldsymbol{i}}_x$, $\vec{\boldsymbol{i}}_y$, and $\vec{\boldsymbol{i}}_z$ are unit vectors parallel to the ox, oy and vertical directions, respectively.

In a non isotropic aquifer, the permeability is a tensor $\bar{\mathrm{k}}$. For a saturated groundwater flow, the flux $\vec{\mathrm{q}}$ is given by the general Darcy's law.

$$\vec{\mathrm{q}} = -\bar{\mathrm{k}}\,\vec{\nabla}\,\mathrm{h_f}$$

$$\bar{\mathrm{k}} = \begin{bmatrix} \mathrm{k_{xx}} & \mathrm{k_{xy}} & \mathrm{k_{xz}} \\ \mathrm{k_{xy}} & \mathrm{k_{yy}} & \mathrm{k_{yz}} \\ \mathrm{k_{xz}} & \mathrm{k_{yz}} & \mathrm{k_{zz}} \end{bmatrix} = \begin{bmatrix} \mathrm{k_{xx}} \\ \mathrm{k_{yy}} \\ \mathrm{k_{zz}} \\ \mathrm{k_{yz}} \\ \mathrm{k_{zx}} \\ \mathrm{k_{xy}} \end{bmatrix} \tag{3.47d}$$

$\bar{\mathrm{k}}$ is the intrinsic permeability tensor (m^2).

The change in the mass of water stored per unit volume of the REV per unit time is then:

$$\frac{\partial\,(\mathrm{n}\rho)}{\partial \mathrm{t}} = \rho\frac{\partial \mathrm{n}}{\partial \mathrm{t}} + \mathrm{n}\frac{\partial \rho}{\partial \mathrm{t}}. \tag{3.48}$$

This states that the mass of water stored decreases due to decreases in the porosity of the medium (**compaction**) or a decrease in the density of the fluid (**fluid compression**).

3.2.3.2 The Hydrology Approach: Suppose That $\rho(p)$; $n(p)$

Rather than attempt to measure both the porosity and water density with time, it is more practical to relate changes in compaction and fluid compression to changes in the pressure of the water:

$$\frac{\partial n}{\partial t} = \frac{\partial n}{\partial P}\frac{\partial P}{\partial t}$$
$$\frac{\partial \rho}{\partial t} = \frac{\partial \rho}{\partial P}\frac{\partial P}{\partial t}. \tag{3.49}$$

From Eq. (3.11), the change in the density of water for a given change in pressure is:

$$\beta = \frac{-1}{V_0}\frac{dV}{dP}. \tag{3.50}$$

The **medium compressibility** can be defined as the change in the volume of the medium in response to a change in the pressure exerted on the medium:

$$\alpha = \frac{1}{V_T}\frac{dV_T}{dP} = \frac{1}{V_T}\frac{d\left(V_s + V_v\right)}{dP}, \tag{3.51a}$$

where V_s is the volume of the matrix solids, V_v is the volume of the voids and V_T is the total original volume of the sample. Generally, it can be assumed that the solid grains are incompressible:

$$\frac{dV_s}{dP} = 0. \tag{3.51b}$$

The compressibility per unit (original) volume of porous medium is then,

$$\alpha = \frac{1}{V_T}\frac{dV_v}{dP} = \frac{d\left(\frac{V_v}{V_T}\right)}{dP} = \frac{dn}{dP}. \tag{3.52}$$

Physically, the assumption of incompressible grains means that **the medium compressibility is provided entirely by rearranging of the grains into a more efficient packing.** α is commonly taken to be constant (see Freeze and Cherry 1979, p. 54, for a further discussion).

Equation (3.39) may be written for each components in space. This leads to:

$$-\rho\frac{\partial \mathrm{q}}{\partial \mathrm{x}}-\rho\frac{\partial \mathrm{q}}{\partial \mathrm{y}}-\rho\frac{\partial \mathrm{q}}{\partial \mathrm{z}}=\rho\left(\alpha+\mathrm{n}\beta\right)\frac{\partial \mathrm{P}}{\partial \mathrm{t}}. \tag{3.53}$$

Considering Eq. (3.53), let us estimate whether it is more likely that water will be added to (or removed from) storage by the expansion/contraction of water or of the aquifer solids. The compressibility of water is $3.88\times10^{-10}\ \mathrm{m}^2\ \mathrm{N}^{-1}$. The compressibility of aquifer materials ranges from $2\times10^{-6}\ \mathrm{m}^2\ \mathrm{N}^{-1}$ for plastic clay, to $2\times10^{-8}\ \mathrm{m}^2\ \mathrm{N}^{-1}$ for dense sand, to $3\times10^{-10}\ \mathrm{m}^2\ \mathrm{N}^{-1}$ for solid rock. The compressibility of water is also multiplied by the porosity, which ranges from 0.24 to 0.60 for unconsolidated media. Therefore, it is clear that most changes in water storage occur due to compaction of the medium, not due to the compression of water.

Substituting for the flux using Darcy's equation gives:

$$-\rho\frac{\partial\left(-K_x\frac{\partial H}{\partial x}\right)}{\partial x}-\rho\frac{\partial\left(-K_y\frac{\partial H}{\partial y}\right)}{\partial y}-\rho\frac{\partial\left(-K_z\frac{\partial H}{\partial z}\right)}{\partial z}=\rho\left(\alpha+n\beta\right)\frac{\partial p}{\partial t}. \tag{3.54a}$$

This can be written in a more convenient form using the Einstein notation as:

$$\frac{\partial}{\partial x_i}\left(K_i\frac{\partial H}{\partial x_i}\right)=\left(\alpha+n\beta\right)\frac{\partial P}{\partial t}. \tag{3.54b}$$

Note that this has been developed for an REV that is aligned with the principal directions of the hydraulic conductivity field. That is, for instance, there is no flow in the x-direction due to a hydraulic head gradient in the y-direction.

The pressure can be related to the hydraulic head as:

$$\mathrm{P}=\rho g\mathrm{H}-\rho g z. \tag{3.55}$$

At each location g and z are constant in time. Assuming that ρ is also time invariant, the expression describing the change in the pressure with time simplifies to:

$$\frac{\partial \mathrm{P}}{\partial \mathrm{t}}=\rho g\frac{\partial \mathrm{H}}{\partial \mathrm{t}}. \tag{3.56}$$

Substituting Eq. (3.55) into the flow equation gives:

$$\frac{\partial}{\partial \mathrm{x_i}}\left(\mathrm{K_{ij}}\frac{\partial \mathrm{H}}{\partial \mathrm{x_j}}\right)=\rho g\left(\alpha+\mathrm{n}\beta\right)\frac{\partial \mathrm{H}}{\partial \mathrm{t}}. \tag{3.57}$$

In a more general form using Eq. (3.48):

$$\frac{\partial}{\partial x_i}\left(K_{ij}\frac{\partial H}{\partial x_j}\right) = \rho\frac{\partial n}{\partial t} + n\frac{\partial \rho}{\partial t} \tag{3.58}$$

Equation (3.58) may also be written in terms of the water pressure P:

$$\frac{\partial}{\partial x_i}\left(\frac{K_{ij}}{\rho g}\frac{\partial P}{\partial x_j}\right) = \rho\frac{\partial n}{\partial t} + n\frac{\partial \rho}{\partial t} \tag{3.59a}$$

This is the general form of the flow equation in the principal directions for a medium saturated with a single fluid. There are some simplifications that can be made to this equation for specific applications.

A **steady-state flow** is defined as a flow that occurs with no change in mass stored with time. That is, the flow into each section is equal to the flow out of the section. For this condition, the right-hand side of Eq. (3.59a) equals zero:

$$\frac{\partial}{\partial x_i}\left(K_{ij}\frac{\partial H}{\partial x_j}\right) = 0. \tag{3.59b}$$

For a homogeneous, isotropic medium, the hydraulic conductivity has a single value, regardless of the direction of flow or the location in the medium. Therefore, K can be removed from the derivative giving:

$$\frac{\partial^2 H}{\partial x_i^2} = \frac{\rho g\,(\alpha + n\beta)}{K}\frac{\partial H}{\partial t}. \tag{3.59c}$$

Steady-state flow in a homogeneous and isotropic medium gives the Laplace equation:

$$\frac{\partial^2 H}{\partial x_i^2} = 0. \tag{3.59d}$$

Most quantitative analyses of subsurface water flow and solute transport make use of **boundary value problems**. This method of analysis is based on a definition of the value of some physical property at one point in space (and time) with respect to the values of that property at neighboring locations (and times).

3.2.3.3 A Heat-Water Problem: Suppose That $\rho(p, T)$

When the temperature of the water T is not constant, the water density $\rho(p, T)$ depends not only on the water pressure but also on the temperature T. A linear dependence is generally assumed:

$$\rho(p, c) = \rho_0(p) + \kappa(T - T_0) \tag{3.60a}$$

Where $\rho_0(p)$ is the fresh water density at a reference depth, κ a regression coefficient and T_0 the temperature of the water at the reference depth.

In this case Eq. (3.48) lead to

$$\begin{aligned} \frac{\partial n}{\partial t} &= \frac{\partial n}{\partial P}\frac{\partial P}{\partial t} \\ \frac{\partial \rho}{\partial t} &= \frac{\partial \rho}{\partial P}\frac{\partial P}{\partial t} + \frac{\partial \rho}{\partial c}\frac{\partial c}{\partial t} = \frac{\partial \rho}{\partial P}\frac{\partial P}{\partial t} + \kappa\frac{\partial T}{\partial t} \end{aligned} \tag{3.61}$$

Then the flow Eq. (3.58) is written as:

$$\frac{\partial}{\partial x_i}\left(K_{ij}\frac{\partial H}{\partial x_j}\right) = \rho g\,(\alpha + n\beta)\,\frac{\partial H}{\partial t} + \kappa\frac{\partial T}{\partial t} \tag{3.62}$$

The solution of the above equation requires a heat equation for the temperature field T(see Gray et al. 1976; Thomas et al. 1998).

3.2.3.4 Boundary Conditions Associated with the Flow Equation

There are two primary types of boundary conditions that are used to analyze subsurface water flow. The first type (Type I), also called a Dirichlet boundary condition, defines the value of the hydraulic head along the boundary. Type II flow boundary conditions define the gradient of the primary variable (hydraulic head) and are known as a Neumann boundary condition. The flux can be defined based on the hydraulic conductivity and the hydraulic head gradient using Darcy's equation. If the boundary condition is defined based on the water flux, then the hydraulic conductivity of the medium must be defined at the boundaries to determine the gradient.

It is important to realize that either the hydraulic head or the water flux may be constrained on a boundary, but not both. If the hydraulic head is fixed, then the water flux (or, more accurately, the hydraulic head gradient) will adjust accordingly. Conversely, if the flux is known, the hydraulic head on the boundary will adjust to produce this flux.

It is important to remember that boundary conditions are used to isolate a region from its surroundings. That is, by assigning a boundary condition, you are claiming that regardless of what happens outside of the domain of interest, the value at the boundary will remain fixed. Often, the most challenging aspect of analyzing flow is the definition of boundary conditions that represent the physical flow system. The definition is not as simple for many real-world conditions. Physical features that are often used to define boundary conditions include rivers, lakes, and impermeable

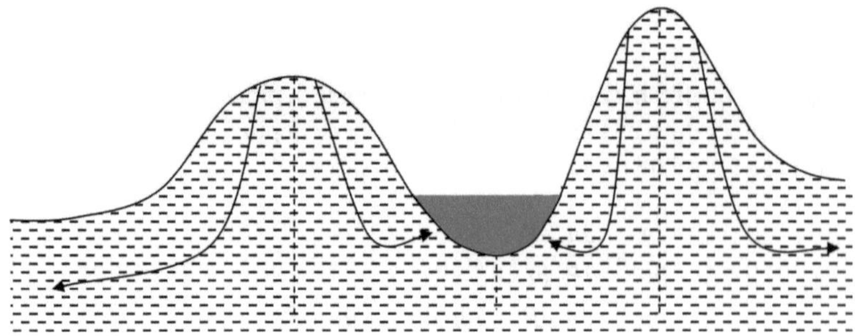

Fig. 3.9 Flow lines in a valley with a lake and topographic divides

boundaries (faults, bedrock contacts). Some of these features represent known flow boundaries, but others are used as best guess alternatives when constraining hydrologic data are incomplete (which they usually are). In some cases, it is not possible to define boundary conditions based on physical features. In these cases, boundaries must be chosen based on an understanding of the effects of such choices on the flow regime being represented. One approach is to model a larger area to define approximate flow lines along boundaries of a smaller domain and then to use these flow lines as no flow boundaries.

A **flow divide** describes a plane in the subsurface across which there is no ground water flow. One example, a **topographic divide**, is shown in Fig. 3.9. Water does not cross this boundary, as shown by the flow lines. As a result, this boundary can be considered to be impermeable. For a plan view analysis, a line along the hilltop can be set as a no flow boundary. A convergent flow boundary beneath a stream at the base of a valley, as shown on Fig. 3.9, can also be represented as a no flow boundary with the assumption that all of the subsurface flow enters the stream; no water flows beneath the stream.

A lake or stream can be used to define a constant head boundary (Fig. 3.10). Commonly, the assumption is made that the water in the lake is under hydrostatic conditions. That is, the hydraulic head throughout the lake is equal to the hydraulic head at the surface of the lake, which is equal to the height of the lake above the datum elevation. Taking the lake-bed to be an equipotential line, the flow lines will be perpendicular to the base of the lake at the top of the hill in Fig. 3.10.

There are some points to clarify concerning boundary conditions. First, a "constant head" or "constant flux" boundary may have a varying value of head or flux along the boundary. Generally, it is clearest to think of a boundary condition being applied at a point along a boundary. The flux may change from point to point, but it has a constant value at each point. In fact, one "boundary" may have Type I and Type II sections. Additionally, the constant value at any point can change in time. Therefore, to be even more precise, a boundary condition is a fixed condition at a point for a fixed time period. In addition, while boundary conditions must be

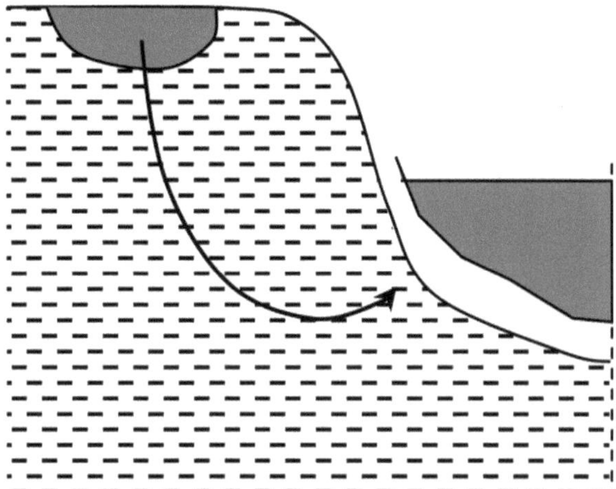

Fig. 3.10 Flow lines between two lakes at different elevations

Table 3.3 Physical attributes used to define flow boundary conditions (after Ferre J (2005))

Boundary type	Physical condition
Impermeable "no-flow"	Bedrock
	Clay layer
	Ground water flow divide
	Geometrical symmetry
	Laboratory column wall
Constant flux	Pumping well
	Precipitation
	Evaporation
	Irrigation
Constant hydraulic head	Surface water body
	Recharge boundary
	Discharge boundary
	Top or bottom of a column

applied around the boundary of a domain, boundary values also may be applied at points that are internal to the domain. For example, when modeling an area in plan view a pumped well can be incorporated at a point within the domain. Specifically, if the pump is operated at a constant rate, a constant flow of water leaves the domain through this internal point, represented as a Type II boundary. Similarly, precipitation and evaporation at the ground surface can be incorporated into plan view models as **distributed sources** to allow for the addition of water to the domain within the domain boundaries.

For reference, Table 3.3 lists some common physical conditions that can be represented using Type I and Type II flow boundary conditions.

3.2.3.5 The Flow Equation for Deformable Soils

The general form of the flow Eq. (3.45) may be written in terms of the water pressure as:

$$-\rho\frac{\partial q_x}{\partial x}-\rho\frac{\partial q_y}{\partial y}-\rho\frac{\partial q_z}{\partial z}=-\rho\nabla\cdot\vec{q}\ =\rho\frac{\partial n}{\partial t}+n\frac{\partial \rho}{\partial t}; \tag{3.63a}$$

$$\vec{q}=-\vec{\nabla}(Kp) \tag{3.63b}$$

Where the operator nabla is used: $\vec{\nabla}=\left(\frac{\partial}{\partial x},\frac{\partial}{\partial y},\frac{\partial}{\partial z}\right)$.

Now, for saturated media, the volume of the REV V_T is equal to the sum of the solid volume and the water volume leading to:

$$V_T=V_s+V_w \tag{3.64}$$

Taking the time derivative of this last relation leads to

$$\frac{\partial V_s}{\partial t}+\frac{\partial V_w}{\partial t}=0 \tag{3.65}$$

Let's define the solid dilation as $\varepsilon_V=\frac{\Delta V_s}{V_s}$; then

$$\frac{\partial \varepsilon_V}{\partial t}=\frac{1}{V_s}\frac{\partial V_s}{\partial t} \tag{3.66}$$

Also the time derivative of the volume of water may be expressed as:

$$n=\frac{V_w}{V_T}\Rightarrow\frac{\partial V_w}{\partial t}=V_T\frac{\partial n}{\partial t}. \tag{3.67}$$

Inserting these last relations into Eq. (3.65) lead to:

$$V_s\frac{\partial \varepsilon_V}{\partial t}+V_T\frac{\partial n}{\partial t}=0\Rightarrow\frac{\partial n}{\partial t}=(1-n)\frac{\partial \varepsilon_V}{\partial t} \tag{3.68}$$

Inserting Eq. (3.68) into the conservation equation leads to another form of the flow conservation equation similar to the Biot equation for water pressure (see Biot 1941).

$$\nabla.\nabla(Kp)=(1-n)\frac{\partial \varepsilon_V}{\partial t}+\frac{n}{\rho}\frac{\partial \rho}{\partial t}; \tag{3.69}$$

The term $\frac{n}{\rho}\frac{\partial \rho}{\partial t}$ is treated as above and in the case where $\rho(p)$ this term is written as: $\frac{n}{\rho}\frac{\partial \rho}{\partial t} = n\left[\frac{1}{\rho}\frac{\partial \rho}{\partial p}\right]\frac{\partial p}{\partial t} = nS_s\frac{\partial p}{\partial t}$ where the parameter $S_s = \left[\frac{1}{\rho}\frac{\partial \rho}{\partial p}\right]$ is called specific storage.

Finally Eq. (3.69) leads to:

$$\frac{1}{n}\nabla.\nabla(Kp) = \frac{(1-n)}{n}\frac{\partial \varepsilon_V}{\partial t} + S_s\frac{\partial p}{\partial t}; \tag{3.70}$$

Similar derivations have been performed by Corapcioglu (1976).

Notice that the derivation of the last equation is based on the Darcy law and the analysis of the matrix deformation. Other approaches have been proposed for the modeling of subsidence and land motion. For example, recently Li and Helm proposed a theory of three-dimensional land motion based on velocity fields of both the solid skeleton and the fluid (Li and Helm 2010).

In order to solve Eq. (3.70), we need two additional equations:

(a) A constitutive equation that relates the strain on the porous structure of the sediment's skeletal frame.
(b) An equation of conservation of the linear momentum that relates the displacement of the matrix to the applied load on it.

These equations and issues will be discussed in the next sections.

Also, Eq. (3.70) cannot be solved for the water pressure. A way of solving this problem is to use a constitutive equation based on linear elasticity:

$\varepsilon_V = G\sigma'_V = G(\sigma_V - p)$, where G is the bulk modulus, σ'_V is the trace of the tensor of effective stress, and σ_V is the trace of the total stress tensor.

Finally Eq. (3.70) leads to:

$$\frac{1}{n}\nabla.\nabla(Kp) = G\frac{(1-n)}{n}\left[\frac{\partial \sigma_V}{\partial t} - \frac{\partial p}{\partial t}\right] + S_s\frac{\partial p}{\partial t}; \tag{3.71}$$

If we assume that the total stress resulting from a loading on the soil is not changed, then Eq. (3.71) is an equation in terms of the water pressure.

$$\frac{1}{n}\nabla.\nabla(Kp) = \left[S_s + G\frac{(n-1)}{n}\right]\frac{\partial p}{\partial t}; \tag{3.72}$$

For secondary consolidation and a Kelvin-Voigt visco-elastic model:

$$\varepsilon_V = G\sigma'_V + \int_{t_0}^{t} K(t-\tau)\sigma'_V(\tau)d\tau = G(\sigma_V - p) + \int_{t_0}^{t} K(t-\tau)(\sigma_V(\tau) - p(\tau))d\tau$$

Then, taking the time derivative of this last expression and inserting it into Eq. (3.71) leads to:

$$\frac{1}{n}\nabla.\nabla(Kp) = G\frac{(1-n)}{n}\left[\frac{\partial \sigma_V}{\partial t} - \frac{\partial p}{\partial t}\right] + K(0)\frac{(1-n)}{n}(\sigma_V(t) - p(t)) + S_s\frac{\partial p}{\partial t}; \tag{3.73}$$

Assume now that $\sigma_V(t) = \sigma_0$ constant, then Eq. (3.73) becomes:

$$\frac{1}{n}\nabla.\nabla(Kp) = \left[S_s + G\frac{(n-1)}{n}\right]\frac{\partial p}{\partial t} + K(0)\frac{(1-n)}{n}(\sigma_0 - p(t)); \tag{3.74}$$

Similar models have been proposed by Miao and Wu (1991).

3.2.4 Stress Conservation of the Soil Matrix

When soil subsidence is involved, the soil matrix is not rigid and it has its own deformation. In most of the classical models the soil matrix is supposed to have a deformation.

In this section, the system of equations of Biot's model is presented. This model results from a series of field equations of linear isotropic poroelastic media. The linear isotropic poroelastic processes may be described by (See Detourney and Cheng (1993); Wang (2000)):

1. The constitutive equations of linear elasticity for the porous solid.
2. The constitutive equations for the fluid presented in the previous section.
3. Darcy's law presented in the previous section.
4. The equilibrium equations for the solid phase, including the compatibility equations to ensure a single valued continuous displacement field.
5. The continuity equation for the fluid phase.

3.2.4.1 Strain and Displacements

The total state of strain in a given material may consist of the following components (see Elsworth and Bai 1992; Plotkowiak 2005).

$$\overline{\varepsilon}_{Total} = \overline{\varepsilon}_{elastic} + \overline{\varepsilon}_{plastic} + \overline{\varepsilon}_{initial} \tag{3.75}$$

In this section, only the elastic strain will be considered, assuming there is no plastic deformation and the initial strain will be discussed later. In a general case, the elastic deformation is a result of the mechanical deformation and the deformation due to temperature changes (see Li and Helm 2010 for the use of thermo-elastic models).

$$\varepsilon_{elastic} = \varepsilon_{mechanical} + \varepsilon_{thermal} \tag{3.76}$$

The relationship between the strain tensor ε, and displacements field $\vec{u} = (u_1(x, y, z), u_2(x, y, z), u_3(x, y, z))$, is defined by

$$\varepsilon_{ij} = \frac{1}{2}\left(\frac{\partial u_i}{\partial x_j} + \frac{\partial u_j}{\partial x_i}\right) \quad i = 1, 2, 3 \tag{3.77}$$

or in short as

$$\overline{\varepsilon} = \begin{bmatrix} \varepsilon_{xx} \\ \varepsilon_{yy} \\ \varepsilon_{zz} \\ \varepsilon_{xy} \\ \varepsilon_{xz} \\ \varepsilon_{yz} \end{bmatrix} = \begin{bmatrix} \frac{\partial}{\partial x} & 0 & 0 \\ 0 & \frac{\partial}{\partial y} & 0 \\ 0 & 0 & \frac{\partial}{\partial z} \\ \frac{1}{2}\frac{\partial}{\partial x} & \frac{1}{2}\frac{\partial}{\partial y} & 0 \\ \frac{1}{2}\frac{\partial}{\partial x} & 0 & \frac{1}{2}\frac{\partial}{\partial z} \\ 0 & \frac{1}{2}\frac{\partial}{\partial y} & \frac{1}{2}\frac{\partial}{\partial z} \end{bmatrix} \begin{bmatrix} u_1 \\ u_2 \\ u_3 \end{bmatrix} = \overline{a}\,\vec{u} \tag{3.78}$$

Where **a** is the partial differential operator matrix.

Thermal strains are defined by:

$$\varepsilon_{thermal} = \alpha\,(T - T_0) \tag{3.79a}$$

where α is the matrix of thermal expansion coefficient and T and T_0 are the temperature and its reference temperature, respectively.

Assuming an isotropic behavior, the thermal expansion coefficient is equal in all directions. The total strain can then be written as

$$\overline{\varepsilon}_{elastic} = \overline{a}\,\vec{u} + \overline{\alpha}(T - T_0) \tag{3.79b}$$

3.2.4.2 Deformation

Hooke's Law states that a general stress, σ and the strain, ε within a deforming material are related by a constant E which is the modulus of elasticity.

$$\sigma = E\,\varepsilon$$

This is the constitutive relations for linearly elastic material, and can be represented in matrix form as:

$$\overline{\sigma} = \overline{\mathrm{D}}\,\overline{\varepsilon}; \quad \overline{\varepsilon} = \overline{\mathrm{C}}\,\overline{\sigma} \tag{3.80a}$$

Where $\overline{\mathrm{D}}$ is the elasticity matrix and $\overline{\mathrm{C}} = \overline{\mathrm{D}^{-1}}$, the compliance matrix. For a three-dimensional isotropic solid used in most problems, $\overline{\mathrm{C}} = \overline{\mathrm{D}^{-1}}$ is given by

$$\overline{\mathrm{C}} = \overline{\mathrm{D}^{-1}} = \begin{bmatrix} \frac{1}{\mathrm{E}} & -\frac{\nu}{\mathrm{E}} & -\frac{\nu}{\mathrm{E}} & 0 & 0 & 0 \\ -\frac{\nu}{E} & \frac{1}{\mathrm{E}} & -\frac{\nu}{\mathrm{E}} & 0 & 0 & 0 \\ -\frac{\nu}{\mathrm{E}} & -\frac{\nu}{\mathrm{E}} & \frac{1}{\mathrm{E}} & 0 & 0 & 0 \\ 0 & 0 & 0 & \frac{1}{\mathrm{G}} & 0 & 0 \\ 0 & 0 & 0 & 0 & \frac{1}{\mathrm{G}} & 0 \\ 0 & 0 & 0 & 0 & 0 & \frac{1}{\mathrm{G}} \end{bmatrix} \tag{3.80b}$$

where ν is the Poisson's ratio, and the shear modulus, $G = E/2\,(1+\nu)$.

3.3 Models Associated with the Causes of Subsidence

3.3.1 The Biot Model for Saturated Soil and Subsidence Due to Water Withdrawal

When water withdrawal is causing subsidence, the Biot model is computing the water pressure and the soil displacement.

The Biot model relies on the following assumptions:

1. At the grains level the Terzaghi's principle of effective stress inter-granular stress holds.
2. The solid grains changes their volume exclusively because the pore or neutral pressure variation P. The volume of the soil changes exclusively due to changes in the volume of voids which is caused by changes in effective stress. (See Herrera et al. 1976).

It is important to note that this last assumption takes into account the variation of solid stress due to water pressure variation which is a part of the sedimentation phenomena.

If we indicate by ε_{zz}, $\varepsilon_{xx}, \varepsilon_{yy}$, the principal components of the incremental volumetric strain ε, and both σ and p are incremental quantities, with the aid of the elastic theory and the three-dimensional form of the Terzaghi's stress principle, governing equations for the fluid flow and the poroelastic stress in a three-dimensional isotropic porous medium are reviewed here in terms of the incremental variables:

$$\begin{cases} (\lambda + G)\dfrac{\partial \varepsilon}{\partial x} + G\nabla^2 u = \chi\dfrac{\partial p}{\partial x} + g_x(x,y,z,t) \\ (\lambda + G)\dfrac{\partial \varepsilon}{\partial y} + G\nabla^2 v = \chi\dfrac{\partial p}{\partial y} + g_y(x,y,z,t) \\ (\lambda + G)\dfrac{\partial \varepsilon}{\partial z} + G\nabla^2 w = \chi\dfrac{\partial p}{\partial z} + g_z(x,y,z,t) \end{cases}$$

$$(x,y,z) \in \Omega \quad ; \ t > 0 \tag{3.81}$$

Where the components of the displacement vector $\vec{d}$ are u, v, w and the pore pressure is $p, \vec{g} = (g_x, g_y, g_z)$ is the loading vector.

Ω is the aquifer domain. The solution $\vec{d}$ belongs to the Sobolev space:

$$\vec{d} \in \{\vec{v} \in H^1(\Omega) \times H^1(\Omega) \times H^1(\Omega);\ \vec{v} = \vec{u}\, on\ \partial\Omega\} \times [0.T] = [H_0^1(\Omega, T)]^3$$

Following Bear (1972, 1979) and Gambolati et al. (2000), the flow equation for a compressible fluid moving in a deformable porous medium (single grains are compressible).

$$\frac{1}{\gamma}\vec{\nabla}.\left(k\vec{\nabla}p\right) = \left[n\beta + c_{gr}(1-n)\right]\frac{\partial p}{\partial t} + \frac{\partial \varepsilon}{\partial t} + Q$$

$$(x,y,z) \in \Omega;\quad t > 0$$

$$\vec{\nabla}\text{is the gradient operator}\ \left(\vec{\nabla} = \left(\frac{\partial}{\partial x}, \frac{\partial}{\partial y}, \frac{\partial}{\partial z}\right)\right) \tag{3.82}$$

where t is the time, γ and β are the specific weight and the volumetric compressibility of water, respectively, k is the hydraulic conductivity tensor, n is the porosity, c_{gr} is the rock grain compressibility and the medium volume strain is defined as $\varepsilon = \vec{\nabla} \cdot \vec{d} = \frac{\partial u}{\partial x} + \frac{\partial v}{\partial y} + \frac{\partial w}{\partial z}$. Q is the pumping term. Denoting by E and ν the Young and Poisson moduli of the bulk porous medium, and by c_b the medium compressibility, the remaining symbols are defined as $\chi = (1 - c_{gr}/c_b)$ the Biot coefficient, $G = E/2(1+\nu)$ the shear modulus, and $\lambda = \nu E/[(1-2\nu)/(1+\nu)]$ the Lamè constant. While the coupled model solves simultaneously equations (1) and (2), the uncoupled approach they are solved separately. First the time evolution of the pressure field p is obtained by the only equation (2), eliminating the dependence on ε with the equation $\varepsilon = \alpha\chi p$, being $\alpha = 1/(\lambda + 2G)$ the vertical medium compressibility, and then p is supplied to the right-hand-side of equation (1) for the calculation of the corresponding displacement field.

A more convenient form of the flow equation is:

$$\vec{\nabla}.\left(K\vec{\nabla}p\right) = \frac{1}{S}\frac{\partial p}{\partial t} + \frac{\partial \varepsilon}{\partial t} + Q:\quad ;(x,y,z) \in \Omega;\quad t > 0$$

$$K = \frac{k}{\gamma};\quad \frac{1}{S} = \left[n\beta + c_{gr}(1-n)\right] \tag{3.83}$$

Boundary conditions:

For the water pressure p:

$$p(x,t) = g_1(x,t)\qquad (x \in \partial\Omega^1,\ t > 0);$$

$$-K\nabla p.\vec{n}_{\partial\Omega} = g_2(x,t)\quad (x \in \partial\Omega^2,\ t > 0); \tag{3.84}$$

where Ω is a bounded, polygonal open set of R^2, $\partial\Omega^1$ and $\partial\Omega^2$ are partitions of the boundary $\partial\Omega$ of Ω corresponding to Dirichlet and Neumann boundary conditions and $\vec{n}_{\partial\Omega}$ the unit outward normal to the boundary $\partial\Omega$.

The solution p belongs to the Sobolev space:

$$p \in \{ f \in H^1(\Omega);\ f(x,t) = g_1(x,t)$$

$$(x \in \partial\Omega^1,\ t > 0); -\mathrm{K}\nabla f\ .\vec{n}_{\partial\Omega} = g_2(x,t)$$

$$(x \in \partial\Omega^2,\ t > 0); \} \times [0,T] = H(div,\Omega) \times [0,T]$$

The boundary conditions for the flow equations have been discussed previously.

For the displacement field, the boundary condition depends on the depth of the geological layer where the subsidence occurs.

At the above layer, the boundary conditions are of Dirichlet types.

$$\begin{aligned} u(x,t) &= u_1(x,t) \quad (x \in \partial\Omega,\ t > 0); \\ v(x,t) &= v_1(x,t) \quad (x \in \partial\Omega,\ t > 0); \\ w(x,t) &= w_1(x,t) \quad (x \in \partial\Omega,\ t > 0); \end{aligned} \tag{3.85}$$

However, when the geological layer is an internal layer, the stress of the upper layer has to be computed. Thus, there is a result with a stress boundary condition.

The solution of the Biot system of Eqs. (3.81), (3.83) and the associated boundary conditions (3.84) and (3.85) is a coupled system of equations for displacement field and pressure $(\vec{d}, p)$ that belongs to: $[H_0^1(\Omega, T)]^3 \times H(div, \Omega) \times [0, T]$. $(\vec{d}, p)$ is associated to the loading vector $\vec{g}$ and the pumping rate function Q.

3.3.2 *The Two-Phase Flow Model for Subsidence Caused by Oil and Gas Withdrawal*

When oil and gas withdrawal are causing soil subsidence, the flow equation has to be replaced by a two-phase flow equation (see Esaki et al. 1991; Gambolati et al. 1991).

- Governing equations for a two-phase flow – Richards' equation:

Water-air systems can be simplified by considering air to be infinitely mobile as we did when considering water rising in a capillary tube previously. The pressure in the air phase is then zero gauge everywhere and spatial and temporal derivatives of the air phase pressure are zero as well. As a result, only the wetting phase equation is necessary to describe the flow system. The remaining wetting phase equation is referred to as Richards' equation using a repeated index convention. Also, land subsidence induced by rainwater infiltration may be described using the Richard's equation (see Martinez et al. 2010; Masoudzade et al. 2010).

$$\frac{\partial}{\partial x_i}\left[\frac{kk_{r_{w}ij}}{\mu_w}\left(\frac{\partial P_w}{\partial x_j}+\rho_w g\frac{\partial z}{\partial x_j}\right)\right]=\phi\frac{\partial S_w}{\partial t}+\frac{\partial \varepsilon}{\partial t}. \tag{3.86}$$

Different applications require different forms of the Richards' equation. Celia et al. (1990) developed the "mixed form" of the Richards' equation.

By replacing the water pressure with the pressure head by substituting the relationship becomes:

$$\frac{\partial}{\partial x_i}\left[\frac{kk_{r_{wij}}\rho_w g}{\mu_w}\left(\frac{\partial \psi_w}{\partial x_j}+\frac{\partial z}{\partial x_j}\right)\right]=\phi\frac{\partial S_w}{\partial t}, \tag{3.87}$$

where ψ is the pressure head of the water phase ($P_w/\rho g$).

For air-water systems, the hydraulic conductivity can be used rather than the permeability:

Substituting $K_s = \frac{k\rho_w g}{\mu_w}$ gives:

$$\frac{\partial}{\partial x_i}\left[K_s k_{r_{wij}}\left(\frac{\partial \psi_w}{\partial x_j}+\frac{\partial z}{\partial x_j}\right)\right]=\phi\frac{\partial S_w}{\partial t}+\frac{\partial \varepsilon}{\partial t}. \tag{3.88}$$

Finally, the water phase saturation can be replaced by the volumetric water content, $\theta = \phi S_w$, to give the final form:

$$\frac{\partial}{\partial x_i}\left[K_s k_{r_{wij}}\left(\frac{\partial \psi_w}{\partial x_j}+\frac{\partial z}{\partial x_j}\right)\right]=\frac{\partial \theta}{\partial t}+\frac{\partial \varepsilon}{\partial t}. \tag{3.89}$$

3.3.2.1 Water Capacity

With the assumption that the air phase pressure is zero gauge everywhere, the capillary pressure can be simplified to:

$$P_c = P_{nw} - P_w = -P_w = -\psi_w \rho_w g. \tag{3.90}$$

Therefore, a plot of the saturation as a function of the negative of the wetting phase pressure head will differ from the capillary pressure – wetting phase saturation curve by a constant multiplier, as shown below on the left.

The water capacity is defined as the change in water content per unit change in the pressure head of the water phase. This is equivalent to the negative inverse slope of the soil characteristic curve (below on the left),

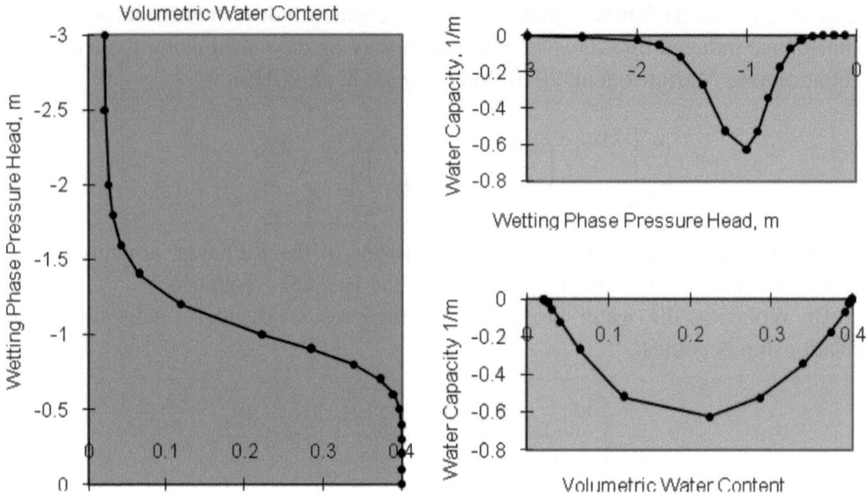

Fig. 3.11 Suction pressure versus Water content in a non saturated soil.

$$C(\psi) = \frac{\partial \theta}{\partial \psi}.$$

The function is plotted against both water content and pressure head (below on the right) (Fig. 3.11).

3.3.2.2 The Pressure Head form of Richards' Equation

Beginning with the mixed form of Richards' equation:

$$\frac{\partial}{\partial x_i}\left[K_s k_{r_{wij}}\left(\frac{\partial \psi_w}{\partial x_j} + \frac{\partial z}{\partial x_j}\right)\right] = \frac{\partial \theta}{\partial t} + \frac{\partial \varepsilon}{\partial t}, \tag{3.91}$$

Substitute the pressure dependent water capacity into the storage term:

$$\frac{\partial \theta}{\partial t} = \frac{\partial \theta}{\partial \psi}\frac{\partial \psi}{\partial t} = C(\psi)\frac{\partial \psi}{\partial t}. \tag{3.92}$$

The final form is:

$$\frac{\partial}{\partial x_i}\left[K_s k_{r_{wij}}(\psi)\left(\frac{\partial \psi}{\partial x_j} + \frac{\partial z}{\partial x_j}\right)\right] = C(\psi)\frac{\partial \psi}{\partial t} + \frac{\partial \varepsilon}{\partial t} \tag{3.93}$$

This pressure head form is difficult to solve because it is highly nonlinear through $k_{rij}(\psi)$ and $C(\psi)$. However, as we have seen above, the pressure varies smoothly even in heterogeneous or layered media, making this form very robust.

To extend this form of the equation for use in saturated media, the saturated storage term, S_s, must be added to the right-hand side. For completeness, the storage mechanisms discussed for a saturated flow could be added to all of the multiphase flow equations. However, their contributions are insignificant compared to the drainage of the pores.

3.3.2.3 A Water Content form of Richards' Equation

Richards' equation can be formulated with the water content as the only dependent variable as well.

Beginning with the mixed form of Richards' equation:

$$\frac{\partial}{\partial x_i}\left[K_s k_{r_{wij}}\left(\frac{\partial \psi_w}{\partial x_j}+\frac{\partial z}{\partial x_j}\right)\right]=\frac{\partial \theta}{\partial t}+\frac{\partial \varepsilon}{\partial t}, \tag{3.94}$$

Substitute the water content dependent water capacity into the gradient term:

$$\frac{\partial \psi}{\partial x_j}=\frac{\partial \psi}{\partial \theta}\frac{\partial \theta}{\partial x_j}=\frac{1}{C(\theta)}\frac{\partial \theta}{\partial x_j}.$$

The final form is:

$$\frac{\partial}{\partial x_i}\left[K_s k_{r_{wij}}(\theta)\left(\frac{1}{C(\theta)}\frac{\partial \theta}{\partial x_j}+\frac{\partial z}{\partial x_j}\right)\right]=\frac{\partial \theta}{\partial t}. \tag{3.95}$$

Defining the soil water diffusivity as $D(\theta)=Kk_r(\theta)/C$ gives:

$$\frac{\partial}{\partial x_i}\left[\left(D_{ij}(\theta)\frac{\partial \theta}{\partial x_j}+K_s k_{r_{wij}}(\theta)\frac{\partial z}{\partial x_j}\right)\right]=\frac{\partial \theta}{\partial t}+\frac{\partial \varepsilon}{\partial t}. \tag{3.96}$$

This diffusion-like form is more amenable to analytical solutions such as those found in Carslaw and Jaeger (1959). The water content form of the unsaturated flow equation is not valid in regions saturated with water because the water capacity becomes zero, making the soil water diffusivity undefined. Furthermore, as we saw above, the water content is discontinuous across material boundaries, also precluding the use of this form. The advantage of this form is that the diffusivity is only weakly nonlinear and is relatively easy to measure in the lab. However, the relative permeability is nonlinear and the diffusivity may be hysteretic.

Also, the soil matrix displacement has to be computed using Eq. (3.81):

$$\begin{cases} (\lambda + G)\dfrac{\partial \varepsilon}{\partial x} + G\nabla^2 u = \chi\dfrac{\partial p_w}{\partial x} + g_x(x, y, z, t) \\ (\lambda + G)\dfrac{\partial \varepsilon}{\partial y} + G\nabla^2 v = \chi\dfrac{\partial p_w}{\partial y} + g_y(x, y, z, t) \\ (\lambda + G)\dfrac{\partial \varepsilon}{\partial z} + G\nabla^2 w = \chi\dfrac{\partial p_w}{\partial z} + g_z(x, y, z, t) \end{cases}$$

$$(x, y, z) \in \Omega; \ t > 0 \tag{3.81}$$

where p_w is the water pressure.

3.3.3 A Model for Subsidence Due to Encroachment of Sea Water Into Coastal and Inland Flooding

For sea water or brine water, the water density $\rho(p, c)$ depends not only on the water pressure but also on the salt concentration. For natural water chemical components exists at a concentration that does not influence the water density (Hem 1970).

For sea water, a linear dependence is generally assumed (see Babu et al. 1993, 1997; Melloul and Zeitoun 1999):

$$\rho(p, c) = \rho_0(p) + \eta(c - c_0)$$

Where $\rho_0(p)$ the fresh water density or a reference density is the compressibility of the water, η is a regression coefficient and c_0 the salt concentration of the fresh water. In practice, $\eta = 1/(c_{\max} - c_{\min})$ is approximated as the inverse of the range of the salt concentration.

In this case Eq. (3.41) lead to

$$\frac{\partial n}{\partial t} = \frac{\partial n}{\partial P}\frac{\partial P}{\partial t}$$

$$\frac{\partial \rho}{\partial t} = \frac{\partial \rho}{\partial P}\frac{\partial P}{\partial t} + \frac{\partial \rho}{\partial c}\frac{\partial c}{\partial t} = \frac{\partial \rho}{\partial P}\frac{\partial P}{\partial t} + \eta\frac{\partial c}{\partial t}$$

Then the flow Eq. (3.41) is written as:

$$\frac{\partial}{\partial x_i}\left(K_{ij}\frac{\partial H}{\partial x_j}\right) = \rho g\left(\alpha + n\beta\right)\frac{\partial H}{\partial t} + \eta\frac{\partial c}{\partial t} \tag{3.97}$$

The solution of the above equation requires a transport equation for the concentration c.

The mass balance of solutes can be expressed as the change in solute mass being equal to the divergence of mass fluxes (adjective, diffusive and dispersive ones) plus sink/source contributions.

For the concentration of salt the basic equation is given by:

$$\frac{\partial (\rho c)}{\partial t} = -\vec{\nabla} \cdot \left(\vec{j_{adv}} + \vec{j_{disp}} \right) + \rho Q \xi^* \tag{3.98}$$

The advective mass flux is straightforward to express and is equal to $\vec{j_{adv}} = \rho c \vec{q}$. The dispersive and diffusive fluxes are treated together as $\vec{j_{dif}} = -\rho D \vec{\nabla} c$ where D is the diffusion and dispersion tensor. According to the Bear-Scheidegger dispersion relationship (Bear 1972, 1979), it is equal to

$$\overline{D} = \phi \left(\overline{D}_d + \overline{D}_m \right) \tag{3.99a}$$

in the case of a saturated flow.

$$\overline{D}_d = \overline{D}_d T \tag{3.99b}$$

$$\overline{D}_m = \alpha_T \left| \vec{q} \right| \bar{I} + (\alpha_L - \alpha_T) \frac{\vec{q} \otimes \vec{q}}{\vec{q}} \tag{3.99c}$$

where D_d is the molecular diffusion tensor and T the tortuosity tensor. α_L and α_T are called longitudinal and transversal dispersivity (m).

$D_d = D_d T$ will not be considered in this work.

The mass balance equation may be written as:

$$\rho \theta \frac{\partial c}{\partial t} = -\rho \vec{q} \cdot \vec{\nabla} c + \vec{\nabla} \cdot \left(\rho \overline{D} \vec{\nabla} c \right) + \rho Q \left(c^* - c \right) \tag{3.100}$$

3.3.4 A Model of Subsidence in Fracture Porous Media

Due to its high permeability, fractured rocks are found to be the largest resource for practical geothermal development (Bowen 1989). It has also been considered for use in underground storage of carbon dioxide (Holloway 1997; Wawersik and Rudnicki 1998; Koutsabeloulis and Hope 1998). The fractured media acts as a means of transport for the fluid or injected fluids where occurs interaction between rock and fluid flow combined with effects of high temperature, and occasionally with chemical reaction. The behavior of the system that constitutes the three interacting processes: thermal, hydrological and mechanical, is generally called the thermo-hydro-mechanical (THM) coupled processes. Mathematical models developed to describe THM coupled processes treat the fractured rock and the matrix as a continuous medium (Gray et al. 1976) and represent the fractured rock mass by an equivalent porous medium (Pritchett et al. 1976). However, general interaction theories are required to describe the effect of the motion of fluid on the motion of matrix. Much of this has been satisfied through the introduction of the general theory of consolidation (Biot 1941) which relates the influence of pore pressure due to the existence of fluid in a porous rock. This study made it possible for engineers and scientists to solve problems in a much wider scope. The addition of the thermal component coupled later in the mid-1980s (Noorishad et al. 1984) formed a basis to a fully coupled THM solution. Then, many computer codes using the finite element method have been developed to simulate the THM behavior in fractured porous media (Pine and Cundall 1985; Ohnishi et al. 1987; Guvanasen and Chan 1995; Kohl et al. 1995; Nguyen and Selvadurai 1995; Bower and Zyvoloski 1997; Swenson et al. 1997), each of which differ slightly in assumptions, constitutive equations, and initial conditions. Furthermore, commercially available codes, such as ABAQUS (Börgesson 1996), FLAC (Israelsson 1996a), and UDEC/3DEC (Israelsson 1996b) have also extended their capabilities to account for modeling coupled THM in various types of medium. Although, there has been extensive research and computer modeling in this area in the past two decades, the porosity within the fracture and the matrix has long been considered as one entity as originally formulated for a continuous body of porous medium (Biot 1941). However, this continuous approach does not truly reflect the influences of fluid pressures and rock stresses on the fracture and the matrix (Wittke 1973).

The concept of "double-porosity" has been introduced in the early 1960s, which considers the fractured rock mass consisting of two porous systems (Barenblatt et al. 1960; Warren and Root 1963), which are the matrix, having high porosity and low permeability, and the fracture, having low porosity and high permeability. In 1992, a dual-porosity finite element model has been developed to describe the behavior of the fractured porous medium in response to fluid flow and stress (Elsworth and Bai 1992), which accounts for two porous systems and the fluid mass transport between each system. In this section we present the full THM model that involves soil matrix displacement and heating process. When the heat transfer of the fluid is negligible, the model may be used for subsidence prediction in mining problems (see Ma et al. (2010).

For a discontinuous rock mass, the overall elastic properties accounts for the deformation characteristics of the intact rock and the fractures. For an orthotropic behavior of rock mass with three normal-sets of fractures, the compliance matrix can be written as (Amadei and Goodman 1981)

$$\overline{C} = \overline{D^{-1}}$$
$$= \begin{bmatrix} \frac{1}{E_x} + \frac{1}{K_{nx} S_x} & -\frac{\nu_{xy}}{E_y} & -\frac{\nu_{xz}}{E_z} & 0 & 0 & 0 \\ -\frac{\nu_{xy}}{E_y} & \frac{1}{E_y} + \frac{1}{K_{ny} S_y} & -\frac{\nu_{yz}}{E_z} & 0 & 0 & 0 \\ -\frac{\nu_{xz}}{E_z} & -\frac{\nu_{yz}}{E_z} & \frac{1}{E_y} + \frac{1}{K_{ny} S_y} & 0 & 0 & 0 \\ 0 & 0 & 0 & \frac{1}{G_{yz}} + \frac{1}{K_{sy} S_y} + \frac{1}{K_{sz} S_z} & 0 & 0 \\ 0 & 0 & 0 & 0 & \frac{1}{G_{yz}} + \frac{1}{K_{sx} S_x} + \frac{1}{K_{sz} S_z} & 0 \\ 0 & 0 & 0 & 0 & 0 & \frac{1}{G_{xz}} + \frac{1}{K_{sx} S_x} + \frac{1}{K_{sy} S_y} \end{bmatrix} \tag{3.101}$$

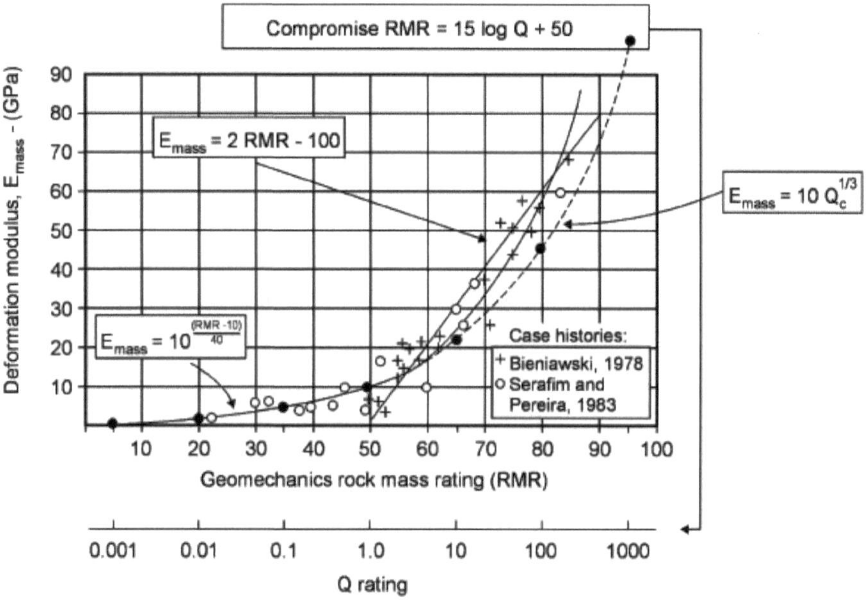

Fig. 3.12 Deformation modulus and rock mass rating (Barton 2002)

where K_n and K_s are the normal and shear stiffness of the rock mass, S is the fracture spacing, and the modulus and poisson's ratio are that of the intact rock. The normal and shear stiffness are generally undetermined; however, they can be approximated from the modulus of the rock mass using the rock mass rating obtained from experimental data (Bieniawski 1973; Bieniawski 1976) as shown in Fig. 3.12.

Accounting for the effective stress, and the differing fluid pressures within the fracture and porous phases, the relationship for stress and strain becomes

$$\sigma'_{p} = D_{p}\,\varepsilon_{p} \tag{3.102a}$$

$$\sigma'_{F} = D_{F}\,\varepsilon_{F} \tag{3.102b}$$

The total strain, ε, which accounts for the fluid pressure becomes:

$$\varepsilon = (C_p + C_F)\sigma - C_p p_p - C_F p_F \tag{3.103}$$

If we also account for thermal expansion, the total strain:

$$\varepsilon = (C_p + C_F)\sigma - C_p p_p - C_F p_F + \alpha_P(T_P - T_0) + \alpha_F(T_F - T_0) \tag{3.104}$$

Thus, rearranging for stress gives

$$\sigma = D_m[au + C_p p_p + C_F p_F - \alpha_P(T_P - T_0) - \alpha_F(T_F - T_0) \tag{3.105}$$

where $D_m = (C_p + C_F)^{-1}$ is the elastic matrix for the dual porosity medium.

3.3.4.1 Conservation of Mass

The governing equation for fluid flow which relates the flux, q, and its potential, is defined by Darcy's law which is expressed in matrix form as:

$$\vec{q}_f = -\frac{k}{\mu}\vec{\nabla}(p + \gamma Z) \tag{3.106}$$

where γ is the specific weight of the fluid, μ the dynamic viscosity, Z the elevation, k the permeability matrix, and ∇ the gradient operator.

The equilibrium equation for fluid transport in an isotropic medium according to Darcy's law is:

$$\beta P - \vec{\nabla} \bullet \left[\frac{k}{\mu}\vec{\nabla}(p + \gamma Z)\right] + Q_f = 0 \tag{3.107a}$$

The term P on the left hand side of Eq. (3.107a) refers to the rate of change of pressure with time, t, due to grain compressibility, β. The second term is Darcy's flux, which is function of the permeability, k, dynamic viscosity, μ, specific weight of the fluid, γ, elevation, z, and fluid pressure, p. The third term, Q_f, is the source term.

However, accounting for the thermal effects and the fluid mass transfer between the porous and fractured phases, the equilibrium equation for the porous phase may be rewritten as:

$$-\vec{\nabla} \bullet \left[\frac{k}{\mu}\vec{\nabla}(p + \gamma Z)\right] - \beta_P P_p - \varphi(p_p - p_F) + \dot{\varepsilon} + [Q_f]_p = 0 \tag{3.107b}$$

where the added component left-hand side of the last equation is a positive component referring to the flow of fluid mass driven by the changes of the volume due to strain, $\dot{\varepsilon}$, and the second term is the fluid mass transfer between the porous and the fractured phase, where the coefficient φ which governs the quasi-steady response (Warren and Root 1963) is a function of the geometry of the porous block and the flow characteristics (permeability and dynamic viscosity).

$$\varphi = \frac{4n(n+2)k_p}{l^2\mu_p} \tag{3.108}$$

The first term is the shape factor, where n is the number of normal set of fractures, and l is the characteristic length which can be approximated from the dimensions a, b, and c, of the porous block as:

$$1 \cong \frac{3abc}{ab + bc + ac};\ n = 3;\ 1 \cong \frac{3ab}{a + b};\ n = 2;\ 1 \cong x;\ n = 1 \tag{3.109}$$

Similarly, the equilibrium equation for the fractured phase may be expressed as:

$$-\vec{\nabla} \bullet \left[\frac{k_F}{\mu_F}\vec{\nabla}(p_F + \gamma Z)\right] - \beta_F P_F - \varphi(p_p - p_F) + \dot{\varepsilon} + [Q_f]_F = 0 \tag{3.110}$$

3.3.4.2 Thermal Behavior

The governing equation for fluid flow which relates the heat flux, $\vec{q_T}$ with the temperature field. It is defined by Fick's law as:

$$\vec{q_T} = -\lambda \vec{\nabla} T \tag{3.111}$$

Where λ is the thermal conductivity.

The general thermal transport equilibrium equation for thermal conduction and convection in an isotropic medium is defined by

$$\rho C_{ap} T - \vec{\nabla} \bullet [\lambda\ \vec{\nabla} T] - \rho C_{ap} \vec{\nabla} \bullet \vec{q_f} + [Q_f]_F = 0 \tag{3.112}$$

Where ρ is the density, C_{ap} is the heat capacity, and Q_T are is the heat source.

References

Amadei B, Goodman RE (1981) A3-D constitutive relation for fractured rock masses. International symposium on the mechanical behavior of structured media, Ottawa

Audiguier M, Cojean R, Geremew Z (2010) Micro-cracking of expansive soils during shrinkage processes: roles of mineralogy and microstructure. In: Proceedings of EISOLS 2010, Queretaro, Mexico, October 2010, IAHS publication 339

Babu DK, Pinder GF, Niemi A, Ahlfeld DP, Stothoff SA, Zeitoun DG (1993) Chemical transport by three-dimensional density-dependent groundwater flows. Princeton University Report 01/93/243

Babu DK, Pinder GF, Niemi A, Ahlfeld DP, Stothoff SA (1997) Chemical transport by three-dimensional groundwater flows, Report 84-WR-3. Princeton University, Princeton, NJ

Balint S, Balint AM (2007) Mathematical models for mass and heat transport in porous media, Part I. Summer University, Vrnjacka Banja, West University of Timisoara, Romania

Barenblatt GE, Zheltov IP et al (1960) Basic concepts in the theory of seepage of homogeneous liquids in fissured rocks. J Appl Math Mech-USS 24:1286–1303

Barton NR (2002) Some new Q-value correlations to assist in site characterisation and tunnel design. Int J Rock Mech Min 39:185–216

Bear J (1972) Dynamics of fluids in porous media. Dover, New York

Bear J (1979) Hydraulics of groundwater. McGraw-Hill, New York

Bieniawski ZT (1973) Engineering classification of jointed rock masses. Trans South Afr Inst Civil Eng 15:335–344

Bieniawski ZT (1976) Rock mass classifications in rock engineering. In: Proceedings of the symposium on exploration for rock engineering, Johannesburg

Biot MA (1941) General theory of three-dimensional consolidation. J Appl Phys 12:155–164

Bishop AW (1959) The principle of effective stress. Tek Ukeblad 39:859–863

Börgesson L (1996) ABAQUS. Coupled thermo-hydro-mechanical processes of fractured media. In: Stephansson O, Jing L, Tsang, CF (eds) Developments in geotechnical engineering 79:565–570. Elsevier, Amsterdam

Bowen R (1989) Geothermal resources. Elsevier, New York

Bower KM, Zyvoloski G (1997) A numerical model for thermo-hydro-mechanical coupling in fractured rock. Int J Rock Mech Min 34:1201–1211

Brenner H (1980) Dispersion resulting from flow through spatially periodic porous media. Philos T R Soc Lond A 297:81–133

Carslaw HS, Jaeger JC (1959) Conduction of heat in solids, 2nd edn. Oxford University Press, London

Celia MA, Bououtas ET, Zarba RL (1990) A general mass-conservative numerical solution for the unsaturated flow equation. Water Resour Res 26:1483–1496

Corapcioglu Y (1976) Mathematical modeling of leaky aquifers with rheological properties. In: Proceedings of the second international symposium on land subsidence, Anaheim, CA, December 1976

Detournay E, Cheng AH-D (1993) Fundamentals of poroelasticity. In: Fairhurst C (ed) Chapter 5 in comprehensive rock engineering: Principle, Practice and Projects, Vol. II, Analysis and Design method. Pergamon Press, pp 113–171

Elsworth D, Bai M (1992) Flow-deformation response of dual-porosity media. J Geotech Eng-ASCE 118:107–124

Esaki T, Shikata K, Aoki K, Kimura T (1991) Surface subsidence in natural gas fields. In: Proceedings of the fourth international symposium on land subsidence, May 1991, IAHS publication – 200

Ferre J (2005) Course on hydrology. http://academic.engr.arizona.edu/HWR/ferre/tyhome.html

Fowler AC (1998) Mathematical models in the applied sciences. Cambridge University Press, Cambridge, UK

Freeze RA, Cherry JA (1979) Groundwater. Prentice-Hall, Englewood Cliffs

Gambolati G, Ricceri G, Bertoni W, Brighenti G, Vuillermin E (1991) Numerical analysis of land subsidence at Ravenna due to water withdrawal and gas removal. In: Proceedings of the fourth international symposium on land subsidence, May 1991. IAHS publication – 200

Gambolati G, Teatini P, Bau D, Ferronato M (2000) The importance of poro-elastic coupling in dynamically active aquifers of the Po River basin, Italy. Water Resour Res 36:2443–2459

Germain Paul et Patrick Muller (1995) Introduction ‘a la mecanique des milieux continus. Edition Masson, 2 ´edition

Gray WG, O'Neill K et al (1976) Simulation of heat transport in fractured, single-phase geothermal reservoirs. Summaries, second workshop on geothermal reservoir engineering. Stanford University, Stanford, pp 222–228

Guvanasen V, Chan T (1995) A new three-dimensional finite-element analysis of hysteresis thermo-hydromechanical deformation of fractured rock mass with dilatance in fractures. The second conference on mechanics of jointed and faulted rocks, Vienna, Austria

Hasimoto H (1959) On the periodic fundamental solution of the Stokes equations and their application to viscous flow passed a cubic array of spheres. J Fluid Mech 5:317–328

Hem JD (1970) Study and interpretation of the chemical characteristics of natural water. US Geological Survey Water-Supply Paper 1473

Herrera I, Alberro J, Graue R, Hanel J, Homsy GM (1976) Development of artificial reservoirs by inducing land subsidence. In: Proceedings of the second international symposium on land subsidence, Anaheim, CA
Holloway S (1997) An overview of the underground disposal of carbon dioxide. Energy Convers Manage 38:193–198
Houlsby GT (1997) The work input to an unsaturated granular material. Geotechnique 47:193–196
Hueckel TA (1992) On the effective stress concepts and deformation in clays subjected to environmental loads: discussion. Can Geotech J 29:1120–1125
Israelsson JI (1996a) Short description of FLAC version 3.2. Coupled thermo-hydro-mechanical processes of fractured media. In: Stephansson O, Jing L, Tsang CF (eds) Developments in geotechnical engineering 79:513–522. Elsevier, Amsterdam
Israelsson JI (1996b) Short description of UDEC and 3DEC. Coupled thermo-hydro-mechanical processes of fractured media. In: Stephansson O, Jing L, Tsang CF (eds) Developments in geotechnical engineering 79:523–528. Elsevier, Amsterdam
Kohl T, Evans KF et al (1995) Coupled hydraulic, thermal and mechanical considerations for the simulation of hot dry rock reservoirs. Geothermics 24:345–359
Koutsabeloulis N, Hope SA (1998) Coupled stress/fluid/thermal multiphase reservoir simulation studies incorporating rock mechanics. SPE/ISRM EUROCK-98 symposium, Norway
Lee CK, Sun CC, Mei CC (1996) Computation of permeability and dispersivities of solute or heat in a periodic porous media. Int J Heat Mass Trans 39:661–676
Li J, Helm DC (2010) A theory of three-dimensional land motion in terms of its velocity field. In: Proceedings of the EISOLS 2010, Queretaro, Mexico, 17–22 Oct 2010. IAHS publication 339
Liepmann HW, Roshko A (1957) Elements of gas dynamics. Wiley, New York
Ma FS, Zhao HJ, Zhang YM, Guo J (2010) Ground subsidence induced by backfill mining of a nickel mine and development forecasts. In: Proceedings of the EISOLS 2010, Queretaro, Mexico, 17–22 Oct 2010. IAHS publication 339
Martinez I, Hinkelmann R, Savidis S (2010) Modeling land subsidence processes induced by fast rainwater infiltration through fractures into the unsaturated zone. In: Proceedings of the EISOLS 2010, Queretaro, Mexico, 17–22 Oct 2010. IAHS publication 339
Masoudzade SA, Toufigh MM, Yazdani H, Rahgozar R (2010) 1-D infiltration, analysis of unsaturated flow and increase in land subsidence. In: Proceedings of the EISOLS 2010, Queretaro, Mexico, 17–22 Oct 2010. IAHS publication 339
Melloul AJ, Zeitoun DG (1999) A semi-empirical approach to seawater intrusion monitoring in Israel's coastal aquifer. In: Bear J et al (eds) Seawater intrusion in coastal aquifers – concept, methods, and practices. Kluwer, Dordrecht
Miao JF, Wu LG (1991) Mathematical modeling of land subsidence due to pumping of a multi-aquifer system with visco-elastic properties. In: Proceedings of the fourth international symposium on land subsidence, May 1991. IAHS publication – 200
Murray EJ (2002) An equation of state for unsaturated soils. Can Geotech J 39:125–140
Nguyen TS, Selvadurai APS (1995) Coupled thermal-hydrological-mechanical processes in sparsely fractured rock. Int J Rock Mech Min 32:465–480
Noorishad J, Tsang CF et al (1984) Coupled thermal-hydraulic-mechanical phenomena in saturated fractured porous rocks: numerical approach. J Geophys Res 89:10365–10373
Ohnishi Y, Shibata H et al (1987) Development of finite element code for the analysis of coupled thermo-hydro-mechanical behavior of a saturated-unsaturated medium. Coupled processes associated with nuclear waste repositories. Academic Press, Orlando, pp 679–696
Pine RJ, Cundall PA (1985) Application of the fluid rock interaction program (FRIP) to the modeling of hot dry rock geothermal energy systems. The international symposium on fundamentals of rock joints, Björkliden, Sweden
Plotkowiak M (2005) Hydro-mechanical coupling in porous media with emphasis on poroelasticity. ISEB publication, University of Hannover
Pritchett JW, Garg SK et al (1976) Geohydrological environmental effects of geothermal power production – phase IIA. Systems, Science and Software, La Jolla

Shanz M (2009) Poroelastodynamics: linear models, analytical solutions and numerical methods. Appl Mech Rev 62:030803-1–030803-15

Snyder LJ, Stewart WA (1966) Velocity and pressure profiles for Newtonian creeping flow in regular packed beds of spheres. AICHE J 12:167–173

Sorensen JP, Stewart WE (1974) Computation of forced convection in slow flow through ducts and packed beds. II. Velocity profile in a simple cubic array of spheres. Chem Eng Sci 29:819–825

Swenson DV, DuTeau R et al (1997) A coupled model of fluid flow in jointed rock applied to simulation of a hot dry rock reservoir (abstract). Int J Rock Mech Min 34:308

Tartar L (1980) Incompressible fluid flow in a porous medium-convergence of the homogenization process in non-homogeneous media and vibration theory. Lect Notes Phys 127:368–377. Springer, Berlin

Terzaghi K (1943) Theoretical soil mechanics. Wiley, New York

Thomas HR, He Y et al (1998) An examination of the validation of a model of the hydro/thermo/mechanical behaviour of engineered clay barriers. Int J Numer Anal Method Geomech 22:49–71

Wang HF (2000) Theory of linear poroelasticity. Princeton University press

Warren JE, Root PJ (1963) The behavior of naturally fractured reservoirs. Soc Petrol Eng J 3: 245–255

Wawersik WR, Rudnicki W (1998) Terrestrial sequestration of CO_2: an assessment of research needs. Invited panelist workshop, Department of Energy, Geoscience Research Program

Whitaker S (1986) Flow in porous media I: a theoretical derivation of Darcy's law. Trans Porous Med 1:3–25

White FM (1991) Viscous fluid flow. McGraw-Hill, New York

Wittke W (1973) Percolation through fissured rock. Int Assoc Eng Geol Bull 7:3–28

Zick AA, Homsy GM (1982) Stokes flow through periodic arrays of spheres. J Fluid Mech 115: 13–26

Chapter 4
Fundamentals of the Consolidation Theory for Soils

Abstract In this chapter, we describe the Terzaghi theory of consolidation as a way of decoupling the general Biot equations using the assumption of constant total stress. We detailed the computation of the water pressure and the soil displacement. The classical one-dimensional theory is presented with solved exercises.

We also discuss the use of the theory of consolidation for large-scale problems such as subsidence due to large-scale groundwater pumping. We review the different approaches that were used in the literature for the numerical modeling of the subsidence due to groundwater pumping. We concentrate on two main points:

1. The mechanical behavior of the clayey soil; and
2. The connection between groundwater movement and stress pressure in the soil.

Then, a critical review of the consolidation theory for large-scale problems is presented in terms of the principle of superposition of loadings.

Also, the three-dimensional theory is described, and we discuss the limitation of the theory in the case of internal loadings.

4.1 The Terzaghi Theory of Consolidation

In the literature, the two following methods of the computation of the consolidation have been proposed:

1. the one-dimensional Terzaghi approach; and
2. the full three-dimensional approach based on the Biot theory.

D.G. Zeitoun and E. Wakshal, *Land Subsidence Analysis in Urban Areas*,
Springer Environmental Science and Engineering, DOI 10.1007/978-94-007-5506-2_4,

4.1.1 Derivation of the Theory of Soil's Consolidation from the Biot Equations

When the displacement field has only a vertical component $w(x, y, z, t)$, the Biot equations reduces to the following system of equations:

$$\begin{cases} \frac{1}{\gamma}\vec{\nabla}.\left(k\vec{\nabla}p\right) = S\frac{\partial p}{\partial t} + \frac{\partial \varepsilon}{\partial t} + Q(x,t) & (4.1)\\ (\lambda + G)\frac{\partial \varepsilon}{\partial z} + G\nabla^2 w = \chi\frac{\partial p}{\partial z} + g_z(x,y,z,t) & (4.2)\\ \varepsilon(x,y,z,t) = \frac{\partial w}{\partial z}(x,y,z,t) & (4.3)\end{cases}$$

Moreover, if we assume that $w(x, y, z, t)$ and $p(x, y, z, t)$ depends only on the depth z and time t, and $Q(x, t) = 0$; $g_z(x, y, z, t) = 0$ then the above system of equations is reduced to:

$$\frac{1}{\gamma}\frac{\partial}{\partial z}\cdot\left(k\frac{\partial}{\partial z}p\right) = S\frac{\partial p}{\partial t} + \frac{\partial \varepsilon}{\partial t} \tag{4.4}$$

$$(\lambda + 2G)\frac{\partial^2 w}{\partial z^2} = \chi\frac{\partial p}{\partial z} \tag{4.5}$$

Integrating Eq. (4.5) with respect to z, one can easily get that:

$$p = \frac{(\lambda + 2G)}{\chi}\varepsilon + c(t) = \frac{(\lambda + 2G)}{\chi}\frac{\partial w}{\partial z} + c(t) \tag{4.6}$$

If we assume that the total stress in the saturated soil is constant over time, then the derivation of the Terzaghi principle leads to:

$$\begin{gathered}\frac{\partial \sigma(t)}{\partial t} = \frac{\partial \sigma'}{\partial t} + \frac{\partial p}{\partial t} = 0;\ \sigma' = a_v\,\varepsilon;\\ \frac{\partial p}{\partial t} = a_v\frac{\partial \varepsilon}{\partial t}\end{gathered} \tag{4.7}$$

A comparison between Eqs. (4.7) and (4.6) leads to the relation $\frac{dc(t)}{dt} = 0$.

Inserting this last equation into Eq. (4.4) and using the relation $\frac{dc(t)}{dt} = 0$ lead for constant permeability to:

$$\frac{k}{\gamma}\frac{\partial}{\partial z}\cdot\left(\frac{\partial}{\partial z}p\right) = \left[S + \frac{\chi}{(\lambda + 2G)}\right]\frac{\partial p}{\partial t} \tag{4.8a}$$

This last equation corresponds to the classical one-dimensional equation of consolidation.

4.1.2 The One-Dimensional Theory of Consolidation

4.1.2.1 The Consolidation Process

We shall use a simple "single cell" analog to examine what happens in a consolidation event.

Starting point: The spring is uncompressed, and the water pressure is zero. There is no external load, and the drainage valve is shut.

At time t = 0: Apply a load P to the piston, keeping the valve shut. The water is incompressible, so no settlement occurs. Therefore, the spring cannot compress, and thus, it carries no load. The load P is completely supported by the water pressure u, and the spring force F is zero:

$$\mathrm{At}\ t = 0 : u = \frac{P}{A} = \sigma_v \text{ and } \sigma'_v = \frac{F}{A} = 0 \tag{4.9a}$$

Open valve: Water starts to leak from the aperture, allowing the settlement to begin, so that the spring starts to compress and to take some of the load. At all the stages, the effective stress principle holds: i.e., $\sigma_v = \sigma'_v + u$ – the applied force is shared between the spring and the water.

As the spring force (σ'_V) increases, the water pressure u decreases by an equal amount. The pressure differential across the drainage value therefore reduces, and this reduces the velocity of the flow. Thus, the *rate of settlement reduces with increasing time*.

4.1.2.2 A Mathematical Analysis of the Problem

One can write down expressions for the process, provided the spring stiffness (E′) is known. Then the relationship between the flow through the aperture and the water pressure is given by:

$$\frac{d\sigma'}{dh/h_0} = E'; \tag{4.9b}$$

for the spring and $\frac{dQ}{dt} = \frac{A.dh}{dt} = k.u$ for flow through the aperture, or

$$\frac{d\sigma'}{dh} = \frac{E'}{h_0} \tag{4.9c}$$

and

$$\frac{dh}{dt} = \frac{k.u}{A} \tag{4.9d}$$

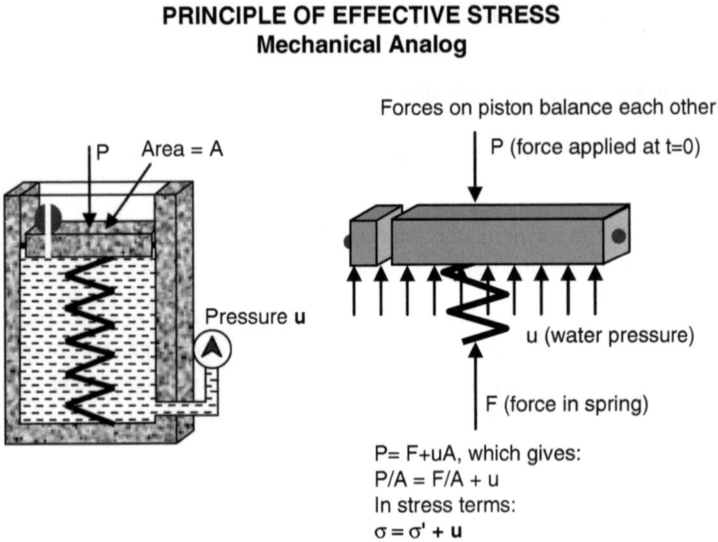

Fig. 4.1 A single-cell mechanical analog for the soil consolidation process

where dQ is the quantity of flow through the aperture in time increment dt, and dh is the compression of the spring due to a stress (force) increment $d\sigma'$. The flow equation states that the rate of flow depends on the pressure differential (note this is ***not*** correct for a small aperture, but we will use this assumption). The solution for this problem can be obtained as:

$$\sigma' = \sigma \cdot \left(1 - e^{-\alpha t}\right) \tag{4.10a}$$

$$u = \sigma \cdot \left(e^{-\alpha t}\right) \tag{4.10b}$$

where the constant $\alpha = \frac{k.E'}{A}$.

These are exponential functions, with σ' increasing exponentially from zero at $t = 0$ to σ at $t \to \infty$ and u reducing from $u = \sigma$ at $t = 0$ to zero at $t \to \infty$.

4.1.2.3 A Solution of the Consolidation Problem: The Terzaghi Solution

When a load is applied to a clay layer (as in Figs. 4.1 and 4.2), a long time may be required for the settlement to be completed. During this stage, the pore pressure decreases gradually, while the effective stress increases at the same rate, and the soil layer compresses (settlement occurs) at an ever-reducing rate.

Terzaghi solved the problem of the time required for this to occur, by making some assumptions:

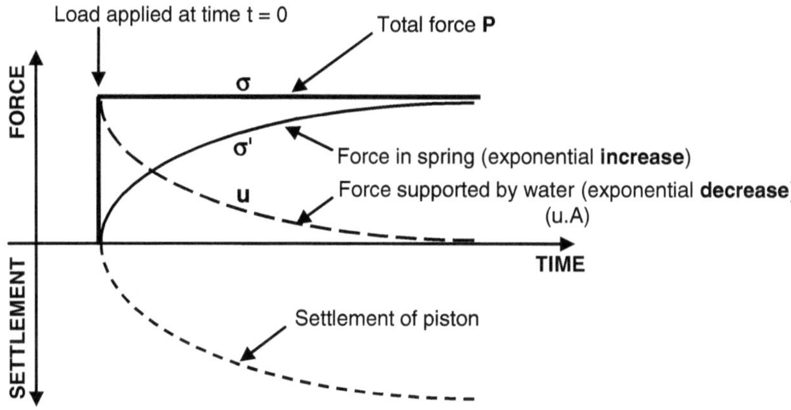

Fig. 4.2 Force versus settlement in a single-cell mechanical analog

- the permeability (k) of the soil does not change as the effective stress increases;
- the compressibility (E'_0) of the soil does not change as the effective stress increases;
- Darcy's Law applies (flow velocity is proportional to hydraulic gradient); and
- The change in thickness of the layer is relatively small compared to the initial thickness.

If the initial state of stress is reasonably high compared to the loading increment, then these assumptions are reasonable (the first effectively assumes that there is not a large change in the voids ratio in the increment; the second assumes that the e–σ'_v curve is linear over the increment of effective stress).

From Eq. (4.8a) and the above assumptions:

$$\frac{k}{\gamma_w}\left(\frac{\partial^2 p}{\partial z^2}\right) = \left[S + \frac{\chi}{(\lambda + 2G)}\right]\frac{\partial p}{\partial t}$$

$$S \approx 0; \quad \frac{\chi}{(\lambda + 2G)} = \frac{1}{E'_0}; \tag{4.11a}$$

$$\frac{kE'_0}{\gamma_w}\left(\frac{\partial^2 p}{\partial z^2}\right) = \frac{\partial p}{\partial t} \tag{4.11b}$$

Terzaghi's solution is written in terms of the permeability of the soil (k), and the compressibility (E'_0), just as our solution to the single cell analog was written. These parameters are combined into a single parameter called the *coefficient of consolidation* (c_v):

$$\mathbf{c_v} = \frac{\mathbf{k\,E'_0}}{\boldsymbol{\gamma_w}} \quad \text{(the units are m}^2\text{/year or m}^2\text{/s)}$$

where γ_w (in kN/m^3) is the unit weight of water ($=\gamma_w \times 9.81$). (This term appears because the solution is in terms of pore pressure (u) rather than the head (h) used in Darcy's Law).

The one dimensional equation of consolidation according to Terzaghi may be derived from Eq. (4.11b):

$$c_V \left(\frac{\partial^2 p}{\partial z^2} \right) = \frac{\partial p}{\partial t} \tag{4.11c}$$

To make the solution general for all problems, Terzaghi used *normalised* parameters.

- The normalized depth **Z** is explained in Fig. 4.3 (Z varies from 0 at a drained boundary to 1 at the maximum distance from a drained boundary).
- The normalized pore pressure dissipation (U_v) at any particular depth Z is:

$$U_v = \frac{u_i - u(t)}{u_i - u_f} \tag{4.12a}$$

  where u(t) is the pore pressure at a particular depth at time t, and u_i and u_f are the initial and final values (the value just after the load is applied, and the value after all consolidation has ceased). **U** varies from 0 at $t = 0$ to 1 (or 100%) at the end of the consolidation.
- The normalized degree of settlement $\mathbf{U_{av}} = \dfrac{\textbf{settlement at time t}}{\textbf{eventual total settlement}}$

  The time required for consolidation depends on:
- the consolidation parameter $\mathbf{c_v}$ (i.e., the combination of permeability and stiffness);
- the maximum distance to a drained boundary, **d** (Fig. 4.3). These are incorporated into a dimensionless *time factor* $\mathbf{T_v}$:

$$T_v = \frac{c_v t}{d^2} \text{ (dimensionless)} \tag{4.12b}$$

The solution is (where $m = 0, 1, 2, \cdots \infty$, and $M = 0.5\,\pi(2\,m + 1)$):

$$U_v = 1 - \sum_{m=0}^{\infty} \frac{2}{M} \sin(MZ)\, e^{-2M^2T_v} \textit{ for the pore pressure dissipation, and}$$

$$\bar{U} \text{ or } U_{av} = 1 - \sum_{m=0}^{\infty} \frac{2}{M^2} e^{M^2T_v} \textit{ for the settlement (or average pore pressure)} \tag{4.13}$$

These equations are used to produce the plots shown in Fig. 4.4. The top figure shows normalized pore pressure isochrones (plots of normalized pore pressure dissipation vs. normalized depth Z), for different time factors. This shows that at

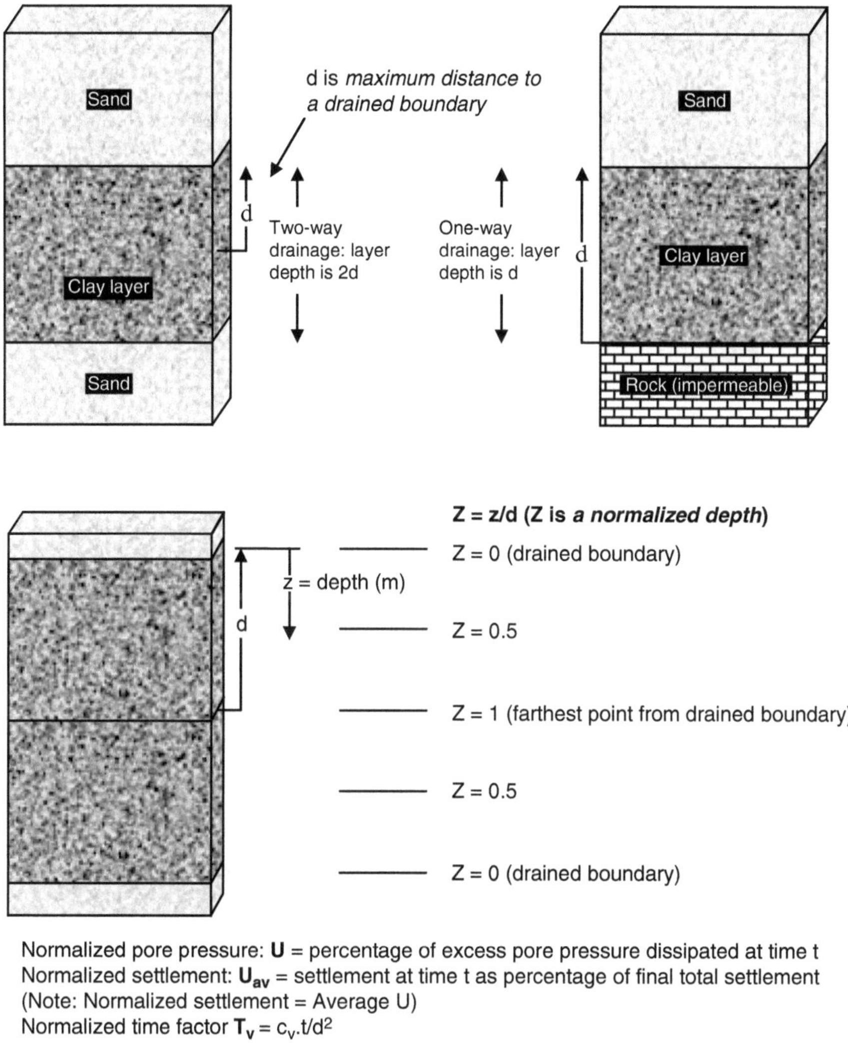

Fig. 4.3 The normalized parameters used in Terzaghi's solution

the drained boundary (Z = 0), the pore pressure dissipates instantaneously but lags behind everywhere else, with the slowest dissipation being at the point farthest from the drained boundary (Z = 1).

The lower plot shows the same information but plotted as pore pressure dissipation vs. the time factor for different points within the mass (different normalised depths Z). The thick black line gives the percentage settlement vs. the time factor (with the settlement axis on the right).

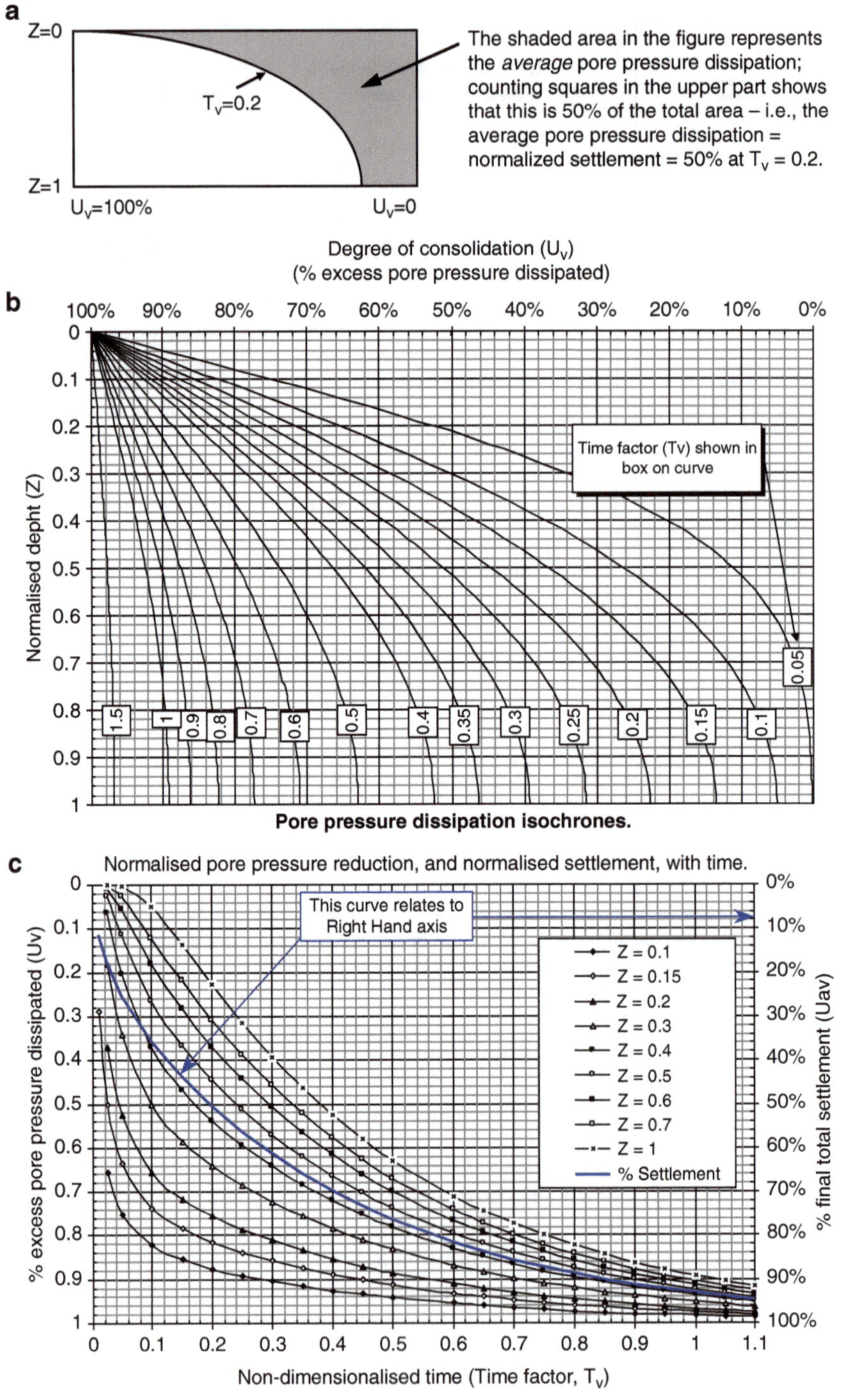

Fig. 4.4 Terzaghi's consolidation solution in graphical form

Looking at the settlement curve (the thick black line in the lower plot), we can see that 50% of the settlement occurs at T_v of 0.2. At this time, only 20% of the pore pressure dissipation has occurred at $Z = 1$, but 88% has occurred at $Z = 0.1$. But as shown below, the average pore pressure dissipation is equal to 50%.

4.1.2.4 Calculating Consolidation Times

We can use Terzaghi's solution (Fig. 4.4a,b,c) to calculate the consolidation time for practical problems. Take the problem discussed previously in Sect. 4.1.2.3.

The layer thickness is 2 m, but it is two-way drained, so **d = 1 m**. The final effective stress increase is 100 kPa. In Fig. 4.6, the initial and final effective stresses in the centre of the layer (48 and 148 kPa) are shown as **points C and D**. The voids ratios are 1.391 and 1.0978. From this, the "one-dimensional stiffness" (E'_0) is:

$$\mathbf{E'_0 = \frac{\Delta\sigma'_v}{\Delta h/h_0} = \frac{\Delta\sigma'_v}{\Delta e/(1+e_0)} = \frac{100}{(1.391 - 1.0978)/(1+1.391)} = 815\ kPa}$$

Assume the permeability $\mathbf{k = 1 \times 10^{-9}\ m/s}$ (low permeability clayey soil). Then the coefficient of consolidation, $\mathbf{c_v}$ is:

$$\mathbf{c_v = \frac{k\,E'_0}{\gamma_w} = \frac{1 \times 10^{-9} \times 815}{9.81} = 8.31 \times 10^{-8}\ m^2/s = 2.62\ m^2/year}$$

So, what is the situation 1 month after the load is applied? This is equivalent to a time factor T_v of:

$$\mathbf{T_v = \frac{c_v\,t}{d^2} = \frac{2.62 \times 1/12}{1^2} = 0.218}$$

From the lower plot in Fig. 4.4, the settlement at this time is 53% completed (i.e., 53% of the 200 mm we calculated earlier has occurred after 1 month). We reach 95% settlement at T_v of 1.125, which is 1.125/0.218 times longer – i.e., 5.16 months.

Note that at 1 month (T_v of 0.218), the pore pressure dissipation at the centre of the layer ($Z = 1$) is about 26%, which means that there is still 74% of the initial excess pore pressure of 100 kPa left, or 74 kPa. If we had a pore-pressure measuring device (a pore pressure *transducer*) located at this point, it would measure this value plus the final hydrostatic value – i.e., 60 + 74 = 134 kPa.

4.1.3 Settlement Computation

This section is taken from the course of the ENPC (Ecole National des Ponts et Chaussees) and the course given by Prof. Martin Fahey at the School of Civil & Resource Engineering at the University of Western Australia.

4.1.3.1 Soil "Density"

The density of the soil particles themselves is denoted γ_s **(the soil particle density)**, and for many soils, this is **between 2.6 and 2.7 t/m³**. The density of water γ_w is **1 t/m³**. The relative density of the soil particles is then $G_s = \frac{\gamma_s}{\gamma_w}$.

As we are generally dealing with soil on the earth's surface (and not on the moon, or in a centrifuge with an elevated g-level), we are usually interested in the *forces* in soil, and thus, we are more interested in weight than density, where weight is simply $\gamma\, g$ (and g, the acceleration due to the earth's gravity, is 9.81 m/s²). The unit weight of water, γ_w, is therefore **9.81 kN/m³**. The unit weight of quartz (the mineral that many sands are comprised of) is about **26 kN/m³**(that is, a solid 1 m³ block of silica would weigh 26 kN.)

If a 1 m³ container is filled with dry silica sand, the weight would be considerably less than the weight of a solid block of silica, due to the air-filled ***voids*** between the particles.

The basic means of expressing the density of packing is to use the ***voids ratio (e):***

$$e = \frac{V_v}{V_s}, \tag{4.11a}$$

where V_v is the volume of the voids, and V_s is the volume of the "solids" (soil particles). Note that **e** can be greater than 1 (it very often is for clay soils).

Thus, for our 1 m³ box of dry sand, the total weight of the soil in this state is the ***dry unit weight*** (γ_d):

$$\gamma_d = \frac{\gamma_s}{1+e} \tag{4.11b}$$

If the voids are now completely filled with water, the box will of course be heavier. The weight would now correspond to the ***saturated unit weight*** (γ_{sat}).

$$\gamma_{sat} = \frac{\gamma_s + e \cdot \gamma_w}{1+e} \tag{4.11c}$$

The "wetness" of a soil is described (in civil engineering) in terms of a weight basis:

$$w = \frac{W_w}{W_s} \quad \text{(often expressed as \%)} \tag{4.11d}$$

(i.e., as the weight of water divided by the weight of the dry soil. This is found by weighing the wet sample first, then drying it in the oven, and then weighing the dry soil).

From these relationships, we can see that (Fig. 4.5):

$$\gamma_d = \frac{\gamma_{sat}}{1+w} \tag{4.11e}$$

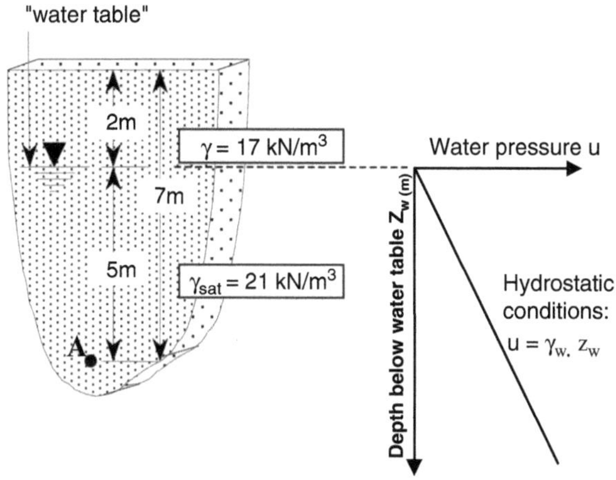

The "**water table**" represents the level that water would eventually be at in a hole dug in the ground, where the water pressure is "hydrostatic" -i.e., where the water pressure increases uniformly with depth in exactly the same way that it increases in an open body of water (i.e., a lake).

"**Hydrostatic**" means exactly that -"static" -i.e., no flow. That means that the total head is equal at all depths (the pressure head increases with depth, but the elevation head reduces with depth). So, it follows that for vertical flow to occur, the water pressure distribution must be different from hydrostatic. This difference is what we will call later ***excess pore pressure.***

For Point A:

Total vertical stress	$\sigma_v = 2 \times 17 + 5 \times 21$	**= 139 kPa**
Pore (water) pressure	$u = 5 \times 10$	**= 50 kPa**
	$(\gamma_w = 9.81\ kN/m^3 \approx 10\ kN/m^3)$	
Therefore, effective vertical stress	σ'_v	**= 89 kPa**

Fig. 4.5 Calculating vertical effective stress in the ground

4.1.3.2 Settlement Calculations

A Normally Consolidated Layer

From the previous section, we can calculate the settlement of a **normally consolidated** soil layer (i.e., where the state is on the initial loading curve) due to any loading increment from:

$$\frac{\Delta h}{h_o} = \frac{\Delta e}{1 + e_o} = \left(\frac{C_c}{1 + e_o}\right) \log_{10}(\sigma'_0/\sigma'_1) \tag{4.12}$$

where σ'_0 is the initial effective stress, and σ'_1 is the final value (i.e., $\Delta\sigma'_V = \sigma'_1 - \sigma'_0$). Therefore, we need to know the initial state (effective stress and voids ratio) and the increase in effective stress due to the applied load (after all excess pore pressure has dissipated).

Example: A 2 m-thick clay layer is located between 5 and 7 m depth at a site. There is dense sand (incompressible in comparison to the clay) above and below the clay. Assume that the water table is starting at the surface with hydrostatic conditions and that the saturated density (γ_{sat} **sat**) of the sand is **18 kN/m³**. This gives the initial effective vertical stress on the top of the clay of:

$$\boldsymbol{\sigma'_v = \sigma_v - u = \gamma_{sat}.z - \gamma_w.z = (18 \times 5) - (10 \times 5) = 90 - 50 = 40\ kPa}$$

where z is depth, and γ_w (the unit weight of water) is taken as 10 kN/m^3, for convenience. If the clay has the same saturated unit weight, the effective stress at the bottom of the layer will be an extra 16 kPa, or **56 kPa**. Thus, the average initial vertical stress in the clay (σ'_o) is **48 kPa**.

Assume that C_c represents the compressibility of the clay and that it is normally consolidated (a natural state lies on the normally consolidated line). For the average vertical stress of 48 kPa, the initial voids ratio (e_o) would be **1.391** (shown as **point C** in Fig. 4.7).

Assume that the large structure placed over a very wide area on the surface imposes a vertical stress of **100 kPa**. Eventually, this will result in an increase in effective stress ($\Delta\sigma'_V$) of 100 kPa (we shall calculate later *how much time* it might take for this to happen). Therefore, the new average effective stress (σ'_1) at the center of the clay layer will be **148 kPa**.

The height change will then be:

$$\boldsymbol{\frac{\Delta h}{h_o} = \frac{\Delta e}{1+e_o} = \left(\frac{C_c}{1+e_o}\right)\log_{10}(\sigma'_o/\sigma'_1)}$$

$$\boldsymbol{= \frac{0.6}{(1+1.391)} \times \log_{10}(48/148) = -0.123}$$

(The result is negative since the height – and the voids ratio – reduces). The result is therefore a height reduction of about 12% of the initial height, or **240** mm.

Note that if the soil was in an *overconsolidated* state, the same stress increase would give a much lower settlement (because the overconsolidated soil is less compressible).

The settlement therefore depends not just on the *increase* in σ'_V, but also on the *starting* value, and whether the starting state is normally consolidated or overconsolidated (Fig. 4.6).

An Over-Consolidated Layer

For soil in an **over-consolidated** state, the process of calculating settlement due to an applied increase in vertical stress is similar to that described in the previous section, except that now some of the compression will be along the re-compression line (until the stress reaches the pre-consolidation stress σ'_c), and then it follows the normal consolidation line.

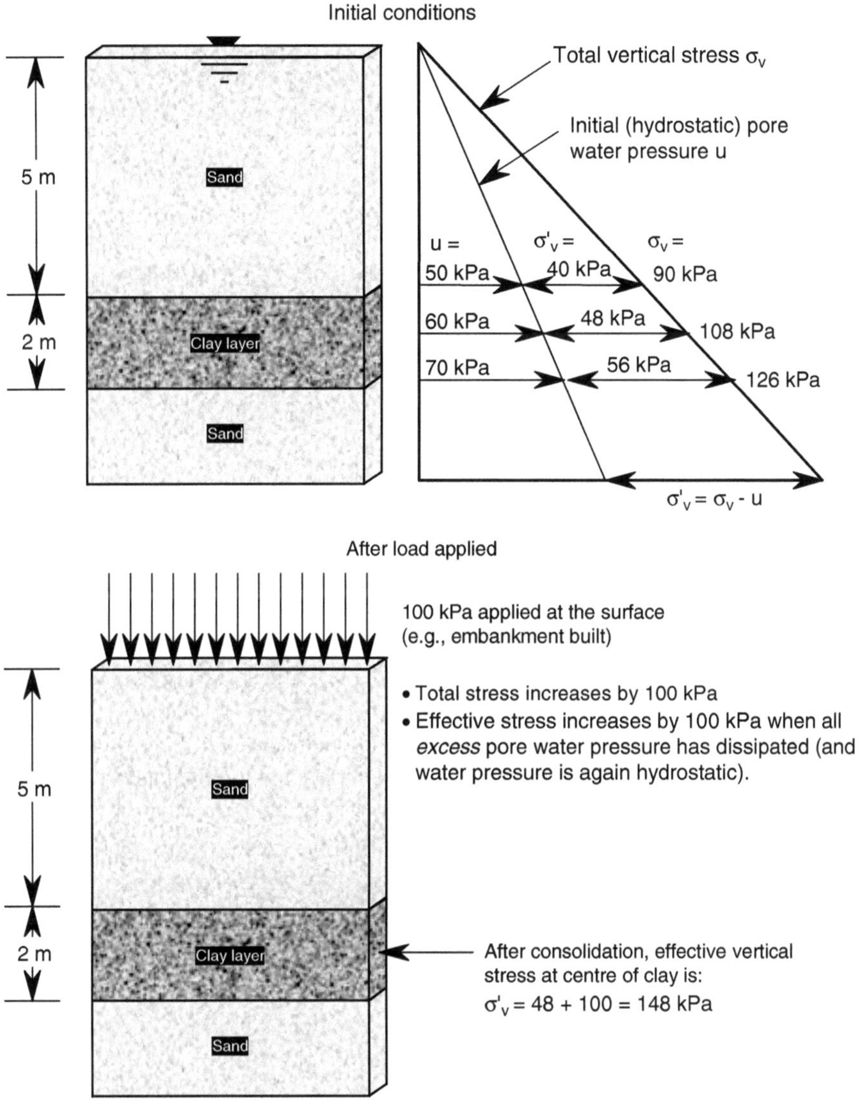

Fig. 4.6 Settlement of a layer of clay due to load (stress) applied at the ground surface

The steps involved are:

1. Determining the *initial* vertical effective stress σ'_{v0} and *initial* voids ratio $\mathbf{e_0}$, as before
2. Knowing the additional effective stress that will result from the surface loading and calculating the *final* vertical effective stress σ'_{vf}
3. There are then three possibilities to consider:

(a) $\sigma'_{v0} = \sigma'_c$ (in which case the soil is actually normally
(b) $\sigma'_{vf} \leq \sigma'_c$: in this case the state remains in the over consolidated state at the end of loading, and the settlement is calculated using the re-compression index only:

$$\Delta e = C_r \log_{10}\left(\frac{\sigma'_{vo}}{\sigma'_{vf}}\right) \Rightarrow \Delta h = h_o \frac{\Delta e}{1+e_o} \tag{4.13a}$$

(c) $\sigma'_{v0} < \sigma'_c$ **and** $\sigma'_{vf} > \sigma'_c$: in this case, some of the loading is calculated using the recompression index, and the rest using the compression index:

$$\Delta e = C_r \log_{10}\left(\frac{\sigma'_{vo}}{\sigma'_c}\right) + C_c \log_{10}\left(\frac{\sigma'_c}{\sigma'_{vf}}\right) \Rightarrow \Delta h = h_o \frac{\Delta e}{1+e_o} \tag{4.13b}$$

An Example of Computation of the Displacement

Assume that the compression diagram shown in Fig. 4.7 represents a **2-m thick layer**, with the initial conditions represented by Point A (i.e., $\mathbf{e_o} = \mathbf{1.178}$ and $\sigma'_{v0} =$ **30 kPa**). Assume that a surface load is applied, resulting in an increase in the vertical stress of **290 kPa**. This would produce a final effective vertical stress of $\sigma'_{vf} =$ **320 kPa**. Assume that the soil is overconsolidated, with preconsolidation pressure (past maximum consolidation pressure) of $\sigma'_c =$ **150 kPa**.

This case therefore corresponds to (c) above, with the initial stress (Point A) being *less than* the preconsolidation stress (Point B), and the final stress (Point C) being *greater than* the preconsolidation stress. Thus, assuming that $\mathbf{C_c} = 0.6$ and $\mathbf{C_r} = 0.12$, as before, the change in the voids ratio is:

$$\Delta e = 0.12\log_{10}\left(\frac{30}{150}\right) + 0.6\log_{10}\left(\frac{150}{320}\right) = -0.0839 - 0.1974$$

$$= -0.2813 \Rightarrow e_f = 1.178 - 0.281 = 0.897$$

$$\Rightarrow \Delta h = h_o \frac{\Delta e}{1+e_o} = 2000\left(\frac{0.281}{2.178}\right) = \underline{258\ mm}$$

Note, if normally consolidated :

$$\Delta e = 0.6\log_{10}\left(\frac{30}{320}\right) = -0.617 \Rightarrow \Delta h = 2000\left(\frac{0.617}{2.178}\right) = \underline{566\, mm}$$

The settlement of a thicker layer can be calculated by dividing the layer into a number of sub-layers as shown in Fig. 4.8. This is necessary because both the initial and final effective stress vary with depth as do the voids ratio and the OCR.

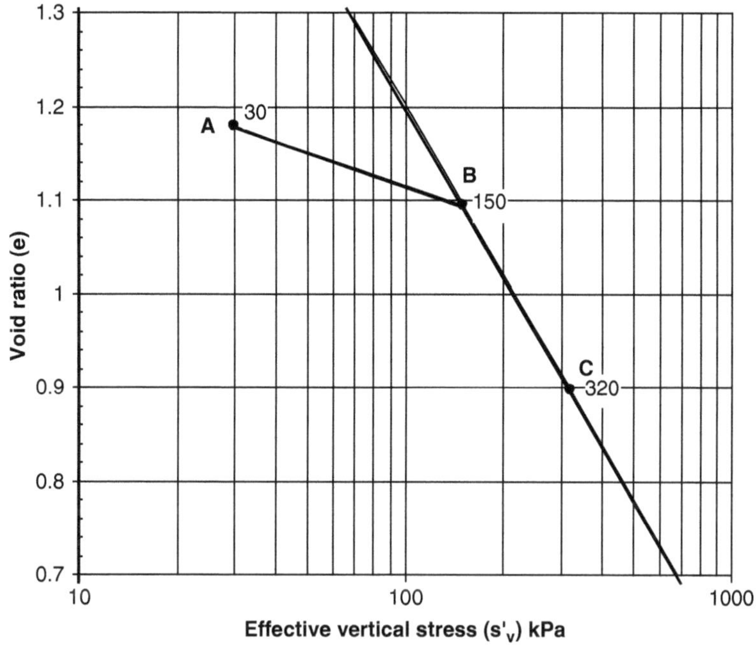

Fig. 4.7 An example of overconsolidated soil and a calculation of the settlement

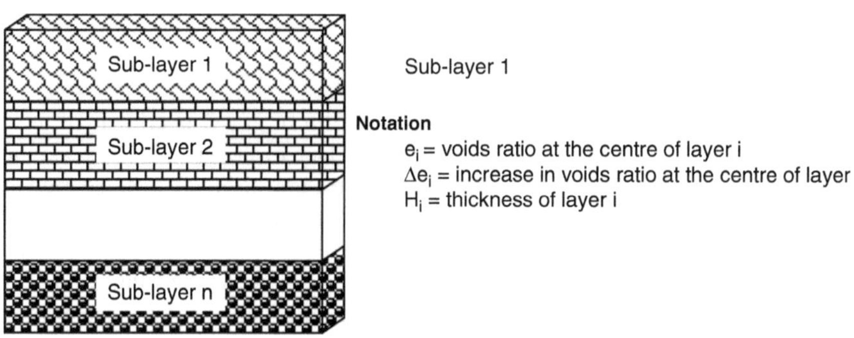

Fig. 4.8 Division of soil layers into sub-layers

The settlement of the soil layer is calculated by calculating the settlement of the individual sub-layers and adding them; in doing this, it is assumed that the voids ratio and the effective stress are constant throughout the sub-layer and equal to their values at the center of thesub-layer.

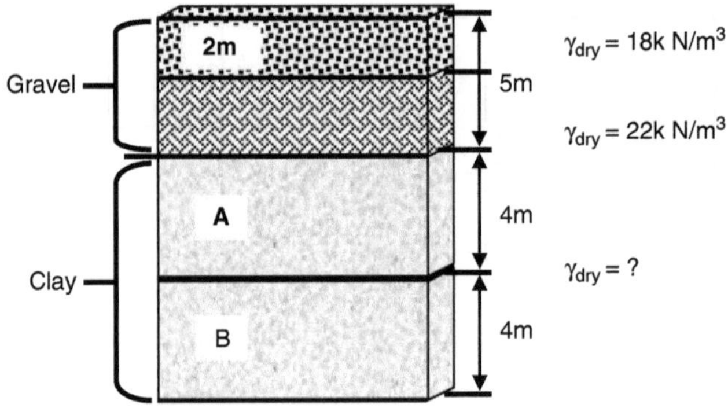

Fig. 4.9 A layered soil deposit

thus

$$\text{For sub-layer i } \Delta S_i = -\frac{\Delta e_i H_i}{1 + e_i}$$

so that

$$\text{Total Settlement S} = \sum_1^n \Delta S_i = \sum_1^n \left[-\frac{\Delta e_i H_i}{1 + e_i} \right]$$

An Example of a Settlement Calculation

A soil deposit, as shown in Fig. 4.8, consists of 5 m of gravel overlaying 8 m of clay. Initially the water table is 2 m below the surface of the gravel. Calculate the settlement if the water table rises to the surface of the gravel slowly over a period of time and surface loading induces an increase of total stress of 100 kPa at the point A and 60 kPa at the point B. The preconsolidation pressure at A is 120 kPa, and the deposit is normally consolidated at B. The gravel has a saturated bulk unit weight of 22 kN/m^3 and a dry unit weight of 18 kN/m^3 and is relatively incompressible when compared to the clay. The voids ratio of the clay is 0.8, and the skeletal particles have a specific gravity of 2.7. The compression index of the clay is 0.2, and the recompression index is 0.05.

In solving this problem, it will be assumed that the gravel is far less compressible than the clay and thus, that the settlement of the gravel can be neglected. The settlement of the clay layer will be calculated by dividing it into two sub-layers.

In order to begin the calculations, it is first necessary to calculate the unit weight of the clay; this is shown schematically in Fig. 4.9.

$$\gamma_{sat} = \frac{W_w + W_s}{V_v + V_s} = \frac{7.84 + 26.46}{0.8 + 1} = 19.06\ \text{kN/m}^3$$

or

$$\gamma_{sat} = \frac{(G_s + e)\gamma_w}{1 + e} = 19.06\ kN/m^3$$

- **The initial state at A**

$$\begin{array}{lll} \text{Total stress} & \sigma_{zz} & = 2 \times 18 + 3 \times 22 + 2 \times 19.06 = 140.12\ kPa \\ \text{Pore water pressure} & u_w & = 5 \times 9.8\ kPa = 49\ kPa \end{array} \qquad (3a)$$

$$\text{Effective stress} \quad \sigma'_{zz} = \sigma_{zz} - u_w = 140.12 - 49 = 91.12\ kPa$$

Notice that the initial effective stress is less than σ'_{pc} =120 kPa; thus, the clay is initially over-consolidated.

- **The final state at A**

$$\begin{array}{lll} \text{Total stress} & \sigma_{zz} & = 100 + 2 \times 22 + 3 \times 22 + 2 \times 19.06 = 248.12\ kPa \\ \text{Pore water pressure} & u_w & = 7 \times 9.8\ kPa = 68.6\ kPa \end{array} \qquad (3b)$$

$$\text{Effective stress} \quad \sigma'_{zz} = \sigma_{zz} - u_w = 248.12 - 68.6 = 179.52\ kPa$$

Notice that the final effective stress exceeds the initial preconsolidation stress and thus, the clay moves from being initially over-consolidated to finally being normally consolidated.

- **The settlement of the first sub-layer**
  The soil in the first sub-layer moves from being overconsolidated to normally consolidated and thus, the calculation of the change in the voids ratio must be made in two stages.

Stage 1 Soil overconsolidated ($\sigma' < \sigma'_{pc\ (initial)}$)

$$\Delta e_1 = -C_r \times \log_{10}(\sigma'_{pc\ (initial)}/\sigma'_I)$$

Stage 2 Soil normally consolidated ($\sigma' = \sigma'_{pc}$)

$$\Delta e_2 = -C_c \times \log_{10}\left(\sigma'_F/\sigma'_{pc(initial)}\right)$$

now

$$\begin{aligned} \Delta S &= -\frac{H\Delta e}{1 + e} \\ &= \frac{4}{1.8}\left[0.05 \times \log_{10}\left(\frac{120.00}{91.12}\right) + 0.2 \times \log_{10}\left(\frac{179.52}{120.00}\right)\right] \\ &= 0.0911\ m \end{aligned}$$

- **The initial state at B**

$$\begin{array}{lll} \text{Total stress} & \sigma_{zz} & = 2 \times 18 + 3 \times 22 + 6 \times 19.06 = 216.36\ \text{kPa} \\ \text{Pore water pressure} & u_w & = 9 \times 9.8\ \text{kPa} = 88.20\ \text{kPa} \end{array} \qquad (4a)$$

$$\text{Effective stress} \quad \sigma'_{zz} = \sigma_{zz} - u_w = 216.36 - 88.20 = 128.16\ \text{kPa}$$

- **The final state at B**

$$\begin{array}{lll} \text{Total stress} & \sigma_{zz} & = 60 + 2 \times 22 + 3 \times 22 + 6 \times 19.06 = 284.36\ \text{kPa} \\ \text{Pore water pressure} & u_w & = 11 \times 9.8\ \text{kPa} = 107.80\ \text{kPa} \end{array} \qquad (4b)$$

$$\text{Effective stress} \quad \sigma'_{zz} = \sigma_{zz} - u_w = 284.36 - 107.80 = 176.56\ \text{kPa}$$

- **The settlement of the second sub-layer**
  The soil in the second sub-layer is normally consolidated and thus:

$$\Delta e_2 = -C_c \times \log_{10}(\sigma'_F/\sigma'_I) \qquad (4c)$$

now

$$\begin{aligned} \Delta S &= -\frac{H\,\Delta e}{1+e} \\ &= \frac{4}{1.8} 0.2 \times \log_{10}\left(\frac{176.56}{128.16}\right). \\ &= 0.0620\ \text{m} \end{aligned}$$

- **Total settlement**

$$\text{Total settlement} = 0.0911 + 0.0620\ \text{m} = 0.1531\ \text{m}$$

4.2 The Three-Dimensional Theory of Consolidation

The derivation of the three-dimensional theory of consolidation may be performed as follows:

Starting from Eq. (3.54) and the basic relations:

$$\frac{\partial}{\partial \boldsymbol{x}_i}\left(\frac{\boldsymbol{K}_{ij}}{\rho g}\frac{\partial \boldsymbol{P}}{\partial \boldsymbol{x}_j}\right) = \rho\frac{\partial \boldsymbol{n}}{\partial \boldsymbol{t}} + \boldsymbol{n}\frac{\partial \rho}{\partial \boldsymbol{t}} \qquad (3.54)$$

$$\frac{\partial n}{\partial t} = \frac{\partial n}{\partial P} \frac{\partial P}{\partial t}$$
$$\frac{\partial \rho}{\partial t} = \frac{\partial \rho}{\partial P} \frac{\partial P}{\partial t}$$

the consolidation process under drained or undrained conditions involves mainly a deformation of the skeleton of the soil. Therefore, $\frac{\partial \rho}{\partial P} \approx 0$ and only the term $\frac{\partial n}{\partial P}$ is considered. The classical theory of consolidation used the void indice $e = \frac{V_v}{V_s}$ (the ratio of the void volume to the solid volume) rather than the porosity n. When the soil is saturated, $e = \frac{V_w}{V_s}$ then

$$n = \frac{V_w}{V_T} = \frac{V_w}{V_w + V_s} = \frac{1}{1 + \frac{1}{e}} = \frac{e}{1+e}. \tag{4.14}$$

The relation between the void ratio and the water pressure depends on the rheological behavior of the soil.

For granular soil:

$$e = e_0 - a_v \sigma' \quad or\ e = e_0 - a_v(\sigma' - \sigma_0{}') \tag{4.15}$$

Where a_v (> 0) is a coefficient of compressibility; e_0 is the void ratio at the effective stress σ'_0.

For cohesive soils, experimental results lead to a relationship of the type:

$$e = e_0 - C_c \log(\sigma' / \sigma'_0)$$

C_c is the compression index. For small changes in e, the coefficient of compressibility is also used for cohesive soils.

When we assume that the total stress, σ due to external load remains unchanged, then using the Terzaghi principle:

$$e = e_0 - a_v(\sigma - p) \Rightarrow \frac{\partial e}{\partial p} = a_v$$
$$\frac{\partial n}{\partial t} = \frac{\partial n}{\partial P} \frac{\partial P}{\partial t} = \frac{\partial n}{\partial e} \frac{\partial e}{\partial p} \frac{\partial p}{\partial t} = \frac{\rho g a_v}{(1+e_0)^2} \frac{\partial H}{\partial t} \cong \frac{\rho g a_v}{(1+e_0)} \frac{\partial H}{\partial t} \tag{4.16}$$

The consolidation equation may be written in terms of the water pressure:

$$\frac{\partial}{\partial x_i} \left(\frac{K_{ij}}{\rho g} \frac{\partial P}{\partial x_j} \right) = \frac{\rho g a_v}{(1+e_0)} \frac{\partial p}{\partial t} \tag{4.17}$$

4.3 The Elastic Model in Compression and Extension

4.3.1 Normally Consolidated and Overconsolidated Soils

This linear relationship is called **Hooke's law**. The classic model of linear elasticity is the perfect spring. Although the general proportionality constant between stress and strain in three dimensions is a fourth-order tensor, when considering simple situations of higher symmetry such as a rod in one-dimensional loading, the relationship may often be reduced to applications of Hooke's law.

Because most materials are only elastic under relatively small deformations, several assumptions are used to linearize the theory. Most importantly, higher order terms are generally discarded based on the small deformation assumption. In certain special cases, such as when considering a rubbery material, these assumptions may not be permissible. However, in general, elasticity refers to the linearized theory of the continuum stresses and strains.

Elasticity is a branch of physics which studies the properties of elastic materials. Tensile stress (or tension) is the stress state leading to expansion, i.e., the length of a material tends to increase in the tensile direction.

For example, the rod has a length L and a cross-sectional area A. Its extension (strain) is linearly proportional to its tensile stress, σ by a constant factor, the inverse of its modulus of elasticity, E.

$$\sigma = E\,\varepsilon$$

or

$$\Delta L = \frac{F}{EA} L = \frac{\sigma}{E} L \tag{4.18}$$

Hooke's law only holds for some materials under certain loading conditions. Granular soils and also clays exhibit linear-elastic behavior in most soil mechanical problems; Hooke's law is valid for it throughout its **elastic range** (i.e., for stresses below the yield strength). For some soils, such as gravel, Hooke's law is only valid for a portion of the elastic range. For these materials, proportional limit stress is defined, below which the errors associated with the linear approximation are negligible.

This model has been used by soil-mechanic scientists using a variant of the linear model called the compressibility model, briefly explained in the following.

After a soil forms by sedimentation from a slurry, the "sedimented voids ratio" may be very high. The exact value depends heavily on the *clay content*, and the *type of clay mineral* present. For instance, the voids ratio at this stage is 3.0. We can imagine the structure of the clay at this stage to be very "open", with the voids volume being three times the solids volume. (This is like a very loosely-packed box of cornflakes).

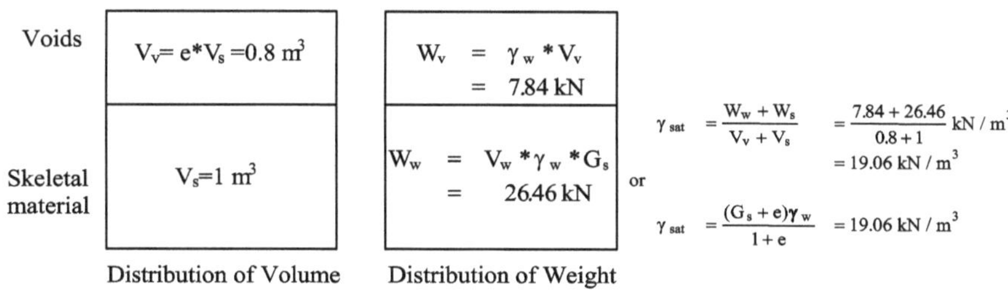

Fig. 4.10 The determination of saturated unit weight

This structure is very *compressible* – i.e., a very small stress applied to the surface will cause a large amount of settlement. However, if further equal increments of stress are applied, the amount of compression under successive increments reduces. The soil becomes progressively *less compressible* with increasing stress. Thus, the compression curve is *highly non-linear*, which is illustrated in the upper part of the following figures (Fig. 4.10).

If the stress is reduced at any stage, some swelling occurs (water is sucked back into the sample), but the initial state is never recovered. Most of the compression is irrecoverable.

If the state lies on the initial compression curve, the soil is said to be ***normally consolidated***. If it has swelled back from a higher stress state, it is ***overconsolidated***. The ratio of the *present* effective stress (σ'_v) to the *past maximum* effective stress (σ'_c) is called the ***overconsolidation ratio*** (**OCR**). Thus, for point A, σ'_v is about 10 kPa, σ'_c is about 100 kPa (point B), and thus, at point A, the OCR is about 10.

When plotted on a *semi-logarithmic* plot, the initial loading points tend to lie on a straight line. This shows in this case that to reduce the voids ratio by 0.6 from 3 to 2.4 requires a stress increase of 0.9 kPa, but to reduce it by a further 0.6 requires a 9 kPa increase, with the next 0.6 reduction requiring an increase of 90 kPa, etc (Fig. 4.11).

The swelling-recompression data also plot on straight lines, which are flatter than the initial loading line and parallel to each other.

The ***Compression Index*** (**C_c**) is defined as the gradient (magnitude) of the initial loading curve:

$$C_c = \left|\frac{\textit{change in voids ratio}}{\textit{change in log of effective stress}}\right| = \left|\frac{(e_o - e_1)}{\log_{10}(\sigma'_o) - \log_{10}(\sigma'_1)}\right|$$

$$= \left|\frac{\Delta e}{\log_{10}(\sigma'_o / \sigma'_1)}\right| \tag{4.19}$$

Similarly, the swelling/recompression index (C_s or C_r) is the gradient of the swelling and recompression lines.

In the following Fig. (4.12), the value of $\mathbf{C_c}$ is **0.6**, and $\mathbf{C_s = C_r = 0.12}$.

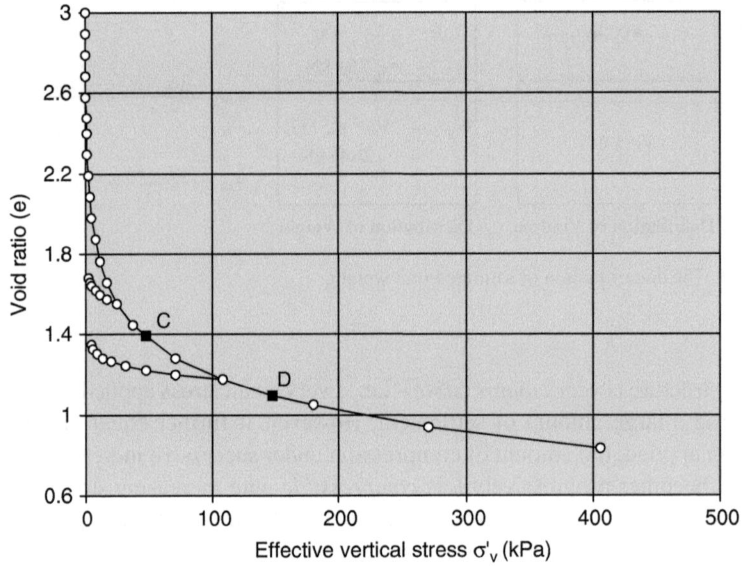

Fig. 4.11 Void ratio versus effective stress curve

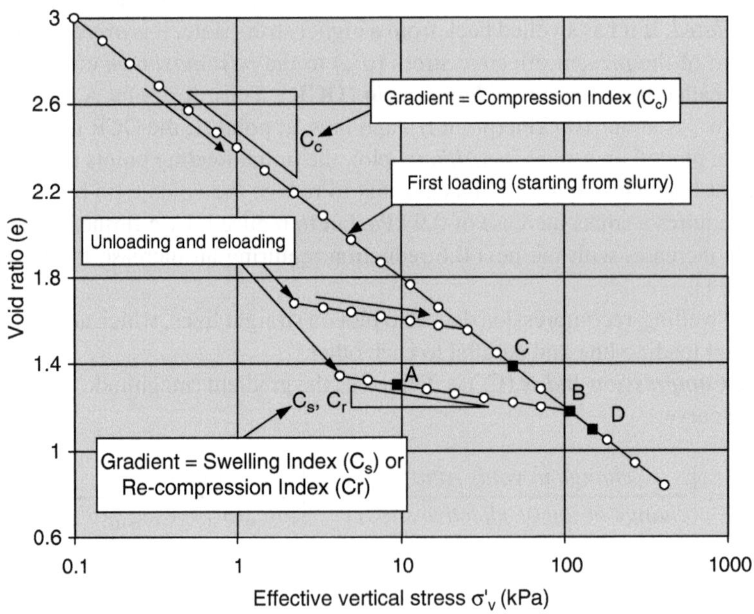

Fig. 4.12 The compression coefficient from a slurry

For soil in an **overconsolidated** state, the process of calculating settlement due to an applied increase in vertical stress is similar to that described in the previous section, except that now, some of the compression will be along the re-compression line (until the stress reaches the pre-consolidation stress σ'_c), and then it follows the normal consolidation line.

The steps involved are:

1. Determining the *initial* vertical effective stress $\boldsymbol{\sigma'_{vo}}$ and the *initial* void ratio $\mathbf{e_o}$, as before
2. Knowing the additional effective stress that will result from the surface loading and calculating the *final* vertical effective stress $\boldsymbol{\sigma'_{vf}}$
3. There are then three possibilities to consider:

   (a) $\sigma_{v0} = \sigma'_c$ (in which case the soil is actually normally consolidated, so proceed as in previous section);

   (b) $\sigma'_{vf} \leq \sigma'_c$: in this case the state remains in the over-consolidated state at the end of loading, and the settlement is calculated using the re-compression index only:

$$\Delta e = C_r \log_{10}\left(\frac{\sigma'_{vo}}{\sigma'_{vf}}\right) \Rightarrow \Delta h = h_o \frac{\Delta e}{1 + e_o}; \tag{4.20}$$

   and

   (c) $\sigma_{v0} < \sigma'_c \leq \sigma'_{vf}$: in this case, some of the loading is calculated using the recompression index, and the rest using the compression index:

4.3.2 *Use of the Consolidation Theory for Computation of Subsidence Due to Pumping*

The Terzaghi theory of consolidation was mainly developed for computing the soil compression due to loading. However, the rheological model of Terzaghi was also used in hydrogeology to compute the subsidence due to water withdrawal due to water pumping. In this section, we summarized this approach presented by Poland and Davis (1969) and implemented by the USGS into the MODFLOW 2000 code (see Hoffmann et al. 2003). This approach was used in several engineering projects (see Bravo et al. 1991a, b; Stanley et al. 2010).

According to Terzaghi, the soil subsidence is caused by the compaction (or consolidation) of the soil matrix, that results from a change in effective stresses.

To incorporate calculations of consolidation into ground-water flow models, the common approach is to use the Terzaghi relation. This relation relates the effective (inter-granular) stress to the total stress and the water pore pressure.

$$\sigma'_{ij} = \sigma_{ij} - \delta_{ij} u \tag{4.21a}$$

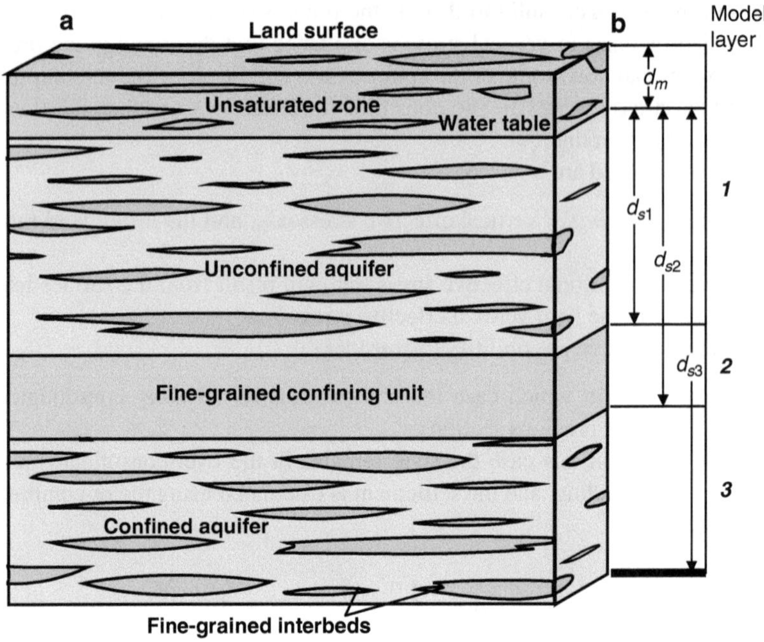

Fig. 4.13 Vertical section of a two-aquifer system with potential for compaction of fine-grained sediments. (**a**) Hydrogeology of the system. (**b**) Representation of the system with three model layers

σ'_{ij}: is the component ij of the effective stress tensor
σ_{ij}: is the component ij of the geostatic total tensor
δ_{ij}: is the Kronecher delta function and
u: is the fluid pore pressure or hydrostatic stress.

The relation (1) shows that changes in effective stress can result from changes in geostatic stress or changes in pore pressure.

The geostatic stress is the sum of the loads due to buildings, the load of the overlying saturated and unsaturated sediments and also the tectonic stress.

The change in water pore pressure results from water level changes. If the interbeds are assumed to be horizontal and laterally extensive with respect to their thickness, changes in pore pressure gradients is primarily vertical (Fig. 4.13).

Assuming that the resulting strains also are primarily vertical and ignoring tectonic stresses, the tensor equation in a vertical direction with one dimensional assumption result with:

$$\sigma' = \sigma - u \tag{4.21b}$$

Where σ' and σ are the vertical components of effective and geostatic stress, respectively.

When no tectonic stresses are present, the geostatic stresses can be expressed as the sum of the loading on the soil surface, the geostatic and the hydrostatic stresses (Poland and Davis 1969).

$$\sigma = \sigma_{load} + \gamma_m d_m + \gamma_s d_s \tag{4.21c}$$

$$\gamma_m = \gamma_g (1 - n) + n_w \gamma_w \tag{4.22d}$$

$$\gamma_s = \gamma_g - n \left(\gamma_g - \gamma_w\right) \tag{4.22e}$$

and

$$u = d_s \gamma_w$$

where

γ_m is the unit weight of moist sediments above the water table;
γ_g is the unit weight of sediments grains;
γ_s is the unit weight of saturated sediments below the water table;
γ_w is the unit weight of water;
n is the porosity;
n_w is the moisture content of sediments in the unsaturated zone, as a fraction of total volume;
d_m is the depth below land surface in the unsaturated-zone interval, land surface ($z = 0$) to the water table ($z = z_{wt}$); and
d_s is the depth of interest in the saturated zone.

4.3.2.1 The Relationship Between the Variation of Effective Stresses and the Change in the Water Table

Given the geostatic and hydrostatic stresses, the effective stress can be calculated by using the Terzaghi equation. A change in effective stress resulting from a given head change generally differs in unconfined and confined aquifers. In an unconfined aquifer, a change in head corresponds to a change in the position of the water table. The draining or re-wetting of pore space in the zone of water table fluctuation results in a change in the geostatic stress on the underlying sediments in the unconfined and confined aquifers. A change in effective stress caused by a head change in an unconfined aquifer can be described as (Poland and Davis 1969):

$$\Delta\sigma' = -\gamma_w (1 - n + n_w) \Delta h_{wt} \tag{4.23}$$

where $\Delta h_{wt} = \Delta u/\gamma_w$ is the change in head of the water table in a confined aquifer, the geostatic stress changes negligibly with changes in head in the confined aquifer

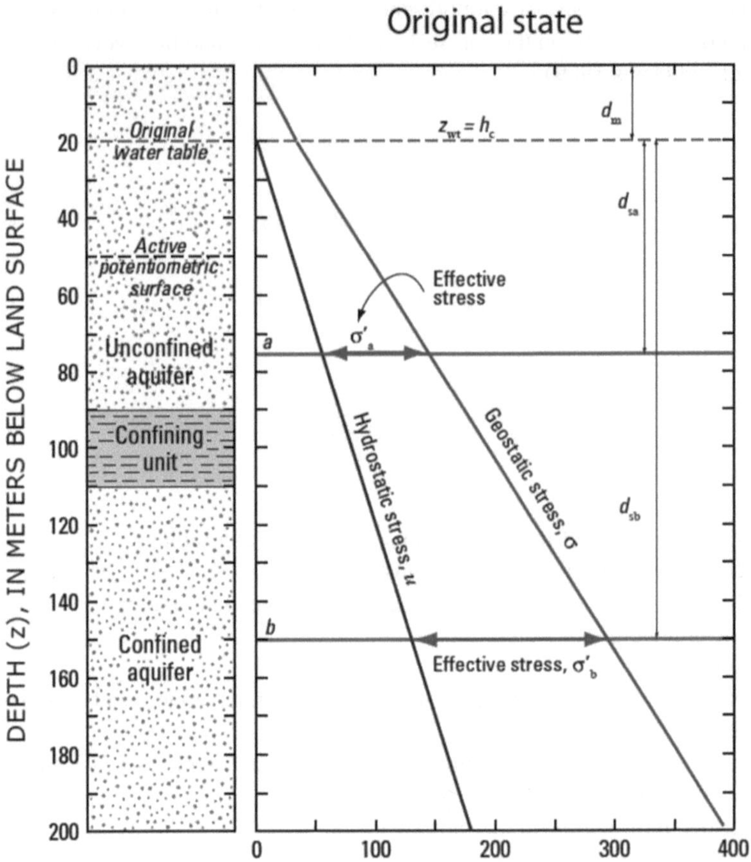

Fig. 4.14 Stress diagrams for water-table decline with a stable underlying confined aquifer,and for stable water-table with head decline in an underlying confined aquifer

owing to the small changes in the unit weight of water associated with the expansion or compression of water. A change in effective stress caused by a head change in a confined aquifer can be described as (Poland and Davis 1969):

$$\Delta\sigma' = -\gamma_w \Delta h_c$$

Where $\Delta h_c = \Delta u/\gamma_w$ is the change in head of the confined aquifer. It is important to note that the change in effective stress caused by a head change in an unconfined aquifer is reduced by a factor of $(1 - n + n_w)$ to that caused by an equivalent head change in a confined aquifer.

Figure 4.14 illustrates the relationships between geostatic, hydrostatic and effective stresses and the changes in these stresses caused by the lowering of

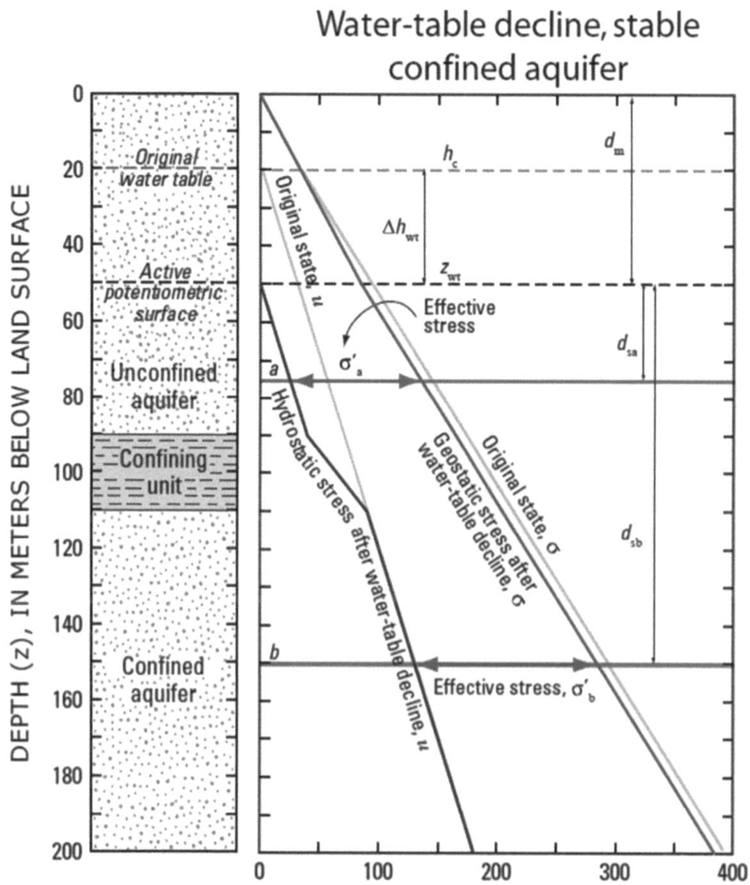

Fig. 4.14 (continued)

hydraulic heads in an idealized unconfined aquifer. The displayed stresses ignore delayed fluid-pressure equilibration in the confined unit. The effective stresses are shown (Fig. 4.14) for two horizons (*a* and *b*) in the aquifer system, one each in the unconfined and confined aquifers. The left-most stress diagram in Fig. 4.14 shows the original state of stresses with the equal head in the confined and unconfined aquifers Table 4.1.

The middle diagram in Fig. 4.14 shows the stresses after a 30-m lowering of the water table. Table 4.2 shows the computed stresses using equations (4.21c–e) at depth *a* and *b* in Fig. 4.15. The geostatic stress decreases slightly below the depth of the original water table. The unconfined aquifer below the depth of the active water table, the confining unit, and the confined aquifer are affected equally by a 9-m decrease in geostatic stress.

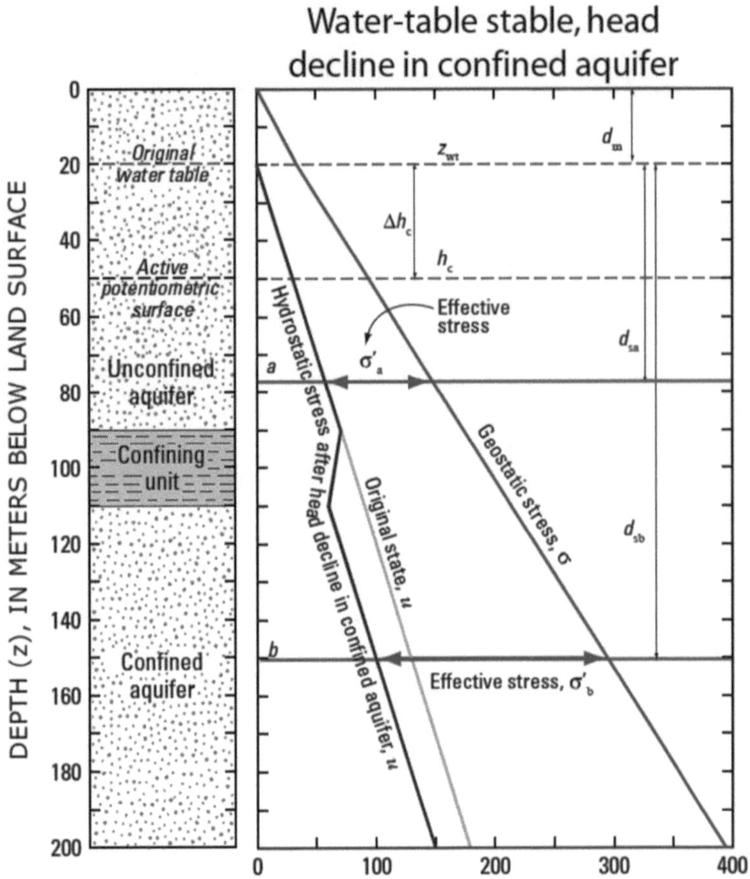

Fig. 4.14 (continued)

Table 4.1 Specified properties of soils used to compute stresses shown in Fig. 4.13

γ_g	2.62×10^4^
γ_w	9.81×10^3^
n	0.40
n_w	0.10

Units for γ terms are newtons per cubic meter; units for n terms are dimensionless

Table 4.2 Computed properties

γ_m	1.67×10^4
γ_s	1.96×10^4
$(1-n+n_w)$	0.70

Units for γ terms are newtons per cubic meter; units for n terms are dimensionless

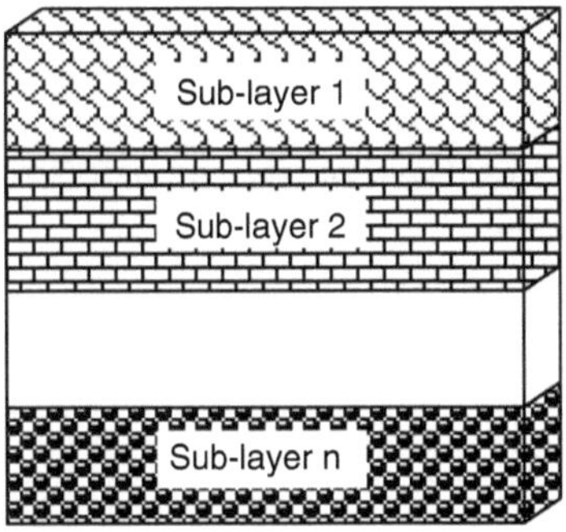

Fig. 4.15 A simplified soil profile

The hydrostatic stress decreases in the unconfined aquifer by 30 m and is unchanged in the confined aquifer.

4.3.2.2 A One-Dimensional Theory

For many types of sediment, the voids ratio, e, decreases linearly with an increase in the logarithm of effective stress.

$$\Delta e = -C_c \Delta \log_{10}\sigma' = -C_c \log_{10}(\sigma' + \Delta\sigma') - \log_{10}(\sigma')$$

$$\Delta e = -C_c \log_{10}\left(\frac{\sigma' + \Delta\sigma'}{\sigma'}\right) = -C_c \log_{10}\left(1 + \frac{\Delta\sigma'}{\sigma'}\right) \qquad \sigma' > \sigma'_c$$

$$\Delta e \approx -C_c \frac{\Delta\sigma'}{\sigma'} \quad \text{for} \left|\frac{\Delta\sigma'}{\sigma'}\right| \ll 1 \tag{4.22a}$$

Similarly

$$\Delta e = -C_r \Delta \log_{10}\sigma' \approx -C_r \frac{\Delta\sigma'}{\sigma'} \quad \sigma' \leq \sigma'_c \tag{4.22b}$$

Where C_c and C_r are the dimensionless compression and recompression indices and σ'_c is the pre-consolidation stress. These relations are valid for both preconsolidated soil and also overconsolidated soil.

It is important to notice that the change in the voids ratio in Eq. (4.22a) is permanent and does not include the instantaneous elastic compaction.

Also the compaction of the layer is related to the changes in the voids ratio by the expression:

$$\Delta b = -\frac{b_0}{1 + e_0}\Delta e$$

b_0: is the initial thickness and e_0 the initial voids ratio. The sign convention for Δb is positive for compaction and negative for expansion.

Elastic compaction Δb_e may be computed by the expression:

$$\Delta b_e = \frac{0.434 b_0 C_c \Delta\sigma'}{(1 + e_0)\,\sigma'} \tag{4.23a}$$

Inelastic compaction Δb_i may be computed by the expression:

$$\Delta b_i = \frac{0.434 b_0 C_c \Delta\sigma'}{(1 + e_0)\,\sigma'} \tag{4.23b}$$

These last relationships permit the introduction of the inelastic (or virgin) S_{skv} and the elastic skeletal specific storage S_{ske} values:

$$S_{skv} = \frac{0.434 C_c \gamma_w}{(1 + e_0)\,\sigma'} \tag{4.24a}$$

$$S_{ske} = \frac{0.434 C_r \gamma_w}{(1 + e_0)\,\sigma'} \tag{4.24b}$$

Generally these parameters are considered constant.

Equations (4.23a) and (4.23b) may be combined in the following expression:

$$\Delta b_{n,n-1} = \frac{0.434 b_0}{(1 + e_0)\,\sigma'}\left[C_n\left(\sigma'_n - \sigma'_{c,n-1}\right) + C_r\left(\sigma'_{c,n-1} - \sigma'_{n-1}\right)\right] \tag{4.25}$$

$$C_c, \sigma'_n > \sigma'_{c,n-1} \tag{4.26a}$$

$$C_n = C_r, \sigma'_n \le \sigma'_{c,n-1} \tag{4.26b}$$

4.3.2.3 A Validation of the Assumptions Used

Please notice that we used the assumption that the ratio between $\Delta\sigma'$ and σ', for example, $\left|\frac{\Delta\sigma'}{\sigma'}\right| \ll 1$. This ratio between two consequent years has an order of magnitude of 0.05–0.1.

In order to analyze the validity of this assumption, let us use a mathematical expansion of the exact integrated formula:

$$de = -C_c \frac{1}{\sigma'} d\sigma$$

$$\Delta e = \int_{e_1}^{e_2} -C_c \frac{1}{\sigma'} d\sigma = -C_c \log_{10}\left(\frac{\sigma' + \Delta\sigma'}{\sigma'}\right)$$

$$\Delta \mathrm{e} = -\mathrm{C_c}\log_{10}\left(\frac{\sigma' + \Delta\sigma'}{\sigma'}\right) = -\mathrm{C_c}\log_{10}\left(1 + \frac{\Delta\sigma'}{\sigma'}\right)$$

$$\text{for } \left|\frac{\Delta\sigma'}{\sigma'}\right| < 1, \ \Delta \mathrm{e} = -\frac{\mathrm{C_c}}{\ln(10)} \ln\left(1 + \frac{\Delta\sigma'}{\sigma'}\right) =$$

$$\Delta \mathrm{e} = -\frac{\mathrm{C_c}}{\ln(10)}\left[\frac{\Delta\sigma'}{\sigma'} - \frac{1}{2}\left(\frac{\Delta\sigma'}{\sigma'}\right)^2 + \frac{1}{3}\left(\frac{\Delta\sigma'}{\sigma'}\right)^3 + \ldots\ldots + (-1)^{\mathrm{n}-1}\left(\frac{\Delta\sigma'}{\sigma'}\right)^{\mathrm{n}}/\mathrm{n}\right] \tag{4.27}$$

Thus, when we are using in the theory the following approximation:

$$\Delta \mathrm{e} = -\mathrm{C_c}\Delta\log_{10}\sigma' \approx -\mathrm{C_c}\frac{\Delta\sigma'}{\sigma'} \tag{4.28}$$

We are neglecting a term with an order of magnitude of

$$\mathrm{app} = \frac{\mathrm{C_c}}{\ln(10)}\left[\frac{1}{2}\left(\frac{\Delta\sigma'}{\sigma'}\right)^2\right]$$

This value is very small compared with the main term $\Delta \mathrm{e} \approx -\mathrm{C_c}\frac{\Delta\sigma'}{\sigma'}$; the ratio of $|\frac{\mathrm{app}}{\Delta \mathrm{e}}| = |\frac{\Delta\sigma'}{\sigma'}|/4.6 \approx 0.02$ corresponds to an error of 2% of the estimation.

4.3.2.4 Computation of the Total Vertical Displacement

To simulate the compaction of a layer in a groundwater flow model, at each time step we must add the computation of the following components:

The compaction for each layer
The total Z displacement
The effective stress field.

Because most of the groundwater-flow models use a hydraulic head $\mathrm{h_n}$ as a dependent variable, the pore pressure $\mathrm{u_n}$ may be expressed as:

$$\mathrm{u_n} = \gamma_\mathrm{w}(\mathrm{h_n} - \mathrm{z_n})$$

At a time t_n, the vertical displacement Δb_n is given by

$$\Delta b_n = \frac{0.434 b_0}{(1+e_0)\,\sigma'_{n-1}}\left[C_n\left(\sigma_n + \gamma_w(z_n - h_n) - \sigma'_{c,n-1}\right) + C_r\left(\sigma'_{c,n-1} - \sigma'_{n-1}\right)\right] \tag{4.29}$$

We may express the vertical displacement in terms of the variation of the loading pressure from the surface $\Delta\sigma_n = \sigma_n - \sigma_{n-1}$ and the variation of the water level head $\Delta h_n = h_n - h_{n-1}$.

This last formula may be expressed as:

4.3.2.5 Normally Consolidated Soils

If $\sigma'_n > \sigma'_{c,n-1}$ then

$$\Delta b_n = \frac{0.434 b_0}{(1+e_0)}\left[C_c\left(\frac{\sigma'_n}{\sigma'_{n-1}} - \frac{\sigma'_{c,n-1}}{\sigma'_{n-1}}\right) + C_r\left(\frac{\sigma'_{c,n-1}}{\sigma'_{n-1}} - 1\right)\right] \tag{4.30a}$$

However $\sigma'_n = \sigma'_{n-1} + \Delta\sigma'_n$ and $\sigma'_n = \sigma_n + \gamma_w(z_n - h_n)$.

So at the same depth z at different times n and n − 1 ($z_n = z_{n-1}$), one can easily find that:

$$\Delta\sigma'_n = \Delta\sigma_n - \gamma_w \Delta h_n$$

Thus, the vertical displacement for an inter-bed system may be written as:

$$\Delta b_n = \frac{0.434 b_0}{(1+e_0)}\left[C_c\left(\frac{\Delta\sigma_n - \gamma_w \Delta h_n}{\sigma'_{n-1}}\right) + (C_r - C_c)\left(\frac{\sigma'_{c,n-1}}{\sigma'_{n-1}} - 1\right)\right] \tag{4.30b}$$

The settlement of the soil layer is calculated by calculating the settlement of the individual sub-layers and adding them; in doing this, it is assumed that the voids ratio and the effective stress are constant throughout the sub-layer and equal to their values at the center of the sub-layer.

thus

$$\textit{For sub-layer } i \quad \Delta S_i = -\frac{\Delta e_i H_i}{1+e_i}$$

so that

$$\textit{Total Settlement} \qquad S = \sum_1^n \Delta S_i = \sum_1^n \left[-\frac{\Delta e_i H_i}{1+e_i}\right] \tag{4.31}$$

The vertical displacement for a multi-layer system of n_b may be written as:

$$\Delta b_n = \sum_{k=1}^{k=n_b} \Delta b_n^k = \sum_{k=1}^{k=n_b} \frac{0.434 {b^{(k)}}_0}{(1 + {e^{(k)}}_0)} \left[\begin{array}{l} {C^{(k)}}_c \left(\frac{\Delta {\sigma^{(k)}}_n - \gamma_w \Delta {h^{(k)}}_n}{{\sigma'^{(k)}}_{n-1}} \right) + \\ ({C^{(k)}}_r - {C^{(k)}}_c) \left(\frac{{\sigma'^{(k)}}_{c,n-1}}{{\sigma'^{(k)}}_{n-1}} - 1 \right) \end{array} \right] \quad (4.32a)$$

4.3.2.6 Preconsolidated Soils

If ${\sigma'}_n \leq {\sigma'}_{c,n-1}$ then

$$\Delta b_n = \frac{0.434 b_0}{(1 + e_0)} \left[C_r \left(\frac{{\sigma'}_n}{{\sigma'}_{n-1}} - 1 \right) \right]$$

Using ${\sigma'}_n = {\sigma'}_{n-1} + \Delta {\sigma'}_n$, $\Delta {\sigma'}_n = \Delta \sigma_n - \gamma_w \Delta h_n$

One can obtain a direct formula for the vertical displacement of a one inter-bed system:

$$\Delta b_n = \frac{0.434 b_0 C_r}{(1 + e_0)} \left(\frac{\Delta \sigma_n}{{\sigma'}_{n-1}} - \frac{\gamma_w \Delta h_n}{{\sigma'}_{n-1}} \right)$$

The vertical displacement for a multi-layer system of n_b may be written as:

$$\Delta b_n = \sum_{k=1}^{k=n_b} \Delta b_n^k = \sum_{k=1}^{k=n_b} \frac{0.434 {b^{(k)}}_0}{(1 + {e^{(k)}}_0)} \left[{C^{(k)}}_r \left(\frac{\Delta {\sigma^{(k)}}_n - \gamma_w \Delta {h^{(k)}}_n}{{\sigma'^{(k)}}_{n-1}} \right) \right] \quad (4.32b)$$

Similar approaches were proposed in the literature, and today, a compaction package SUB-T exists for the MODFLOW software (Hoffmann et al. 2003).

4.4 Limitation of the Theory of Consolidation

4.4.1 Mechanical Behavior of the Soil Matrix

The relationship between the stress applied on a clayey soil and the resulting deformation has been studied for more than three decades.

Most of the researchers have proposed classical compartmental models such as elastic behavior taken from solid mechanics.

In the 1970s, elastic models and elasto-plastic models were commonly used in soil mechanics (see Cours de mecanique des sols 1980; Hill 1998). In the1980s, researchers tried to derive the mechanical behavior of the clay from the microscopic plates existing in the structure of the clay matrix (see Hueckel 1992 for a review).

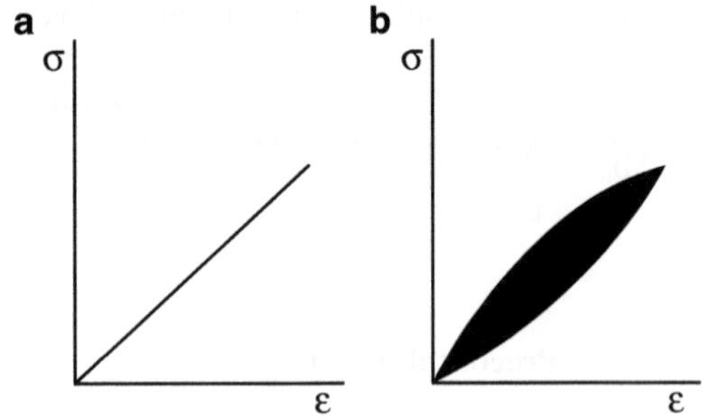

Fig. 4.16 A comparison between elastic and visco-elastic behavior

Today, we may summarize the various rheological behaviors used in the modeling of subsidence as follows:

(a) the elastic model in compression and extension;
(b) the visco-elastic model;
(c) the elasto-plastic model of rupture;
(d) the random elastic model; and
(e) The non-linear large deformation incremental model.

4.4.1.1 Plastic and Visco-Elastic Behavior of the Soil Matrix

Visco-Elastic Material

Some examples of visco-elastic materials include amorphous polymers, semi-crystalline polymers, and biopolymers. Cracking occurs when the strain is applied quickly and outside of the elastic limit.

A **visco-elastic** material has the following properties:

- Hysteresis is seen in the stress–strain curve;
- Stress relaxation occurs: step constant strain causes decreasing stress; and
- Creep occurs: step constant stress causes increasing strain.

Hysteresis is a property of systems (usually physical systems) that do not instantly follow the forces applied to them but react slowly, or do not return completely to their original state.

4.4.1.2 Elastic Behavior versus Visco-Elastic Behavior

Stress-strain curves for a purely elastic material (a) and a visco-elastic material (b) (see Fig. 4.16). The black area is a hysteresis loop and shows the amount of energy lost (as heat) in a loading and unloading cycle. It is equal to $\oint \sigma \, d\varepsilon$, where σ is stress and ε is strain.

Unlike purely elastic substances, a visco-elastic substance has an elastic component and a viscous component. The viscosity of a visco-elastic substance gives the substance a strain rate dependent on time. Purely elastic materials do not dissipate energy (heat) when a load is applied and then removed. However, a visco-elastic substance loses energy when a load is applied and then removed. This can be observed in the stress-strain curve, with the area of the loop being equal to the energy lost during the loading cycle. Since viscosity is the resistance to thermally-activated plastic deformation, a viscous material will lose energy through a loading cycle.

Different types of linear visco-elasticity models are frequently used:

The linear visco-elasticity is when the function is separable in both the creep response and load. All linear visco-elastic models can be represented by a Volterra equation connecting stress and strain:

$$\varepsilon(t) = \sigma(t)/E_{inst,creep} + \int_0^t K(t - t')\sigma(t')dt' \tag{4.33a}$$

or

$$\sigma(t) = E_{inst,relax}\,\varepsilon(t) + \int_0^t F(t - t')\varepsilon(t')dt' \tag{4.33b}$$

where

- t is time;
- $\sigma(t)$ is stress;
- $\varepsilon(t)$ is strain;
- $E_{inst,creep}$ and $E_{inst,relax}$ are instantaneous elastic moduli for creep and relaxation;
- $K(t)$ is the creep function; and
- $F(t)$ is the relaxation function.

Linear visco-elasticity is usually applicable only for small deformations (see Van Vliet 2006).

4.4.1.3 The Maxwell Model

The Maxwell model can be represented by a purely viscous damper and a purely elastic spring connected in a series, as shown in the diagram. The model can be represented by the following equation:

$$\frac{d\varepsilon_{Total}}{dt} = \frac{d\varepsilon_{Damper}}{dt} + \frac{d\varepsilon_{Spring}}{dt} = \frac{\sigma}{\eta} + \frac{1}{E}\frac{d\sigma}{dt} \tag{4.34}$$

A Maxwell material is a visco-elastic material having the properties both of elasticity and viscosity.

The model represents a liquid (ability to have irreversible deformations) with some additional reversible (elastic) deformations.

4.4.1.4 The Kelvin-Voigt Model

The Kelvin-Voigt model, also known as the Voigt model, consists of a Newtonian damper and Hookean elastic spring connected in parallel, as shown in the picture. It is used to explain the stress relaxation behaviors of polymers.

The constitutive relation is expressed as a linear first-order differential equation:

$$\sigma(t) = E\,\varepsilon(t) + \eta\,\frac{d\varepsilon(t)}{dt} \tag{4.35}$$

This model represents a solid undergoing reversible, visco-elastic strain. Upon application of a constant stress, the material deforms at a decreasing rate, asymptotically approaching the steady-state strain. When the stress is released, the material gradually relaxes to its undeformed state. At constant stress (creep), the model is quite realistic as it predicts strain to tend to σ/E as time continues to infinity. Similar to the Maxwell model, the Kelvin-Voigt model also has limitations. The model is extremely good with modelling creep in materials, but with regards to relaxation, the model is much less accurate.

Visco-Elastic Creep

Applied stress (a) and induced strain (b) as functions of time for a visco-elastic material.

At a time t0, a visco-elastic material is loaded with a constant stress that is maintained for a sufficiently long time period. The material responds to the stress with a strain that increases until the material ultimately fails. When the stress is maintained for a shorter time period, the material undergoes an initial strain until a time t1, after which the strain immediately decreases (discontinuity), then gradually decreasing at time $t > t1$ to a residual strain (see Fig. 4.17).

Visco-elastic creep data can be presented by plotting the creep modulus (constant applied stress divided by total strain at a particular time) as a function of time. Below its critical stress, the visco-elastic creep modulus is independent of stress applied. A family of curves describing strain versus time response to various applied stress may be represented by a single visco-elastic creep modulus vs a time curve if the applied stresses are below the material's critical stress value.

Visco-elastic creep is important when considering long-term structural design. Given loading and temperature conditions, designers can choose materials that best suit component lifetimes.

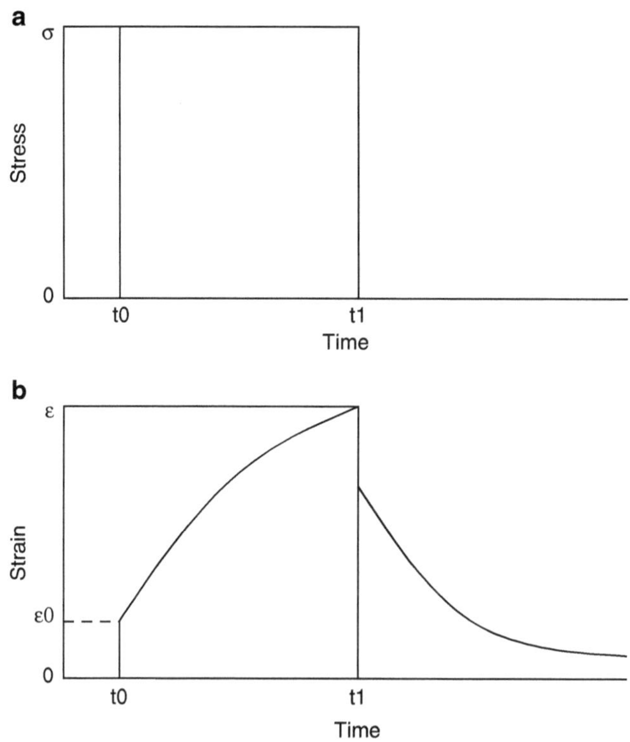

Fig. 4.17 Creep behavior of a visco-elastic model

Incremental Non-linear Visco-Elastic Material

Nonlinear viscoelasticity occurs when the function is not separable. It usually happens when the deformations are large or if the material changes its properties under deformations. In engineering mechanics, deformation is a change in shape due to an applied force. An incremental visco-elastic model for soil has been discussed (see Durban et al. 1990). Large deformation theory has been used in the analysis of mining subsidence by Manchao and Zhida (1991).

Elasto-Plastic Models

If the stress exceeds a critical value, as was mentioned above, the material will undergo plastic, or irreversible, deformation. This critical stress can be tensile or compressive. Such models have been used by Wheeler and Sivakumar (1995, 2000) for the modeling of compaction in non-saturated soils.

1. **Tresca criterion**
   This criterion is based on the notion that when a material fails, it does so in shear, which is a relatively good assumption when considering metals. Given the

principal stress state, we can use Mohr's Circle to solve for the maximum shear stresses our material will experience and conclude that the material will fail if:

$$\sigma_1 - \sigma_3 \geq \sigma_0$$

Where σ_1 is the maximum normal stress, σ_3 is the minimum normal stress, and σ_0 is the stress under which the material fails in uniaxial loading. A yield surface may be constructed, which provides a visual representation of this concept. Inside of the yield surface, the deformation is elastic.

2. **The Von Mises criterion**

   This criterion is based on the Tresca criterion but takes into account the assumption that hydrostatic stresses do not contribute to material failure. Von Mises solves for an effective stress under uniaxial loading, subtracting the hydrostatic stresses, and claims that all effective stresses greater than that which causes material failure in uniaxial loading will result in plastic deformation.

$$\sigma_{\text{effective}}^2 = 1/2\left((\sigma_{11} - \sigma_{22})^2 + (\sigma_{22} - \sigma_{33})^2 + (\sigma_{11} - \sigma_{33})^2\right) + 3\left(\sigma_{12}^2 + \sigma_{13}^2 - \sigma_{23}\right)^2 \tag{4.36}$$

Again, a visual representation of the yield surface may be constructed using the above equation, which takes the shape of an ellipse. Inside the surface, materials undergo elastic deformation. Outside of the surface they undergo plastic deformation (see Hill 1998).

4.4.1.5 The Random Elastic Model

Models where the modulus of elasticity is a random function of the space have been used for modeling the irregularity of the data measurements. Zhang and Guo (2006) presented a subsidence model using a probabilistic approach. Dagan and Zeitoun (1998), and Zeitoun and Braester (1991), Zeitoun and Baker 1992) have presented an extensively used type of these models for soil mechanics computations. Also, computation of the displacement fields and its variability has been presented by Zeitoun and Uzan (1993).

In most of the studies done in the Bangkok region, the one-dimensional theory of Terzaghi has been chosen, mainly because of the simplicity of the computation. In this section, we will analyze the validity of the assumptions of the consolidation theory of Terzaghi.

As discussed above, we may summarize the Terzaghi assumptions as:

The soil displacement is vertical only, and it depends only on the depth and time.
The initial stress distribution in the soil is hydrostatic.
The effective stress acting on the soil matrix in a saturated soil is the difference between the total stress and the water pressure.

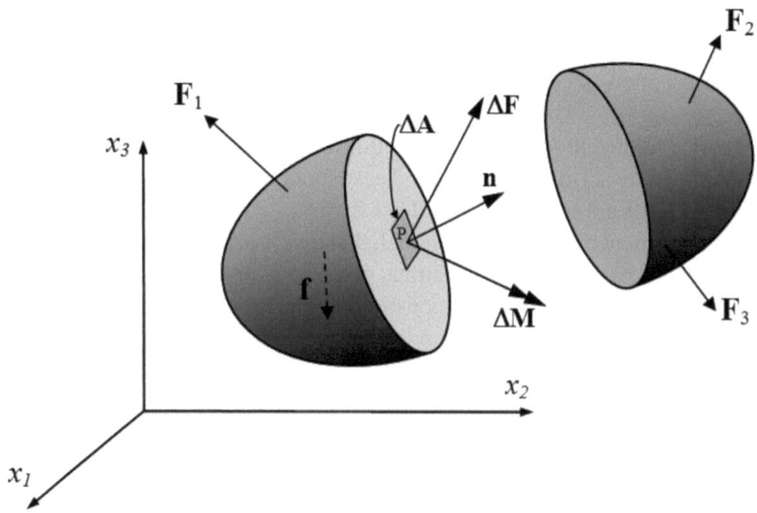

Fig. 4.18 Internal forces in the soil matrix

The water pressure is a function of the depth only.
The soil is loaded from above and not in depth.

Below, we will discuss these assumptions for the Bangkok area:

4.4.2 The Assumption of Vertical Displacement and Constant Total Stress

In most of the analysis performed for the analysis of the subsidence in the Bangkok region, only the effect of water pressure has been considered on a regional scale. The analysis has been performed separately for each subsidence station (see Buapeng 2008). In such an analysis, it is common to assume one-dimensional behavior of the vertical displacement. The soil displacement is assumed to be a function of the vertical component z and the time.

This result with an effective stress tensor with a unique component:

$$\sigma'_{xx} = \sigma'_{yy} = \sigma'_{xy} = \sigma'_{yx} = 0;$$
$$\sigma'_{zx} = \sigma'_{zy} = \sigma'_{xz} = \sigma'_{yz} = 0;$$
$$\sigma'_{zz}(z) \neq 0; \tag{4.37}$$

These different components of the effective stress enable the computation of the internal force acting on an element of the soil matrix. Figure 4.18 illustrates this point. Each solid element has an area ΔA and an internal force ΔF inducing a state of stresses.

This assumption is correct at a local scale, for the analysis of a single station analysis.

However, regional modeling of the groundwater flow in the Bangkok vicinity has been performed by several institutions. These models work on a regional scale with a resolution of the numerical grid until $2 \times 2\ \mathrm{km}^2$.

The development of a numerical model, that involves both groundwater modeling and the mechanical behavior of the strata, needs to address the scale problem.

The correct assumptions on the constitutive equations are fundamental in order to understand the subsidence problem in Bangkok.

In Fig. 4.19, a contour map of the cumulative subsidence in the Bangkok region is presented (see Teparaksa W. Effect of groundwater exploitation on land subsidence in Bangkok City, Personal communication, 2009). The vertical displacement of the soil is not constant over the whole region. In view of the subsidence data in Bangkok, a realistic assumption on the displacement components of the soil should be that the vertical displacement is a function of all the spatial components $w(x,y,z) \neq 0$. We also may assume that the other components are zero.

In such a case, when we assume an elastic behavior of the soil skeleton, the assumption on the behavior results in a shear strain and stress (see Figs. 4.19 and 4.20).

The initial stress distribution in the soil is hydrostatic.

The hydrostatic pressure assumption is generally applied to unconfined formations where the water table is a point of atmospheric pressure. In a confined aquifer; the reference water pressure is not necessarily the atmospheric pressure. Therefore, the assumption of initial hydrostatic pressure may be questionable.

In the Bangkok basin, the different clayey aquifers are confined in such a way that the initial effective pressure is not necessarily hydrostatic.

The effective stress acting on the soil matrix in a saturated soil is the difference between the total stress and the water pressure.

This principle, called the stress principle of Terzaghi, has been used also by Biot and is accepted for saturated soil. For unsaturated soil, a more sophisticated formula has been used.

The water pressure is a function of the depth only.

This is quite difficult to accept from a hydrological point of view. Contour maps of the DGR show clearly that the water pressure is a function of all the dimensions (x,y,z) and the time.

Thus, using the assumption that the water pressure depends only on the depth is problematic.

The soil is loaded from above and not in depth.

This assumption is also very restrictive for our study when we wanted to estimate the effect of loading due to buildings. These buildings were built on piles that are acting in depth on the soil.

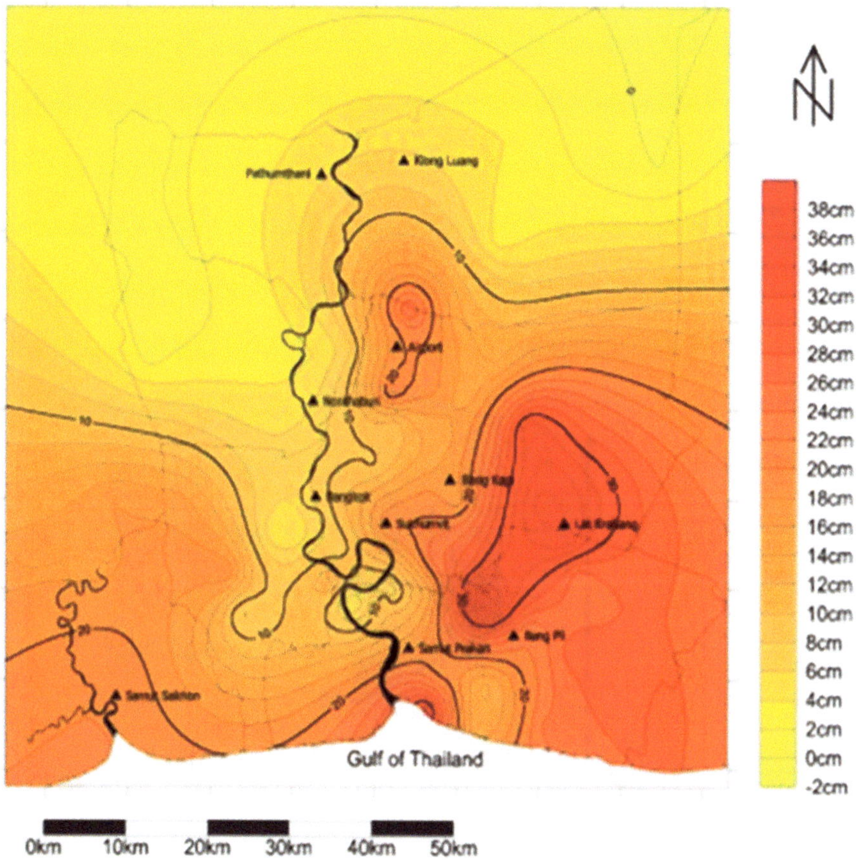

Fig. 4.19 Cumulative subsidence from 1992 to 2003

Following the discussion in the last sections, we adopted the following modeling approach:

1. The use of the Biot's three-dimension consolidation theory.
2. The elastic compressibility model as the rheological model of the soil matrix.

Biot's three-dimension consolidation theory may be seen as a generalization of the Terzaghi consolidation theory. It is the main calculation method of land subsidence caused by groundwater exploitation at the present (see Zhang and Guo 2006).

However, there are some difficulties in calculating the land subsidence through the theory. On the one hand, it is too complicated to obtain the analytical solution. On the other hand, there are some difficulties in obtaining the numerical solutions, too.

$$w(x,y,z,t) \neq w(z,t);\ u = v = 0$$

$$\varepsilon_{xz} = \varepsilon_{zx} = \frac{1}{2}\frac{\partial u_z}{\partial x};\ \sigma'_{xz} = \mu\frac{\partial u_z}{\partial x} \neq 0$$

$$\varepsilon_{yz} = \varepsilon_{yx} = \frac{1}{2}\frac{\partial u_z}{\partial y};\ \sigma'_{yz} = \mu\frac{\partial u_z}{\partial y} \neq 0$$

$$\varepsilon_{zz} = \frac{\partial u_z}{\partial z};\ \sigma'_{zz} = (\lambda + 2\mu)\frac{\partial u_z}{\partial z}$$

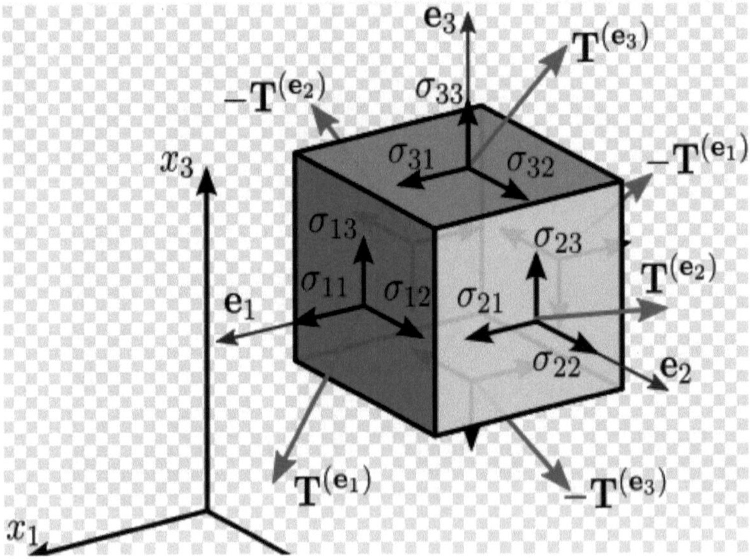

Fig. 4.20 The state of effective stress resulting from a spatial vertical displacement.

First, it is difficult to set up or choose the constitutive model of the soil which can accurately reflect the elasticity, plasticity, and viscidity of the soil.

Second, accurately measuring the parameters of the soil is difficult. In this work, adopting the method of mechanics reasoning, we developed a Biot theory that uses the constitutive parameters of the consolidation theory.

References

Bravo R, Rogers JR, Cleveland TG (1991a) A new three dimensional finite difference model of groundwater flow and land subsidence in the Houston area. In: Proceedings of the fourth international symposium on land subsidence, May 1991. IAHS Publication – 200

Bravo R, Rogers JR, Cleveland TG (1991b) On the determination of the compressible soil properties required to model subsidence in the area of Houston, Texas. In: Proceedings of the fourth international symposium on land subsidence, May 1991. IAHS Publication – 200

Buapeng S (2008) Groundwater situation and land subsidence mitigation in Bangkok and its vicinity. http://siteresources.worldbank.org/EXTWAT/Resources/4602122-1213366294492/5106220-1213649450319/2.7.4_Groundwater_in_Bangkok.pdf

Cours de mecanique des sols (1980) – enseignement specialise – Ecole des Ponts et Chaussees – Les fondations

Dagan G, Zeitoun DG (1998) Seawater-freshwater interface in a stratified aquifer of random permeability distribution. J Contam Hydrol 29:185–203

Durban D, Zeitoun DG, Benaim HE (1990) Finite linear viscoelasticity. J Eng Mech-ASCE 116:2449–2462

Hill R (1998) The mathematical theory of plasticity. Oxford University Press. ISBN-10:0198503679

Hoffmann J, Leake SA, Galloway DL, Wilson AM (2003) MODFLOW-2000 ground-water model –user guide to the subsidence and aquifer-system compaction (SUB) package. US Geological Survey Open-File Report 03–233

Hueckel TA (1992) On the effective stress concepts and deformation in clays subjected to environmental loads: discussion. Can Geotech J 29:1120–1125

Manchao H, Zhida C (1991) Analysis of mining subsidence using large deformation theory. In: Proceedings of the fourth international symposium on land subsidence, May 1991. IAHS Publication – 200

Poland JF, Davis GH (1969) Land subsidence due to withdrawal of fluids. In: Varnes DJ, Kiersch G (eds) Reviews in engineering geology II. Geological Society of America Inc, Boulder

Stanley A, Leake and Galloway DL, Burbey TJ (2010) Use of the SUB-WT package for MODFLOW to simulate aquifer-system compaction in Antelope Valley, CA, USA. In: Proceedings of the EISOLS 2010, Queretaro, Mexico, October 2010. IAHS Publication 339

Van Vliet KJ (2006) 3.032 Mechanical behavior of materials. Massachusetts Institute of Technology, Cambridge, MA

Wheeler SJ, Sivakumar V (1995) An elasto-plastic critical state framework for unsaturated soils. Geotechnique 45:35–53

Wheeler SJ, Sivakumar V (2000) Influence of compaction procedure on the mechanical behaviour of an unsaturated compacted clay. Part 2: Shearing and constitutive modelling. Geotechnique 50:369–376

Zeitoun DG, Baker R (1992) A stochastic approach for settlement predictions of shallow foundations. Geotechnique 42:617–629

Zeitoun DG, Braester C (1991) A Neumann expansion approach to flow through heterogenous formations. Stoch Hydrol Hydraul 5:207–226

Zeitoun DG, Uzan J (1993) Derivation of material variability from settlement measurements. Int J Numer Anal Met 17:303–322

Zhang C, Guo H (2006) Time-space coupled predicting model of land subsidence caused by groundwater exploitation based on stochastic medium theory. International conference on safety science and technology, Shanghai

Chapter 5
Biot's Theory of Consolidation

Abstract In this chapter, we present the constitutive equations of the Biot model. The main difficulties of this model are the coupling of the equations and the application of the superposition principle. We present the general solution of the coupled problem resulting from the Biot model. This solution is based on a new method for decoupling the equations. This decoupling approach permits to derive a superposition principle of the loadings. Several types of loadings and boundary conditions for the water pressure and the displacement field are discussed. Analytical solutions are also derived and discussed.

Also, the problem of the computation of the different causes and their relative contribution to the total subsidence and/or settlement is analyzed in terms of two main principles of the partial differential equations, as follows:

- the coupling/decoupling of the system of equations; and
- the principle of superposition of loadings.

For soft clays and/or loose sand, the ratio between the maximum vertical displacements using the coupled model may be 170 times larger than the maximum vertical displacement computed with the decoupled model.

Also, for large values of the shear modulus and a standard value of the Young modulus (gravel sand), the consolidation ratio of the coupled solution and the decoupled solution are similar. This ratio deceases rapidly with the radial distance to the pumping well. However, for small values of the shear modulus, the consolidation ratio is decreasing slowly with the radial distance to the pumping well.

5.1 The Coupling of the Biot Model

A major debate involving the scientific community in the last decades concerns whether to treat the prediction of land subsidence induced by reservoir exploitation with coupled (Hsieh 1996) or decoupled (Geertsma 1973) models. The former approach supplies the time evolution of both fluid pressures and the soil displacements

D.G. Zeitoun and E. Wakshal, *Land Subsidence Analysis in Urban Areas*,
Springer Environmental Science and Engineering, DOI 10.1007/978-94-007-5506-2_5,

by solving simultaneously the equations governing the mechanical equilibrium and the groundwater flow. The decoupled approach, instead, first calculates the time evolution of the pressure field using the groundwater flow equation and then solves separately the equilibrium equation for the soil displacements at the times of interest. From a physical and mathematical point of view the coupled model represents the more rigorous approach to land subsidence prediction, but it is very computationally expensive, and, in addition, its numerical solutions are often affected by numerical instability and convergence problems (Hsieh 1996; Ferronato et al. 2009). On the other hand, the approximate decoupled model, although theoretically less accurate, requires less computational resources allowing a finer description of the medium properties.

In this chapter we derive a general method of decoupling the three dimensional time dependent Biot equations. The method uses the Helmholtz theorem for a mathematical representation of the displacement vector field of the soil. Thus lead to a series of three non coupled time dependent, parabolic and elliptic partial differential equations.

The method is general and may be used for numerical solution of the Biot equations.

In the present work we derive a closed form solution of the effect of pumping in a deformable porous media. We also compared, the solution with previous solution derived under the decoupled assumption.

The axial symmetric closed-form solutions of horizontal and vertical displacements of the ground surface are obtained by using the image method and the use of the Green theorem. Also the decoupling assumption used by most of the researcher is discussed. It is shown that the ratio between the maximum vertical displacement computed with full coupled solution and the maximum vertical displacement computed with the classical non-coupled solution may be more than 50%. The elastic deformations of the stratum caused by groundwater extraction will fully recover after the excess pore water pressure disappears. The study concludes that horizontal displacement is significant and should be considered in prediction of the transient deformation induced by impulsive groundwater withdrawal.

5.1.1 The Biot Model

The general settlement u_{Total} of the soil will be considered as a sum of three different settlements:

$$u_{Total} = u_{Building} + u_{Pumping} + u_{Natural}$$

Where the settlement due to the load of the building is $u_{Building}$.

The settlement due to the pumping is denoted by $u_{Pumping}$, and the settlement due to the natural consolidation is denoted by $u_{Natural}$.

The natural subsidence is the displacement due to geologic sedimentation of the clay. The weights of the soil particles are causing a displacement of the soil matrix independently of any actions such as loading and/or pumping.

The model of these phenomena is a difficult task and involves large time-scale considerations.

In most of the common theories of consolidation, the soil matrix has no volume change so this phenomenon is not taken into account. It is also considered that on a scale of tens of years, this phenomenon has no influence on the stress field inside the soil matrix.

In the development of the groundwater flow equations used in most of the groundwater code such as MODFLOW (Harbaugh et al. 2000; Harbaugh 2005), the soil matrix is considered as a rigid body. In such model, no deformation exists in the soil matrix. The water pressure is causing the water movement but also it is changing the specific stress acting on the soil matrix. In granular soils, this stress does not induce an important deformation of the soil, and therefore the only movement that is modeled is the groundwater flow movement.

In non consolidated or slightly over consolidated clayey soils, the displacement and the deformation of the soil is not negligible. Moreover, it is the main causes of subsidence.

So all the models of subsidence are coupled models of fluid flow with deformation of the soil matrix (See Bear 1972).

In such a physical model, we have to specify basically two different assumptions:

(a) The mechanical behavior of the soil matrix. This, also called the rheological behavior of the porous media, corresponds to a relationship between the stresses applied to the resulting strain behavior.
(b) The coupling between the water pressure and the stress applied to the soil matrix. Such a model permits to understand the stress transfer that occurs between the water and the soil matrix.

In the literature, two methods of the computation of the consolidation have been proposed:

1. The one dimensional Terzaghi approach.
2. The full three dimensional approach based on the Biot theory.

In most of the studies done on the Bangkok region, the one dimensional theory of Terzaghi has been chosen. This is mainly because of the simplicity of the computation. In this section, we analyze the validity of the assumptions of the consolidation theory of Terzaghi.

As discussed above we may summarize the Terzaghi assumptions as:

1. The soil displacement is vertical only and it depends only on the depth and time.
2. The initial stress distribution in the soil is hydrostatic.
3. The effective stress acting on the soil matrix in a saturated soil is the difference between the total stress and the water pressure.

4. The water pressure is a function of the depth only.
5. The soil is loaded from above and not in depth.

1. The initial stress distribution in the soil is hydrostatic.
   The hydrostatic pressure assumption is generally applied to unconfined formations where the water table is a point of atmospheric pressure. In a confined aquifer, the reference water pressure is not necessarily the atmospheric pressure therefore the assumption of initial hydrostatic pressure may be questionable.
   In the Bangkok basin the different clayey aquifers are confined in such a way that the initial effective pressure is not necessarily hydrostatic.
2. The effective stress acting on the soil matrix in a saturated soil is the difference between the total stress and the water pressure.
   This principle called the stress principle of Terzaghi has been used also by Biot and is accepted for saturated soil. For unsaturated soil more sophisticated formula has been used.
3. The water pressure is a function of the depth only.
   This is quite difficult to accept on an hydrological point of view. Contour maps of the DGR shows clearly that the water pressure is a function of all the dimensions (x,y,z) and the time.
   So using the assumption that the water pressure depends only on the depth is problematic.
4. The soil is loaded from above and not in depth.
   This assumption is also very restrictive for our study when we want to estimate the effect of loading due to buildings. These buildings were built on piles that are acting in depth on the soil.
5. Assumptions of the present model.
   In large metropolitan area built on flood plain, such as Bangkok in Thailand, the main causes of the land subsidence phenomenon are:

   (a) The natural compaction and young sediments of the soil.
   (b) The effects of overpumping from the groundwater aquifers.
   (c) The load of building and infrastructure on the ground surface and the pressure of deep foundations.

Following the discussion in the last chapters, we adopt the following modeling approach:

1. The use of the Biot's three-dimension consolidation theory.
2. The elastic compressibility model as the rheological model of the soil matrix.

Biot's three-dimension consolidation theory may be seen as a generalization of the Terzaghi consolidation theory. It is the main calculation method of land subsidence caused by groundwater exploitation at present (see Zhang and Guo 2006).

However there are some difficulties in calculating the land subsidence through the theory. On one hand, it is too complicated to obtain the analytical solution. On the other hand, there are some difficulties in obtaining the numerical solutions, too.

First, it is difficult to set up or choose the constitutive model of the soil which can accurately reflect the elasticity, plasticity and viscidity of the soil.

Second, accurately measuring the parameters of the soil is difficult. In this work, adopting the method of mechanics reasoning, we develop a Biot theory that uses the constitutive parameters of the consolidation theory.

The new model has many advantages than Biot's method such as simplicity, less calculation parameters and accurate results.

Today, there is not a software on the market that solve the combined equations of groundwater flow and soil (and loose deposits) matrix deformation. Also, there are many models developed combining the flow model with consolidation (or deformation using the Biot theory) model and applied in practice in different geologic environments in different countries. These models are not available in the public domain.

Therefore, we develop software based on the Biot theory of consolidation for the present project. The executable files and the data files will be implemented at the DGR office. Also, in the future the software will be maintained and developed by our team.

In our approach, the coupled system of equations is approximated using a volume cell approach and then solved iteratively. In this advanced cell model (ACM), the cells are built close to the hydrogeology of the aquifer and the stated variables are both the water level, the pumping rate in the cells and the vertical displacement in the cell. A Cells model is simple, easy to use, and allows easy checking of different management paradigms.

Land subsidence due to groundwater withdrawal is a well-known phenomenon (Poland, 1969, 1976). As water pumps from an aquifer, the pore water pressure is reduced in the withdrawal region. It leads to increase in the effective stress between the soil particles and subsidence of ground surface.

The three-dimensional consolidation theory presented by Biot (1941, 1955) is generally regarded as the fundamental theory for modeling land subsidence.

Methods of solutions for the Biot's equation were presented by Detournay and Cheng (1993). Methods of potentials such as Biot's decomposition (similar to the Helmholtz decomposition) and Biot functions permit to solve the Biot' equations in infinite and/or semi-infinite domains. Based on Biot's theory, Booker and Carter (1986a, 1986b, 1987a, 1987b), and Chen (2002, 2005) presented analytical solutions for the steady-state responses of displacements and stresses in a half space subjected to a point sink of constant pumping rate. Wang (2000) presented closed form solution for classical geotechnical problems. Tarn and Lu (1991) presented solutions of subsidence by a point sink embedded in saturated elastic half space at a constant rate. In the studies of Booker and Carter (1986a, 1986b, 1987a, 1987b), the flow properties are considered as isotropic or cross-anisotropic whereas the elastic properties of the soil are treated as isotropic with pervious half space boundary. Tarn and Lu (1991) found that groundwater withdrawal from an impervious half space induces a larger amount of consolidation settlement than a pervious one. The anisotropic permeability was proved to have significant effects on the land subsidence due to fluid extraction.

The effects of impulse pumping on land subsidence are very important. Wilson and Gorelick (1996) found that pulsed pumping may cause a greater amount of subsidence as compared with constant pumping rate. In the studies of Zhang, Baray and Hocking (1999), the steady withdrawal and pulsed pumping of a phreatic aquifer were examined. Lu and Lin (2005, 2006, 2007) displayed transient displacements of the pervious half space due to steady pumping rate. Nevertheless, transient closed-form solutions of the ground surface displacements due to impulsive pumping were not obtained in the above studies.

Closed form solutions of the Biot system of equations for specific domains and boundary conditions may be found in Belotserkovets and Prevost (2011); Kaynia and Banerjee (2003); Jian-Fei Lu and Sheng-Leng (2006); Castelletto et al. (2010), Lu and Lin, (2007).

In the following section, we present the mathematical foundation of the Biot model used.

The general settlement u_{Total} of the soil will be considered as a sum of three different settlements:

$$\mathrm{u}_{\mathrm{Total}} = \mathrm{u}_{\mathrm{Building}} + \mathrm{u}_{\mathrm{Pumping}} + \mathrm{u}_{\mathrm{Natural}} \tag{5.1}$$

Where the settlement due to the load of the building is $u_{Building}$

The settlement due the pumping is denoted by $u_{Pumping}$ and the settlement due to the natural consolidation is denoted by $u_{Natural}$

The natural subsidence is the displacement due to geologic sedimentation of the clay. The weights of the soil particles are causing a displacement of the soil matrix independently any actions such as loading and/or pumping.

The model of these phenomena is a difficult task and involves large time scale considerations.

In most of the common theory of consolidation the soil matrix has no volume change so this phenomenon is not taking into account. It is also consider that at a scale of tens of years this phenomenon has no influence on the stress field inside the soil matrix.

The Biot model has been presented in Chap. 3 and may be summarized as:

$$\begin{cases} (\lambda + G)\dfrac{\partial \varepsilon}{\partial x} + G\nabla^2 u = \chi\dfrac{\partial p}{\partial x} + g_x(x, y, z, t) \\ (\lambda + G)\dfrac{\partial \varepsilon}{\partial y} + G\nabla^2 v = \chi\dfrac{\partial p}{\partial y} + g_y(x, y, z, t) \\ (\lambda + G)\dfrac{\partial \varepsilon}{\partial z} + G\nabla^2 w = \chi\dfrac{\partial p}{\partial z} + g_z(x, y, z, t) \end{cases} \tag{5.2}$$

$$(x, y, z) \in \Omega \quad ; t > 0$$

Where the components of the displacement vector $\vec{d}$ are u, v, w and the pore pressure is $p, \vec{g} = (g_x, g_y, g_z)$ is the loading vector.

Ω is the aquifer domain. The solution $\vec{d}$ belongs to the Sobolev space:

$$\vec{d} \in \{\vec{v} \in H^1(\Omega) \times H^1(\Omega) \times H^1(\Omega);\ \vec{v}=\vec{u}\, on\ \partial\Omega\} \times [0.T] = [H_0^1(\Omega, T)]^3$$

A more convenient form of the flow equation is:

$$\vec{\nabla}.\left(\mathrm{K}\vec{\nabla}\mathrm{p}\right) = \frac{1}{\mathrm{S}}\frac{\partial \mathrm{p}}{\partial \mathrm{t}} + \frac{\partial \varepsilon}{\partial \mathrm{t}} + \mathrm{Q}: \quad ;(\mathrm{x,y,z}) \in \Omega;\ \ \mathrm{t} > 0$$

$$\mathrm{K} = \frac{\mathrm{k}}{\gamma};\quad \frac{1}{\mathrm{S}} = \left[\mathrm{n}\beta + \mathrm{c_{gr}}\,(1-\mathrm{n})\right] \tag{5.3a}$$

Boundary conditions:
For the water pressure p:

$$p(x,t) = g_1(x,t) \qquad (x \in \partial\Omega^1,\ t > 0);$$

$$-\mathbf{K}\nabla\, p\, .\vec{n}_{\partial\Omega} = g_2(x,t) \quad (x \in \partial\Omega^2,\ t > 0); \tag{5.3b}$$

where Ω is a bounded, polygonal open set of R^2, $\partial\Omega^1$ and $\partial\Omega^2$ are partitions of the boundary $\partial\Omega$ of Ω corresponding to Dirichlet and Neumann boundary conditions and $\vec{n}_{\partial\Omega}$ the unit outward normal to the boundary $\partial\Omega$.

The solution p belongs to the Sobolev space:

$$p \in \{\, f \in H^1(\Omega);\ f(x,t) = g_1(x,t)$$

$$(x \in \partial\Omega^1,\ t > 0); -\mathrm{K}\nabla\, f\, .\vec{n}_{\partial\Omega} = g_2(x,t)$$

$$(x \in \partial\Omega^2,\ t > 0);\} \times [0,T] = H(div,\Omega) \times [0,T] \tag{5.4}$$

For the displacement field:

$$u(x,t) = u_1(x,t) \qquad (x \in \partial\Omega,\ t > 0);$$

$$v(x,t) = v_1(x,t) \qquad (x \in \partial\Omega,\ t > 0);$$

$$w(x,t) = w_1(x,t) \qquad (x \in \partial\Omega,\ t > 0); \tag{5.5}$$

The solution of the Biot system of Eqs. (5.2) and (5.3a, 5.3b) and the associated boundary conditions (5.4) and (5.5) is the couple of displacement field and pressure $(\vec{d}, p)$ that belongs to $[H_0^1(\Omega, T)]^3 \times H(div,\Omega) \times [0,T]$. $(\vec{d}, p)$ is associated to the loading vector $\vec{g}$ and the pumping rate function Q.

5.2 General Properties of the Biot Equations

5.2.1 *The Principle of Superposition of Loadings and Pumping*

In this section, we analyze the principle of superposition of loading and pumping. This principle permits to analyze separately the different causes of subsidence. For the Biot system of equations, three different principles of superposition are derived in the following theorems:

The superposition theorem 1 (SP1) Consider now two different solutions of the Biot equations $(\vec{d}_1, p_1)$ associated with a loading vector $\vec{g}_1 \neq \vec{0}$ and a pumping term $Q_1 = 0$ and a second solution $(\vec{d}_2, p_2)$ associated with a loading vector $\vec{g}_2 = \vec{0}$ and $Q_2 \neq 0$, then the solution $(\vec{d}, p)$ is associated with the loading vector $\vec{g}_1$ and the pumping rate function $Q_2 \neq 0$ is given by $\begin{matrix} \vec{d} = \vec{d}_1 + \vec{d}_2 \\ p = p_1 + p_2 \end{matrix}$.

Proof. $(\vec{d}_1, p_1)$ is the solution of the Biot Eqs. (5.1), (5.3a, b), (5.4), and (5.5) with $\vec{g}_1 \neq \vec{0}$ and a pumping term $Q_1 = 0$. Also the boundary conditions are prescribed. $(\vec{d}_2, p_2)$ is the solution of the Biot Eqs. (5.1), (5.3a, b), (5.4), and (5.5) with $\vec{g}_2 = \vec{0}$ and $Q_2 \neq 0$ with prescribed boundary conditions. Using the linearity property of differential operators, it is clear that $\begin{matrix} \vec{d} = \vec{d}_1 + \vec{d}_2 \\ p = p_1 + p_2 \end{matrix}$ is the solution of the Biot equations associated with the loading vector $\vec{g}_1$ and the pumping rate function $Q_2 \neq 0$. The boundary conditions correspond to the sum of the boundary condition associated with each separated solution. Involving the uniqueness of the solution of the Biot equations, the theorem is proved.

The superposition theorem 2 (SP2) Consider now two different solutions of the Biot equations $(\vec{d}_1, p_1)$ associated with a loading vector $\vec{g}_1 \neq \vec{0}$ and a pumping term $Q_1 = 0$ and a second solution $(\vec{d}_2, p_2)$ associated with a loading vector $\vec{g}_2 \neq \vec{0}$ and a pumping term $Q_2 = 0$, then the solution $(\vec{d}, p)$ is associated with the loading vector $\alpha_1\vec{g}_1 + \alpha_2\vec{g}_2$ and the pumping rate function $Q = 0$ is given by $\begin{matrix} \vec{d} = \alpha_1\vec{d}_1 + \alpha_2\vec{d}_2 \\ p = \alpha_1 p_1 + \alpha_2 p_2 \end{matrix}$, where α_1 and α_2 are constant.

Proof. $(\vec{d}_1, p_1)$ is the solution of the Biot Eqs. (5.1), (5.3a, b), (5.4), and (5.5) with $\vec{g}_1 \neq \vec{0}$ and a pumping term $Q_1 = 0$. Also the boundary conditions are prescribed.

$(\vec{d}_2, p_2)$ is the solution of the Biot Eqs. (5.1), (5.3a, b), (5.4), and (5.5) with $\vec{g}_2 \neq \vec{0}$ and $Q_2 = 0$ with prescribed boundary conditions. Using the linearity property of differential operators, it is clear that $\vec{d} = \alpha_1\vec{d}_1 + \alpha_2\vec{d}_2$ is the system of solutions (5.1) associated with the loading vector $\alpha_1\vec{g}_1 + \alpha_2\vec{g}_2$. In this case the deformation $\varepsilon(x, y, z, t)$ is equals to $\varepsilon(x, y, z, t) = \alpha_1\,\varepsilon_1 + \alpha_2\,\varepsilon_2$; $\varepsilon_1 = \vec{\nabla}.\vec{d}_1$; $\varepsilon_2 = \vec{\nabla}.\vec{d}_2$. The boundary conditions correspond to the pondered sum of the boundary condition associated with each separated solution. Because $Q = \alpha_1 Q_1 + \alpha_2 Q_2 = 0$, it is clear that $\begin{matrix} \vec{d} = \alpha_1 \vec{d}_1 + \alpha_2 \vec{d}_2 \\ p = \alpha_1 p_1 + \alpha_2 p_2 \end{matrix}$ is the solution of the Biot equations associated with the loading vector $\alpha_1 \vec{g}_1 + \alpha_2 \vec{g}_2$ and the pumping rate function $Q = 0$.

Involving the uniqueness of the solution of the Biot equations, the theorem is proved.

The superposition theorem 3 (SP3) Consider now two different solutions of the Biot equations $(\vec{d}_1, p_1)$ associated with a loading vector $\vec{g}_1 = \vec{0}$ and a pumping term $Q_1 \neq 0$ and a second solution $(\vec{d}_2, p_2)$ associated with a loading vector $\vec{g}_2 = \vec{0}$ and a pumping term $Q_2 \neq 0$, then the solution $(\vec{d}, p)$ is associated with the loading vector $\vec{g} = \vec{0}$ and the pumping rate function $Q = \alpha_1 Q_1 + \alpha_2 Q_2$ is given by $\begin{matrix} \vec{d} = \alpha_1 \vec{d}_1 + \alpha_2 \vec{d}_2 \\ p = \alpha_1 p_1 + \alpha_2 p_2 \end{matrix}$, where α_1 and α_2 are constant.

Proof. $(\vec{d}_1, p_1)$ is the solution of the Biot Eqs. (5.1), (5.3a, b), (5.4), and (5.5) with $\vec{g}_1 = \vec{0}$ and a pumping term $Q_1 \neq 0$. Also the boundary conditions are prescribed.

$(\vec{d}_2, p_2)$ is the solution of the Biot Eqs. (5.1), (5.3a, b), (5.4), and (5.5) with $\vec{g}_2 = \vec{0}$ and $Q_2 \neq 0$ with prescribed boundary conditions. Using the linearity property of differential operators, it is clear that $\vec{d} = \alpha_1 \vec{d}_1 + \alpha_2 \vec{d}_2$ is the system of solutions (5.1) associated with the loading vector $\alpha_1 \vec{g}_1 + \alpha_2 \vec{g}_2 = \vec{0}$. The boundary conditions correspond to the pondered sum of the boundary condition associated with each separated solution. Because $Q = \alpha_1 Q_1 + \alpha_2 Q_2$, it is clear that $\begin{matrix} \vec{d} = \alpha_1 \vec{d}_1 + \alpha_2 \vec{d}_2 \\ p = \alpha_1 p_1 + \alpha_2 p_2 \end{matrix}$ is the solution of the Biot equations associated with the loading vector $\vec{g} = \vec{0}$ and the pumping rate function $Q = \alpha_1 Q_1 + \alpha_2 Q_2$.

Involving the uniqueness of the solution of the Biot equations, the theorem is proved.

The three superposition theorem will be used to derive the closed form solutions of the Biot system of equations. Also, we will use them for the computation of the contribution of each effect on the total displacement.

5.2.2 *The Decoupling of the Biot System of Equations*

The system of Eqs. (5.1), (5.3a, b), (5.4), and (5.5) is a couple system of equations because we cannot solve separately Eqs. (5.1) and (5.3a, b) for the water pressure p and Eqs. (5.3a, b) and (5.5) for the displacement field $\vec{d}$.

In the previous chapter, we have seen that under the assumption of constant total stress used in the theory of consolidation, the system of Biot equations may be decoupled.

In this section, we present another way of decoupling the Biot system of equations. The basic idea of this new approach is to notice that the term $\frac{\partial \varepsilon}{\partial t}$ appearing in the flow equation is the only term dependent on the displacement field $\varepsilon = \vec{\nabla} \cdot \vec{d} = \frac{\partial u}{\partial x} + \frac{\partial v}{\partial y} + \frac{\partial w}{\partial z}$. This term corresponds to the volume extension of the soil's matrix. Then a decomposition of the displacement field into a rotational part and a non rotational part permit to separate the dilatation of the solid matrix and the rotation of the solid matrix.

Theorem 1. *Consider the displacement vector* $\vec{d}$ *of components* (*u*, *v*, *w*); *let us assume that*

$g_z(x, y, z, t) = g_z(z, t)$; $g_x(x, y, z, t) = g_y(x, y, z, t) = 0$ and $\vec{\nabla} \times \vec{d} = \vec{0}$; then $\varepsilon(x, y, z, t) = \frac{\chi}{\lambda+2G} p(x, y, z, t) + \int_0^z g_z(l, t)\, dl + r(t)$

Proof. Using the theory of potential function (See Appendix A at the end of this chapter), when $\vec{\nabla} \times \vec{d} = \vec{0}$; then there exist a potential function $\phi(x, y, z, t)$ such that

$$\vec{d} = (u, v, w) = \vec{\nabla}(\phi(x, y, z, t)) = \left(\frac{\partial \phi}{\partial x}; \frac{\partial \phi}{\partial y}; \frac{\partial \phi}{\partial z}\right);$$

Therefore

$$\varepsilon = \vec{\nabla} \cdot \vec{d} = \frac{\partial u}{\partial x} + \frac{\partial v}{\partial y} + \frac{\partial w}{\partial z} = \nabla^2 \phi$$

Consider the system (1):

$$\begin{cases} (\lambda + G)\dfrac{\partial \varepsilon}{\partial x} + G\nabla^2 u = \chi \dfrac{\partial p}{\partial x} \\ (\lambda + G)\dfrac{\partial \varepsilon}{\partial y} + G\nabla^2 v = \chi \dfrac{\partial p}{\partial y} \\ (\lambda + G)\dfrac{\partial \varepsilon}{\partial z} + G\nabla^2 w = \chi \dfrac{\partial p}{\partial z} + g_z(x, y, z, t) \end{cases} \tag{5.6}$$

Let us rewrite (1) in terms of $\phi(x, y, z, t)$; $p(x, y, z, t)$ then

$$\begin{cases} \dfrac{\partial}{\partial x}[(\lambda + 2G)\,\nabla^2 \phi] = \chi \dfrac{\partial p}{\partial x} \\ \dfrac{\partial}{\partial y}[(\lambda + 2G)\,\nabla^2 \phi] = \chi \dfrac{\partial p}{\partial y} \\ \dfrac{\partial}{\partial z}[(\lambda + 2G)\,\nabla^2 \phi] = \chi \dfrac{\partial p}{\partial z} + g_z(z, t) \end{cases} \tag{5.7}$$

This last system may be simplified as:

$$\begin{cases} \dfrac{\partial}{\partial x}[(\lambda + 2G)\,\nabla^2\phi - \chi p] = 0 \\ \dfrac{\partial}{\partial y}[(\lambda + 2G)\,\nabla^2\phi - \chi p] = 0 \\ \dfrac{\partial}{\partial z}[(\lambda + 2G)\,\nabla^2\phi - \chi p] = g_z(z,t) \end{cases} \tag{5.8}$$

The system (5.8) has a unique solution given by:

$$\nabla^2\phi = \varepsilon(x,y,z,t) = \frac{\chi}{\lambda + 2G}\,p(x,y,z,t) + \int_0^z g_z(l,t)\,dl + r(t)$$

CQFD.

A consequence of theorem 1 is a characterization of the elastic displacement field $\vec{d}$.

When $\vec{d}$ is a non-rotational field, the displacement $\vec{d}$ is given by the formula: $\vec{d} = \vec{\nabla}.\phi$ where ϕ is the solution of the partial differential equation:

$$\nabla^2\phi = \varepsilon(x,y,z,t) = \frac{\chi}{\lambda + 2G}\,p(x,y,z,t) + \int_0^z g_z(l,t)\,dl + r(t)$$

$$(x,y,z) \in \Omega;\; t > 0 \tag{5.9a}$$

$$\begin{aligned} u(x,t) &= \frac{\partial\phi}{\partial x} \quad (x \in \partial\Omega,\, t > 0); \\ v(x,t) &= \frac{\partial\phi}{\partial y} \quad (x \in \partial\Omega,\, t > 0); \\ w(x,t) &= \frac{\partial\phi}{\partial z} \quad (x \in \partial\Omega,\, t > 0); \end{aligned}$$

From equation (5.9a), it is clear that if both the volume strain and the water pressure vanishes on a part of the boundary and the there is no body forces applied on this boundary, $r(t) = 0$. When $g_z(l,t)$ and also $r(t)$ are zero, then

$$\varepsilon(x,y,z,t) = \frac{\chi}{\lambda + 2G}\,p(x,y,z,t) \tag{5.9b}$$

and Eq. (5.6) reduce to:

$$\begin{cases} G\nabla^2 u = \dfrac{\chi G}{\lambda + 2G}\dfrac{\partial p}{\partial x} \\ G\nabla^2 v = \dfrac{\chi G}{\lambda + 2G}\dfrac{\partial p}{\partial y} \\ G\nabla^2 w = \dfrac{\chi G}{\lambda + 2G}\dfrac{\partial p}{\partial z} \end{cases} \tag{5.9c}$$

These last expressions are similar to the Terzaghi model. Theorem 1.0 generalizes the Terzaghi model and permits a decoupling of the flow and displacement equations.

It is important to notice that other important and practical cases respect the rotational condition (See Appendix A (Sect. 5.9) at the end of this chapter).

5.2.3 Non-rotational Elastic Displacement

Using the generalized Helmholtz theorem (see Sect. 5.9), any displacement field $\vec{d}$ may be written as the sum of two displacements; defined on the domain Ω (Zhou, 2006)

$$\vec{d} = \vec{d}_1 + \vec{d}_2$$

Where $\vec{d}_1$ is a non- rotational field and $\vec{d}_2$ a rotational field. $\vec{d}_1 = \vec{\nabla}\varphi;\ \vec{d}_2 = \vec{\nabla} \times \vec{A};$

Notice that $\varepsilon = \vec{\nabla}.\vec{d} = \vec{\nabla}.\vec{d}_1 = \Delta^2\varphi;\ \vec{\nabla}.\vec{d}_2 = 0.$

The equations incorporating a three-dimensional strain mechanical equilibrium in Cartesian co-ordinates for a homogenous, linearly elastic continuum as described by Biot are:

$$\begin{aligned}
&G\,\nabla^2 u + \frac{G}{1-2\nu}\frac{\partial \varepsilon}{\partial x} = \chi\frac{\partial p}{\partial x} + g_x;\\
&G\,\nabla^2 v + \frac{G}{1-2\nu}\frac{\partial \varepsilon}{\partial y} = \chi\frac{\partial p}{\partial y} + g_y;\\
&G\,\nabla^2 w + \frac{G}{1-2\nu}\frac{\partial \varepsilon}{\partial z} = \chi\frac{\partial p}{\partial z} + \gamma_z;\\
&\nabla.(k\nabla\,p) - \alpha\frac{\partial\,\varepsilon}{\partial t} - \frac{1}{S}\frac{\partial\,p}{\partial\,t} = Q;\\
&p \in \{\,f \in H^1(\Omega);\ f(x,t) = g_1(x,t)\\
&(x \in \partial\Omega^1,\ t > 0); -\mathrm{K}\nabla\,f\,.\vec{n}_{\partial\Omega} = g_2(x,t)\\
&\quad(x \in \partial\Omega^2,\ t > 0);\} \times [0,T] = H(div,\Omega) \times [0,T]
\end{aligned} \tag{5.10}$$

where G is shear modulus, ν is Poisson's ratio, α is the ratio of water volume extracted to the volume change of the soil, assuming the soil is being compressed and independent variables, u; v; w are the x; y and z displacements in the Cartesian plane, p is the pore fluid pressure and ε is the volumetric strain:

$$\varepsilon = \vec{\nabla}.\vec{u} = \frac{\partial u}{\partial x} + \frac{\partial v}{\partial y} + \frac{\partial w}{\partial z}$$

$\vec{\nabla}$ is the Nabla operator and $\vec{u} = (u, v, w)$ is the displacement vector.

k is the coefficient of hydraulic conductivity and 1/S as the amount of water which can be forced into the soil under pressure while the volume of soil is kept constant.

Equation (5.10) can be written more compactly in terms of the unknown vector displacement $\vec{d} = (u, v, w)$, the pressure p and the known vector $\vec{g} = (g_x, g_y, \gamma_z)$.

$$G\,\vec{\nabla}.\vec{\nabla}\,\vec{d} + \frac{G}{1-2\nu}\,\vec{\nabla}(\vec{\nabla}.\vec{d}) = \chi\,\vec{\nabla}\,p + \vec{g} \tag{5.11}$$

$$\frac{\partial}{\partial t}(\vec{\nabla}.\vec{d}) + \frac{1}{S}\frac{\partial p}{\partial t} + Q - \vec{\nabla}.(K\vec{\nabla}\,p) = 0 \tag{5.12}$$

$$p \in \{f \in H^1(\Omega);\ f(x,t) = g_1(x,t) \quad (x \in \partial\Omega^1, t > 0); -\mathbf{K}\nabla f.\vec{n}_{\partial\Omega} = g_2(x,t)$$
$$(x \in \partial\Omega^2,\ t > 0);\} \times [0,T] = H(div,\Omega) \times [0,T]$$

Theorem 2. *Consider the solution* $\vec{d} = \vec{d}_1 + \vec{d}_2$ *where* $\vec{d}_1$ *is the solution of*

$$\begin{aligned}
&\vec{d}_1 = \vec{\nabla}\phi;\\
&\nabla^2\phi = \frac{\chi(1-2\nu)}{2G(1-\nu)}p(x,y,z,t) + \varepsilon_0(t) \quad (x,y,z) \in \Omega;\ t > 0\\
&d_{1x}(x,t) = \frac{\partial\phi}{\partial x} = 0 \qquad (x \in \partial\Omega,\ t > 0);\\
&d_{1y} = (x,t) = \frac{\partial\phi}{\partial y} = 0 \qquad (x \in \partial\Omega,\ t > 0);\\
&d_{1z} = (x,t) = \frac{\partial\phi}{\partial z} = 0 \qquad (x \in \partial\Omega,\ t > 0);
\end{aligned} \tag{5.13}$$

And $\vec{d}_2$ is the solution of:

$$\begin{aligned}
G\,\vec{\nabla}.\vec{\nabla}\,\vec{d}_2 &= \vec{g}\\
d_{2x}(x,t) &= u_1(x,t) \quad (x \in \partial\Omega,\ t > 0);\\
d_{2y} &= v_1(x,t) \quad (x \in \partial\Omega,\ t > 0);\\
d_{2z} &= w_1(x,t) \quad (x \in \partial\Omega,\ t > 0);
\end{aligned} \tag{5.14}$$

And
Then $\vec{d} = \vec{d}_1 + \vec{d}_2$ is the unique solution of Eq. (5.10).

Proof. From Eq. (5.11), one can easily see that:

$$\nabla^2\phi = \vec{\nabla}.\vec{d}_1 = \frac{\chi(1-2\nu)}{2G(1-\nu)}p(x,y,z,t) + \varepsilon_0(t) \tag{5.15}$$

From equation (5.15), it is clear that if both the volume strain and the water pressure vanishes on a part of the boundary and the there is no body forces applied on this boundary, $\varepsilon_0(t) = 0$. This assumption will be used in the following.

Inserting Eq. (5.13) in the left side of Eq. (5.10) leads to

$$G\,\vec{\nabla}.\vec{\nabla}\,\vec{d}_1 + \frac{G}{1-2\nu}\,\vec{\nabla}(\vec{\nabla}.\vec{d}_1) = G\,\vec{\nabla}\left[\frac{\chi(1-2\nu)}{2G(1-\nu)}p\right] + \frac{G}{1-2\nu}\,\vec{\nabla}\left[\frac{\chi(1-2\nu)}{2G(1-\nu)}p\right] = \chi\,\vec{\nabla}\,p \tag{5.16}$$

Subtracting Eq. (5.10) from Eq. (5.15) and using $\vec{d}_2 = \vec{d} - \vec{d}_1$, one can get:

$$G\vec{\nabla}.\vec{\nabla}\vec{d} + \frac{G}{1-2\nu}\vec{\nabla}(\vec{\nabla}.\vec{d}) - \left[G\vec{\nabla}.\vec{\nabla}\vec{d}_1 + \frac{G}{1-2\nu}\vec{\nabla}(\vec{\nabla}.\vec{d}_1)\right] = \chi\vec{\nabla}p + \vec{g} - \chi\,\vec{\nabla}\,p$$

$$\begin{aligned}
&\frac{G}{1-2\nu}\vec{\nabla}(\vec{\nabla}.\vec{d}_2) + G\,\vec{\nabla}.\vec{\nabla}\,\vec{d}_2 = \vec{g}\\
&d_{2x}(x,t) = u_1(x,t) \quad (x \in \partial\Omega,\ t > 0);\\
&d_{2y} = v_1(x,t) \quad (x \in \partial\Omega,\ t > 0);\\
&d_{2z} = w_1(x,t) \quad (x \in \partial\Omega,\ t > 0);
\end{aligned} \tag{5.17}$$

Now, the vector $\vec{d}_2$ is rotational then:

$$\vec{\nabla}\,.\vec{d}_2 = 0, \tag{5.18}$$

therefore Eq. (5.12) leads to:

$$\begin{aligned}
&G\,\vec{\nabla}.\vec{\nabla}\,\vec{d}_2 = \vec{g}\\
&d_{2x}(x,t) = u_1(x,t) \quad (x \in \partial\Omega,\ t > 0);\\
&d_{2y} = v_1(x,t) \quad (x \in \partial\Omega,\ t > 0);\\
&d_{2z} = w_1(x,t) \quad (x \in \partial\Omega,\ t > 0);
\end{aligned} \tag{5.19}$$

The sum of Eqs. (5.17) and (5.19) leads to the identity:

$$\begin{aligned}
&\vec{d} = \vec{d}_1 + \vec{d}_2;\\
&G\,\vec{\nabla}.\vec{\nabla}\,\vec{d} + \frac{G}{1-2\nu}\,\vec{\nabla}(\vec{\nabla}.\vec{d}) = \chi\,\vec{\nabla}\,p
\end{aligned} \tag{5.20}$$

Also the boundary conditions of $\vec{d} = \vec{d}_1 + \vec{d}_2$ respects:

$$u(x,t) = u_1(x,t) \quad (x \in \partial\Omega,\ t > 0);$$

$$v(x,t) = v_1(x,t) \quad (x \in \partial\Omega,\ t > 0);$$

$$w(x,t) = w_1(x,t) \quad (x \in \partial\Omega,\ t > 0);$$

Invoking the uniqueness theorem of the solution $(\vec{d}, p)$ **in the Sobolev spaces** $[H_0^1(\Omega, T)]^3 \times H(div, \Omega) \times [0, T]$ **of Eq.** (5.10) for a given initial condition and boundary conditions, there exists a unique couple of solutions $(\vec{d}, p)$ of Eq. (5.10). Then $\vec{d} = \vec{d}_1 + \vec{d}_2$ is a solution of (5.10).

Notice, that the decoupling of the Biot equations has been proved only for the Dirichlet conditions (Eqs. 5.5). For Neumann boundary conditions and/or mixed boundary conditions, the boundary conditions are coupled.

5.3 Boundary Conditions and Initial Stress Distribution

In this section, we present classical initial and boundary conditions used in most of the hydro-geological modeling. A typical set of geological layers may be viewed in the following cross-section of the Bangkok area (see Phienwej et al. 2006) (Fig. 5.1).

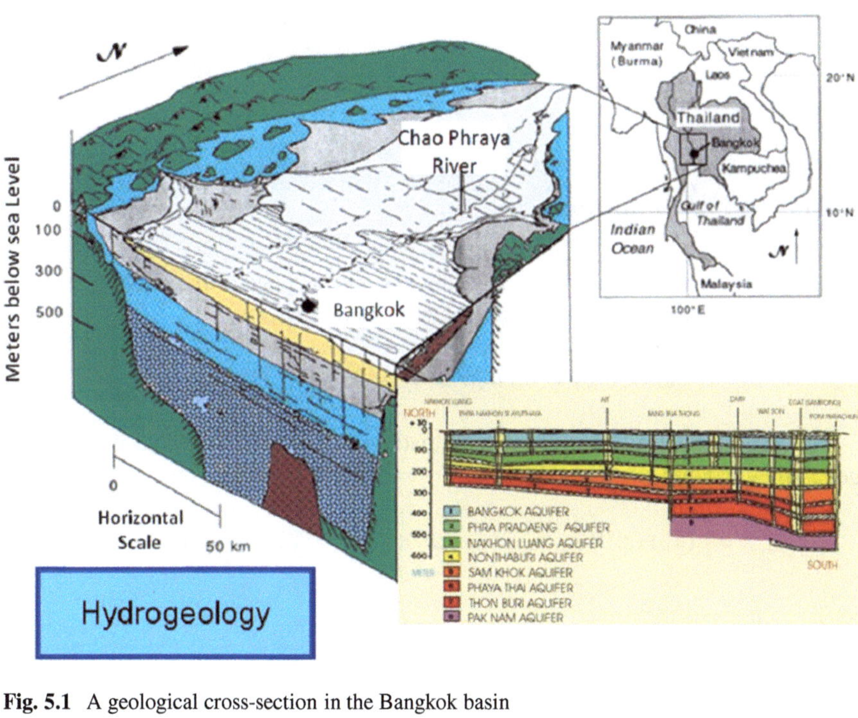

Fig. 5.1 A geological cross-section in the Bangkok basin

In the various modeling that we propose, the system of aquifers is modeled as a series of geological layers where $z = 0$ corresponds to the ground level.

At the initial time t_0, we assume an initial distribution of the vertical displacement $w(x, y, z, t_0) = w_0(x, y, z)$; $u(x, y, z, t_0) = v(x, y, z, t_0) = 0$.

A continuous condition is imposed between the different geological layers.

At $z = 0$, we assume a distribution of the vertical displacement. $w(x, y, 0, t) = w_1(x, y, t)$

At the bottom, we assume $w(x, y, L, t) = 0$, (L is the depth of the whole layers).

Also, we assume that at the boundary $w(x, y, z, t)_{Boundary} = 0$.

The initial stress distribution is considered as the hydrostatic pressure acting in the Bangkok aquifers.

An additional condition is provided by considering the half space as pervious. The mathematical statement of the pervious half space of flow condition at the boundary $z = 0$ is given as below:

$$p\,(r, 0, t) = 0. \tag{5.21a}$$

The boundary conditions at the far boundary $z \to \infty$ due to the effect of an impulsive point sink must vanish at any time. Thus,

$$\lim_{z\to\infty} u_r\,(r, z, t) = 0, \tag{5.21b}$$

$$\lim_{z\to\infty} u_z\,(r, z, t) = 0, \tag{5.21c}$$

$$\lim_{z\to\infty} p\,(r, z, t) = 0. \tag{5.21d}$$

5.3.1 *Initial Conditions*

Assuming no initial changes in displacements and seepage of the aquifer, then the initial conditions at time $t = 0$ of the mathematical model is:

$$u_r\,(r, z, 0) = 0,\ u_z\,(r, z, 0) = 0,\ p\,(r, z, 0) = 0. \tag{5.22}$$

However, when initial stress due to loading is present, the initial condition is generally given by:

$$\sigma_{ij}\,(x, y, z, 0) = \sigma_{0ij}\,(x, y, z, 0);\quad i = 1, 2, 3\quad j = 1, 2, 3$$

5.4 A New System of Equations for Solving the Biot Equations

In summary, one can say that the general solution of the Biot equations may be computed by solving three partial differential systems:

(a) The solution for the water pressure p:

$$\left[\frac{\chi(1-2\nu)}{2G(1-\nu)}+\frac{1}{S}\right]\frac{\partial p}{\partial t}+Q-\vec{\nabla}.(K\vec{\nabla}\,p)=0 \tag{5.22a}$$

(b) The solution of $\vec{d}_1$ via Eq. (5.10)

$$\vec{d}_1=\vec{\nabla}\phi;$$

$$\nabla^2\phi=\frac{\chi(1-2\nu)}{2G(1-\nu)}p(x,y,z,t)+\varepsilon_0 \quad (x,y,z)\in\Omega;\ t>0 \tag{5.22b}$$

$$d_{x1}(x,t)=\frac{\partial\phi}{\partial x}=0 \quad (x\in\partial\Omega,\ t>0); \tag{5.22c}$$

$$d_{y1}(x,t)=\frac{\partial\phi}{\partial y}=0 \quad (x\in\partial\Omega,\ t>0); \tag{5.22d}$$

$$d_{z1}(x,t)=\frac{\partial\phi}{\partial z}=0 \quad (x\in\partial\Omega,\ t>0); \tag{5.22e}$$

Notice that one can solve these last Eq. (5.10) directly in terms of $\vec{d}_1$ without introducing the potential function ϕ. In this case, $\vec{d}_1$ is obtained by solving:

$$G\,\vec{\nabla}.\vec{\nabla}\,\vec{d}_1=\frac{\chi(1-2\upsilon)}{2(1-\upsilon)}\vec{\nabla}p \tag{5.23a}$$

$$G\,\nabla^2u_1=\frac{\chi(1-2\upsilon)}{2(1-\upsilon)}\frac{\partial p}{\partial x}; \tag{5.23b}$$

$$G\,\nabla^2v_1=\frac{\chi(1-2\upsilon)}{2(1-\upsilon)}\frac{\partial p}{\partial y} \tag{5.23c}$$

$$G\,\nabla^2w_1=\frac{\chi(1-2\upsilon)}{2(1-\upsilon)}\frac{\partial p}{\partial z} \tag{5.23d}$$

$$u_1(x,t)=0 \quad (x\in\partial\Omega,\ t>0); \tag{5.23e}$$

$$v_1(x,t)=0 \quad (x\in\partial\Omega,\ t>0); \tag{5.23f}$$

$$w_1(x,t) = 0 \quad (x \in \partial\Omega,\, t > 0); \tag{5.23g}$$

(c) The solution of $\vec{d}_2$ via Eq. (5.11)

$$G\,\vec{\nabla}.\vec{\nabla}\,\vec{d}_2 = \vec{g} \tag{5.24a}$$

$$G\,\nabla^2 d_{2x} = g_1; \tag{5.24b}$$

$$G\,\nabla^2 d_{2y} = g_2 \tag{5.24c}$$

$$G\,\nabla^2 d_{2z} = g_3 \tag{5.24d}$$

$$d_{2x}(x,t) = u_1(x,t) \quad (x \in \partial\Omega,\, t > 0); \tag{5.24e}$$

$$d_{2y} = v_1(x,t) \quad (x \in \partial\Omega,\, t > 0); \tag{5.24f}$$

$$d_{2z} = w_1(x,t) \quad (x \in \partial\Omega,\, t > 0); \tag{5.24g}$$

The solution $\vec{d} = \vec{d}_1 + \vec{d}_2$ is the solution of (5.10).

The system of Eqs. (5.22), (5.23a), and (5.24a) is decoupled.

This system of equations allows the separate treatment of the different loading cases.

In the next Sect. (5.4) the computation of the initial stress $\sigma_0(x,y,z)$ and its corresponding strain $\varepsilon_0(x,y,z)$ is discussed.

Also the practical computation of the loading vector $\vec{g} = (g_1, g_2, g_3)$ for different types of foundations is discussed in Sect. (5.5).

Finally in Sect. (5.6), a closed-form solution of the coupled Biot system of equations is derived.

The present method permits to derive closed-form solutions of the Biot system of equations. Also, it will be used to derive a numerical scheme for the numerical solution of the Biot equations.

5.5 Computation of the Initial Load Using Static Elasticity

There are classic static elasticity problem frequently used in soil and rock mechanics that generates a static deformation. Then, for these problems $\left|\frac{\partial \varepsilon_2}{\partial t}\right| = 0$ and the Biot system of equations is decoupled. In the following, we present several examples

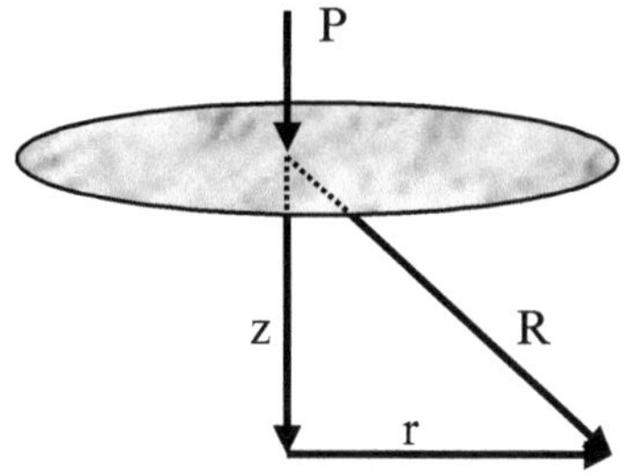

Fig. 5.2 A point load acting on a half space

for the case of a semi-infinite medium with pumping wells and point forces inside and/or at the boundary of the solid. Closed-form solutions for these problems have been presented by Poulos and Davis (1974).

5.5.1 *The Boussinesq Problem*

A fundamental solution of the equations of elasticity is Boussinesq's solution, which relates to a point load applied to the surface of a half-space (very deep layer) and is shown schematically in Fig. 5.2.

Boussinesq found that:

$$\Delta\sigma_{zz} = \frac{3Pz^3}{2\pi R^5}$$

$$\Delta\sigma_{xx} + \Delta\sigma_{yy} + \Delta\sigma_{zz} = \frac{(1+\nu)Pz}{\pi R^3}$$

$$\Delta u_z = \frac{(1+\nu)P}{2\pi ER}\left[2(1-\nu) + \frac{z^2}{R^2}\right]$$

$$\Delta u_r = \frac{(1+\nu)P}{2\pi ER}\left[\frac{rz}{R^2} - \frac{(1-2\nu)r}{R+z}\right]$$

where

$$R = \sqrt{x^2 + y^2 + z^2}$$

and

$$E = \textit{Young's}\ \text{modulus}$$

$$\nu = \textit{Poisson's ratio} \tag{5.25}$$

Δu_z = vertical displacement due to the load.

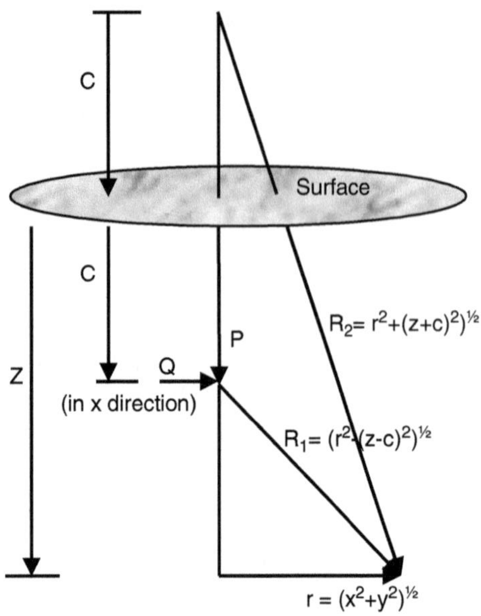

Fig. 5.3 Point loads acting inside a semi-infinite mass

The symbol Δ is used to indicate that each of the quantities in Eq. (5.22) represents the increase in the particular quantity, due to the applied load. u_z is the vertical displacement and u_r the radial displacement ($u_\theta = 0$).

The solution for a point load is important because it can be used to develop solutions for distributed loads by integration.

5.5.2 *The Mindlin's Problem*

Two others classical problems are the Miindlin's problem:

1. A vertical point load P acting beneath the surface of a semi-infinite elastic soil at a depth c (see Fig. 5.3).
2. A horizontal point load Q acting beneath the surface of a semi-infinite elastic soil at a depth c (see Fig. 5.3).

The displacement fields for the Mindlin's problem are given in terms of the vertical displacement u_z and the radial displacement u_r only:

1. A vertical load

$$\Delta u_z = \frac{\mathrm{Pr}}{16\pi G(1-\nu)}\left[\frac{3-4\nu}{R_1} + \frac{(z-c)^2}{{R_1}^3} + \frac{8(1-\nu)^2-(3-4\nu)}{R_2}\right.$$
$$\left.+\frac{(3-4\nu)(z+c)^2-2cz}{{R_2}^3} + \frac{6cz(z+c)^2}{{R_2}^5}\right]$$

$$\begin{aligned}
\Delta u_r &= \frac{\Pr}{16\pi G(1-\nu)}\left[\frac{z-c}{{R_1}^3}+\frac{(3-4\nu)(z-c)}{{R_2}^3}\right.\\
&\quad\left.-\frac{4(1-\nu)(1-2\nu)}{R_2(R_2+z+c)}+\frac{6cz(z+c)}{{R_2}^5}\right]\\
R_1 &= \sqrt{x^2+y^2+(z-c)^2}\\
R_2 &= \sqrt{x^2+y^2+(z+c)^2}\\
G &= \mathit{Shear}\ \text{modulus}\\
\nu &= \mathit{Poisson's\ ratio}
\end{aligned} \tag{5.23}$$

2. A horizontal load

$$\begin{aligned}
\Delta u_z &= \frac{Qx}{16\pi G(1-\nu)}\left[\frac{(3-4\nu)(z-c)}{{R_2}^3}+\frac{(z-c)^2}{{R_1}^3}+\frac{4(1-\nu)(1-2\nu)}{R_2(R_2+z+c)}\right.\\
&\quad\left.-\frac{6cz(z+c)}{{R_2}^5}\right]\\
\Delta u_y &= \frac{Qxy}{16\pi G(1-\nu)}\left[\frac{(3-4\nu)}{{R_2}^3}+\frac{1}{{R_1}^3}-\frac{4(1-\nu)(1-2\nu)}{R_2(R_2+z+c)^2}-\frac{6cz}{{R_2}^5}\right]\\
\Delta u_x &= \frac{Qxy}{16\pi G(1-\nu)}\left[\frac{(3-4\nu)}{R_1}+\frac{1}{R_2}+\frac{x^2}{{R_1}^3}+\frac{(3-4\nu)x^2}{{R_2}^3}\right.\\
&\quad+\frac{4(1-\nu)(1-2\nu)}{(R_2+z+c)}\left(1-\frac{x^2}{R_2(R_2+z+c)}\right)+\frac{2cz}{{R_2}^3}\left(1-\frac{3x^2}{{R_2}^2}\right)\\
R_1 &= \sqrt{x^2+y^2+(z-c)^2}\\
R_2 &= \sqrt{x^2+y^2+(z+c)^2}\\
G &= \mathit{Shear}\ \text{modulus}\\
\nu &= \mathit{Poisson's\ ratio}
\end{aligned} \tag{5.24}$$

These solutions may be used for computing analytically the solution of the Biot problem with a pumping well and an external loading. Using the principle of superposition, the solution of the problem is simply the sum of the displacements for each problem.

5.6 The Model for Loads Due to Building

5.6.1 Methods of Computation

According to the previous development, two loading parameters must be computed, as follows:

1. The initial loading σ_{load} appearing in the loading equation must be evaluated in order to estimate for each layer the total stress $\sigma = \sigma_{load} + \gamma_m d_m + \gamma_s d_s$ and
2. The variation in the total stress $\Delta\sigma_n$ for each year must be evaluated.

In order to estimate both loading for various time steps and for each region of the studied area, we are working in collaboration with the Department of Urbanization of the University of Chulalongkorn.

- **The initial loading:**
  The estimation of the initial loading for each region was based on an assumption of uniformity. The initial year was taken as 1975.
  The computation of the spatial loading for a given region was based on two separate methods:

  (a) A method for computing the density of different types of building. The GIS team provides maps of building types, and we select four types of building, as follows:

  - embankments (F1);
  - residential buildings (5 floors);
  - small towers (less than 15 floors); and
  - large towers (more than 30 floors).

  (b) For each type of building, we developed a method of computation of the initial load.

  The initial loading pressure was assumed to be a function of several parameters, as follows:

  1.1 the type of foundations;
  1.2 the type of buildings; and
  1.3 the point load application.

5.6.2 Classification of Loadings

The various types of buildings were associated with a type of foundations, according to the geological data and the type of pile foundations existing in the Bangkok area.

In Figs. 5.4 and 5.5, the different sub-layers of stiff sand appearing in the top layer of the study are described. In Fig. 5.6, a typical pile foundation in Bangkok

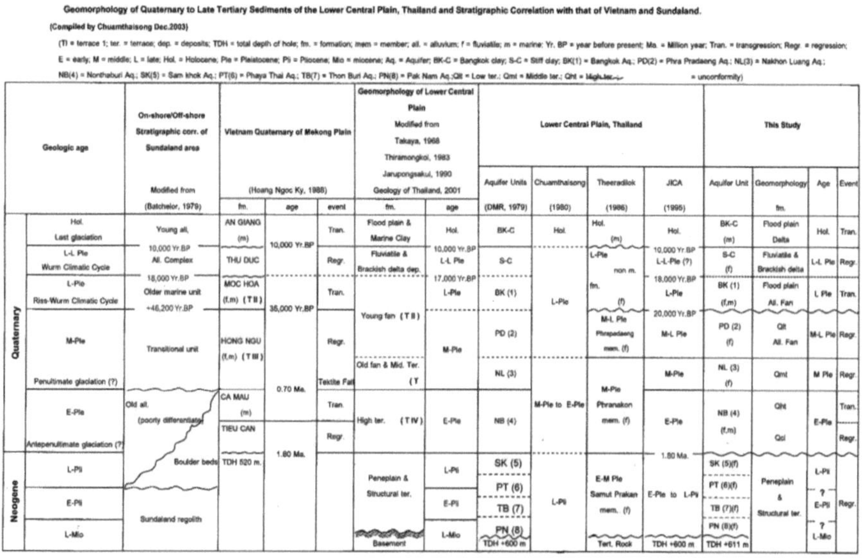

Fig. 5.4 Geological data of the Bangkok clay

is described. Basically, the embankments such as bridges and surface loading are applied on the surface of the ground. However, residential buildings and large towers have pile foundations with different depths. In Fig. 5.7, the type and depth of different piles are described.

The loading effects of the buildings to the soil depend on the type of foundations.

The method of computation differs between shallow foundations and deep ones.

In Figs. 5.6 and 5.7, the different types of foundations are described. Also, the loading effect on the soil is different. The calculation is done using the Boussinesq solution.

In Fig. 5.8, the types of loading due to the weight of a building are described.

There are several technical difficulties in the computation of the stress acting on a soil resulting from the loading of a building:

- Depending on the type of soil, the maximum stress will be at the center of the applied load (sand, sandy clay). However, for soft soils, the stress at the center will be the minimum stress.
- Also, when pile foundations are built close one to the other, the stress acting on the soil is much larger than the sum of the three loadings taken separately.
- Also, the bulb effect shown in Figs. 5.9 and 5.10 may be much larger than the pressure due to a shallow foundation.

In the compartmental model, the point of application of the vertical loads was considered, as follows:

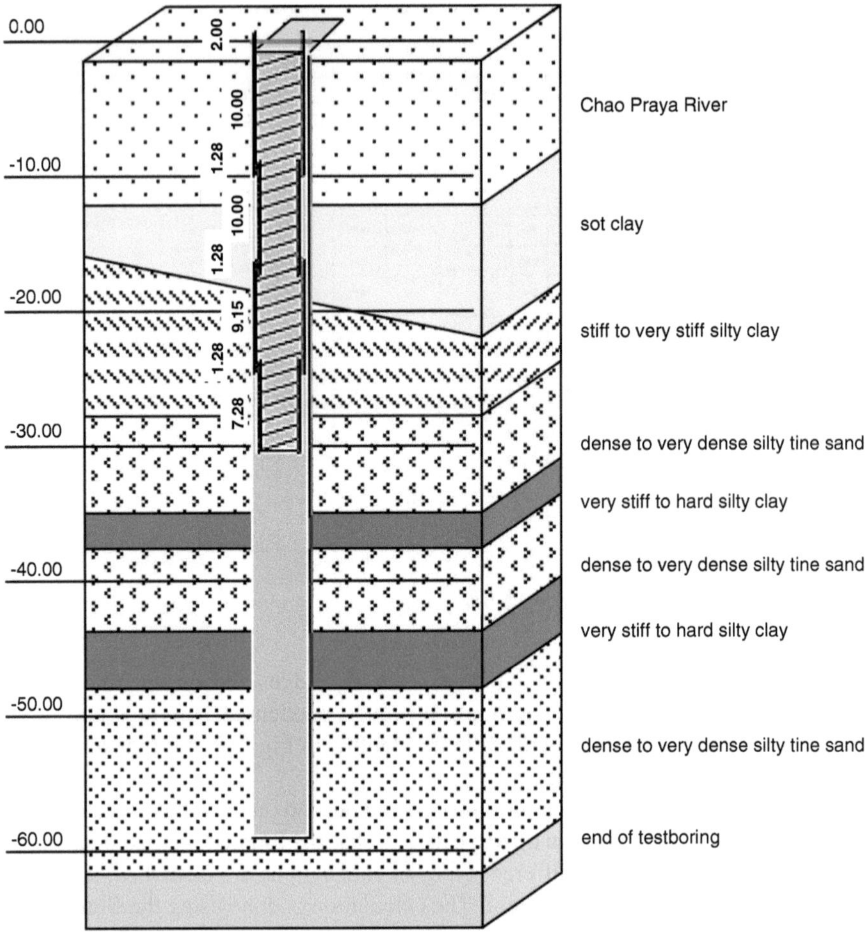

Fig. 5.5 A typical pile perforation in the Bangkok region

1. For embankments buildings, the load effect is applied on top of the Bangkok clay layer.
2. For residential buildings, the pile foundations are acting along the piles at a depth of 13 m; therefore, the integrated load is applied at two-thirds (8 m) in the first layer (see Magnan 2001).
3. For small tower buildings, the pile foundations are acting along the piles at a depth of 24 m; therefore, the integrated load is applied at two-thirds in the first layer.
4. For large tower buildings, the pile foundations are acting along the piles at a depth of 60 m and more; therefore, the integrated load is applied at the middle of the first layer and also at the top of layer 2.

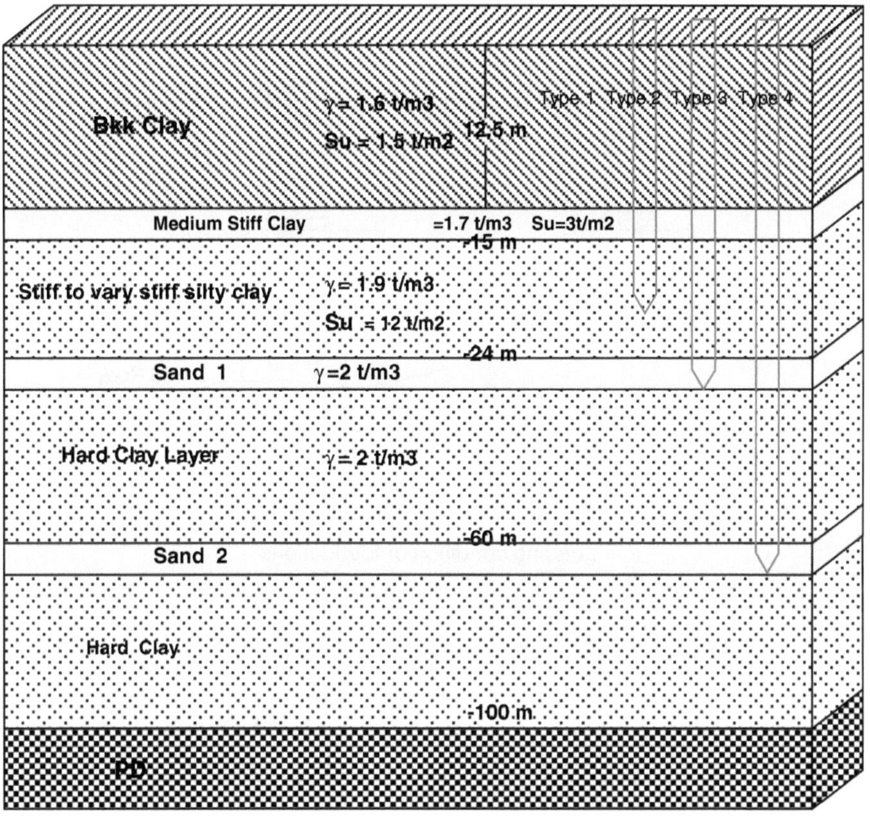

Fig. 5.6 Different types of foundations

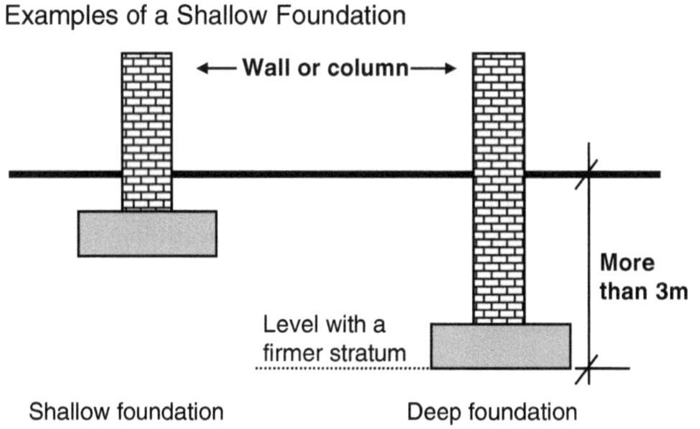

Fig. 5.7 Examples of a shallow foundation

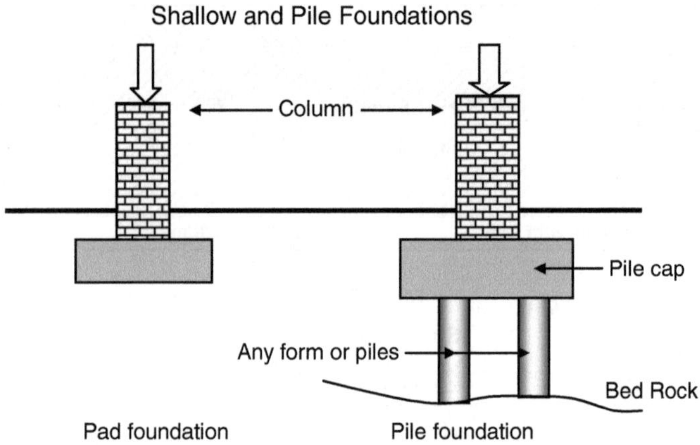

Fig. 5.8 Shallow and pile foundations

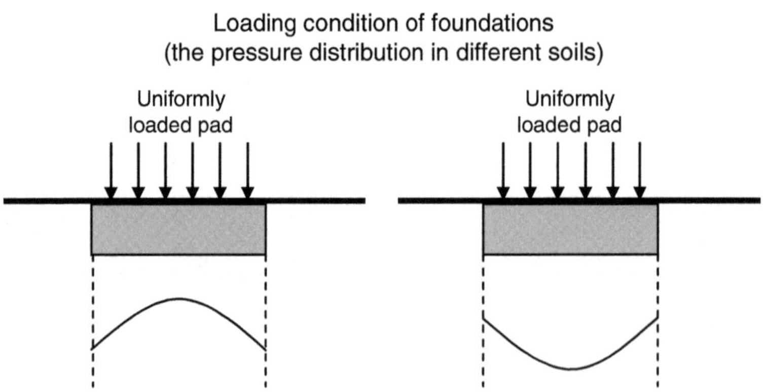

Fig. 5.9 Pressure distribution under a foundation

Also, the value of the above pressure was estimated according to the two following criteria:

(a) an average pressure per floor of building of 0.8 T/m^2; and
(b) the repartition of loads among a group of piles (according to the specification – Magnan 2001).

The method of calculation of the stress is explained in the following sections.

- **Calculation of stress changes acting on the soil**
  The calculation of settlement depends upon knowing the initial and final effective stress within each sub-layer of the deposit. The initial effective stress state can be determined from knowing the bulk unit and the position of the water table. The increase in total stress can be estimated using the theory of elasticity. The

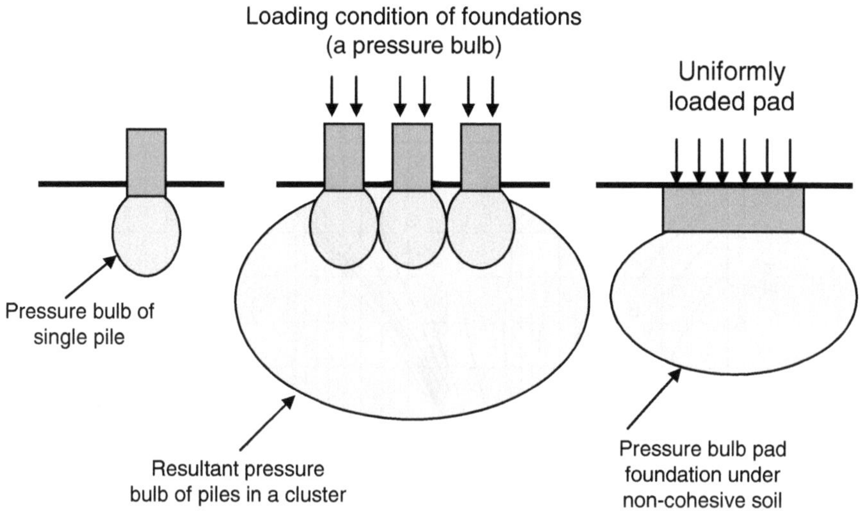

Fig. 5.10 Stress repartition along the pile

Fig. 5.11 A circular-loaded area on a deep elastic layer

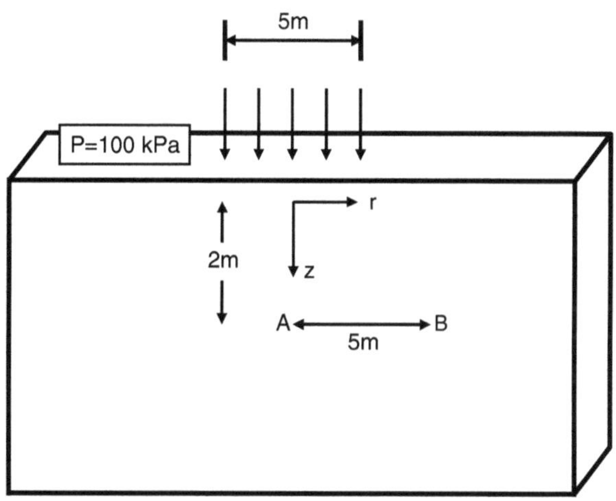

soil is in general not really elastic. However, in the working-stress range, this assumption provides reasonably accurate estimates of the stress increases due to the applied loads.

- **Calculation of stress changes for different types of shallow foundations**

- *Stresses due to* ***circular*** *foundation loads applied at the ground surface*

  A circular foundation of a diameter of 5 m, subjected to an average applied stress of 100 kPa, is shown in Fig. 5.11.

  (a) Calculate the increase in vertical stress at point A

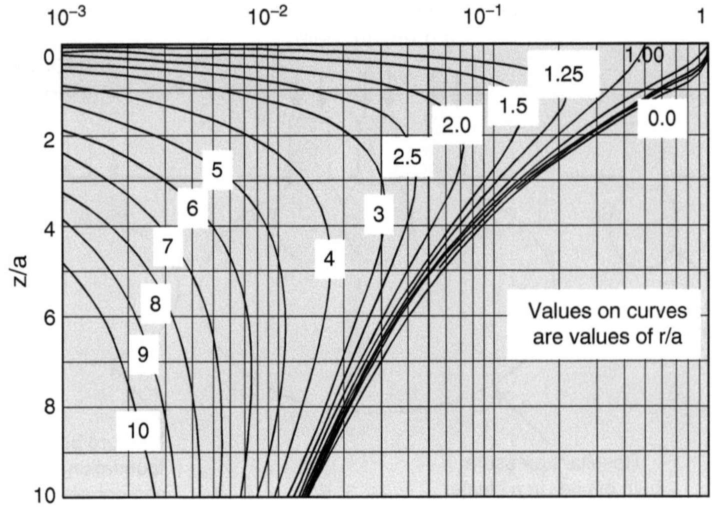

Fig. 5.12 Influence factors for a uniformly-loaded circular area of radius a

There is a simple analytic expression for points on the center line under a circular load:

$$\Delta\sigma_{zz} = p\left(1 - \left[1 + \frac{a^2}{z^2}\right]^{-3/2}\right) \tag{7a}$$

where

p = the surface stress = 100 kPa
a = the radius of the loaded area = 2.5 m
z = the depth of interest = 2 m

$$\Delta\sigma_{zz} = 100 \times \left(1 - [1 + (1.25)^2]^{-3/2}\right) = 75.6\ \text{kPa} \tag{7b}$$

(b) Calculate the increase in vertical stress at point B.
In this case, there is no simple analytic expression, and the solution must be found by using the influence charts given generally in data sheets and reproduced in part in Fig. 5.12. Note that this chart can also be used for points on the center line for which r = 0 (Fig. 5.13).
Now

$$z/a = 2/2.5 = 0.8$$
$$r/a = 5/2.5 = 2 \tag{5.25}$$

using Fig. 5.12, $\Delta\sigma_{zz}/p = 0.03$ and so $\Delta\sigma_{zz} = 3.0$ kPa

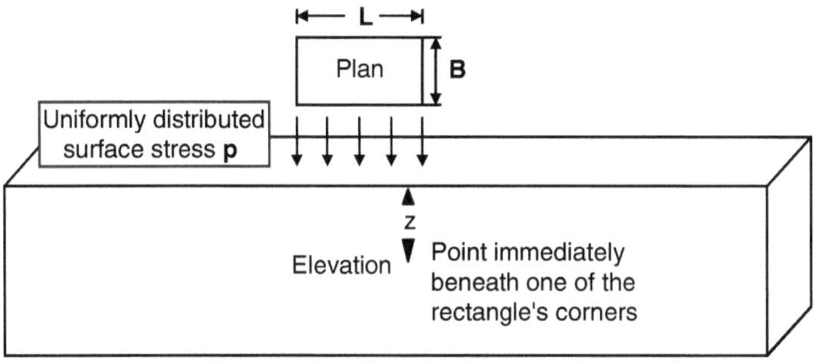

Fig. 5.13 Rectangular uniform loading on a deep elastic layer

- stresses due to **rectangular** foundation loads applied at the ground surface

Many loads which occur in practice are applied to foundations that may be considered to consist of a number of rectangular regions. It is, thus, of interest to be able to calculate the vertical stress increases due to a uniformly-distributed load acting on a rectangular-loaded area. This is shown schematically in Fig. 5.14.

The vertical stress change at a distance z below one of the corners of the rectangular load may be determined from a chart reproduced in Fig. 5.14.

This chart (Fig. 5.14) can be used to determine the value of a stress increase at any point in an elastic layer; the method for doing this is illustrated below.

Calculation of stress below an interior point of the loaded area

This situation is shown schematically in Fig. 5.15a The stress change is required at a depth z below point O.

The first step in using the influence charts is to break the rectangular loading up into a number of components each having a corner at O; this is relatively simple as can be seen in Fig. 5.15a.

It thus follows that at the point of interest, the stress increase $\Delta\sigma_{zz}(\text{ABCD})$ is given by:

$$\Delta\sigma_{zz}(\text{ABCD}) = \Delta\sigma_{zz}(\text{OXAY}) + \Delta\sigma_{zz}(\text{OYBZ}) + \Delta\sigma_{zz}(0\text{ZCT}) + \Delta\sigma_{zz}(\text{OTDX}) \tag{5.26}$$

An example of computation Suppose we wish to evaluate the increase in stress at a depth of 2 m below the point O due to the rectangular loading shown in shown in Fig. 5.15b, when the applied stress over ABCD is 100 kPa.

For rectangular loading OZCT

$$\text{m} = \text{L}/\text{z} = 1$$

$$\text{n} = \text{B}/\text{z} = 1$$

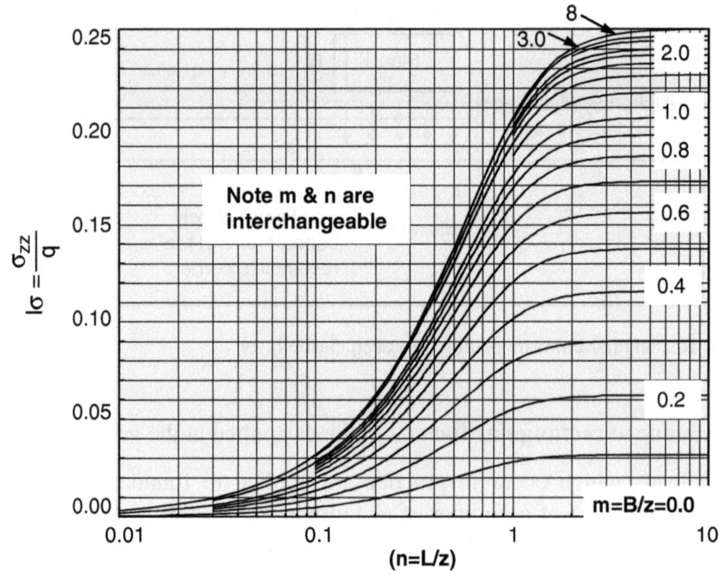

Fig. 5.14 Influence factors for uniformly-loaded *rectangular* areas

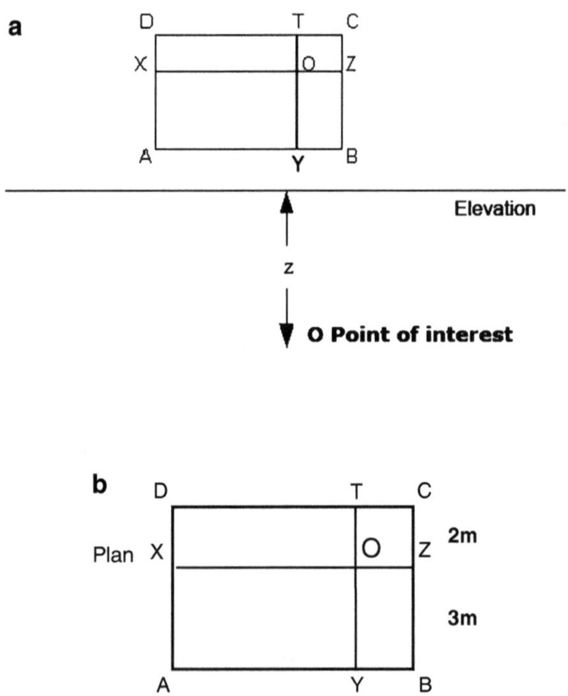

Fig. 5.15 (**a**) The stress increase at a point *below* a loaded *rectangular* region. (**b**) Dimensions of a rectangular-loaded area. (**c**) A rectangular-loaded area ABCD and a point of interest O. (**d**) Decomposition of loading over a *rectangular* area (for stress at an external point)

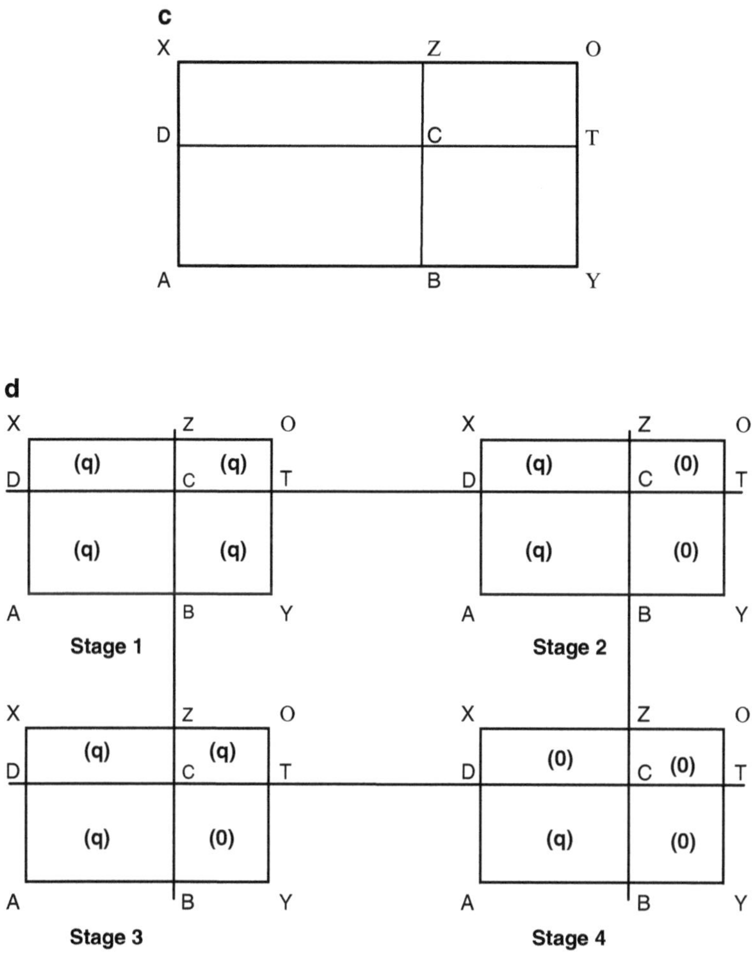

Fig. 5.15 (continued)

thus

$$I_\sigma = 0.175$$

and so

$$\Delta\sigma_{zz} = p\, I_\sigma = 100 \times 0.175 = 17.5 \text{ kPa} \tag{5.27a}$$

For rectangular loading OTDX

$$m = L/z = 1.5$$

$$n = B/z = 1$$

thus

$$I_\sigma = 0.194$$

and so

$$\Delta\sigma_{zz} = p\, I_\sigma = 100 \times 0.195 = 19.4 \text{ kPa} \tag{5.27b}$$

For rectangular loading OXAY

$$m = L/z = 1.5$$

$$n = B/z = 1.5$$

thus

$$I_\sigma = 0.216$$

and so

$$\Delta\sigma_{zz} = p\, I_\sigma = 100 \times 0.216 = 21.6 \text{ kPa} \tag{5.27c}$$

For rectangular loading OYBZ

$$m = L/z = 1.5$$

$$n = B/z = 1$$

thus

$$I_\sigma = 0.194$$

and so

$$\Delta\sigma_{zz} = p\, I_\sigma = 100 \times 0.194 = 19.4 \text{ kPa} \tag{5.27d}$$

Thus, the increase in stress

$$\Delta\sigma_{zz} = 17.5 + 19.4 + 21.6 + 19.4 = 78.9 \text{ kPa} \tag{5.27e}$$

This must, of course, be added to the existing stress state prior to loading to obtain the actual stress σ_{zz}.

Calculation of stress below a point outside the loaded area
The stress increase at a point vertically below a point O which is outside the loaded area can also be found using the influence charts shown in Fig. 5.15c.

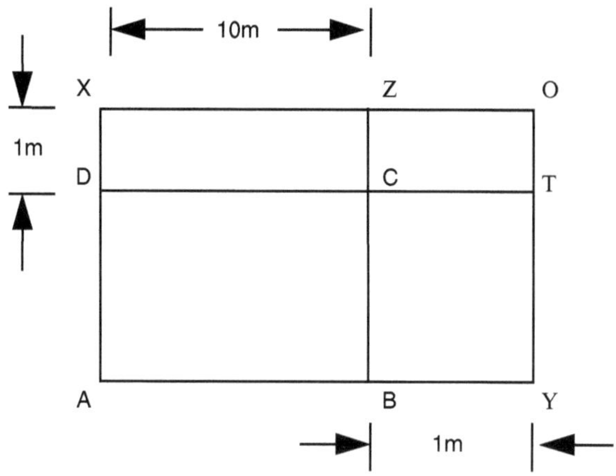

Fig. 5.16 Dimensions of a rectangular-loaded area

This is achieved by considering the stress q acting on ABCD to consist of the following:

1. a stress + q acting over OXAY;
2. a stress + q acting over OZCT;
3. a stress -q acting over OZBY; and
4. a stress -q acting over OXDT

This is illustrated in Fig. 5.15d.

It thus follows that at the point O, the stress increase $\Delta\sigma_{zz}(ABCD)$ is given by:

$$\Delta\sigma_{zz}(ABCD) = \Delta\sigma_{zz}(OXAY) - \Delta\sigma_{zz}(OYBZ) + \Delta\sigma_{zz}(OZCT) - \Delta\sigma_{zz}(OTDX)$$

and thus (10)

$$\sigma_{zz}(ABCD) = q[I_\sigma(OXAY) - I_\sigma(OYBZ) + I_\sigma(OZCT) - I_\sigma(OTDX)]$$

Example. Suppose that the rectangular area ABCD shown in Fig. 5.16 is subjected to a surface stress of 100 kPa AND it is required to calculate the vertical stress increase at a point 1.5 m below the point O.

For rectangular loading OZCT

$$m = L/z = 0.67$$

$$n = B/z = 0.67$$

thus

$$I_\sigma = 0.121$$

and so

$$\Delta\sigma_{zz} = p\, I_\sigma = +100 \times 0.121 = +12.1\ \text{kPa} \tag{5.28a}$$

For rectangular-loading OTDX

$$m = L/z = 7.67$$
$$n = B/z = 0.67$$

thus

$$I_\sigma = 0.167$$

and so

$$\Delta\sigma_{zz} = p\, I_\sigma = -100 \times 0.167 = -16.7\ \text{kPa} \tag{5.28b}$$

For rectangular-loading OXAY

$$m = L/z = 7.67$$
$$n = B/z = 2.00$$

thus

$$I_\sigma = 0.240$$

and so

$$\Delta\sigma_{zz} = p\, I_\sigma = +100 \times 0.240 = +25.0\ \text{kPa} \tag{5.28c}$$

For rectangular-loading OYBZ

$$m = L/z = 2$$
$$n = B/z = 0.67$$

thus

$$I_\sigma = 0.164$$

and so

$$\Delta\sigma_{zz} = p\, I_\sigma = -100 \times 0.164 = -16.4\ \text{kPa} \tag{5.28d}$$

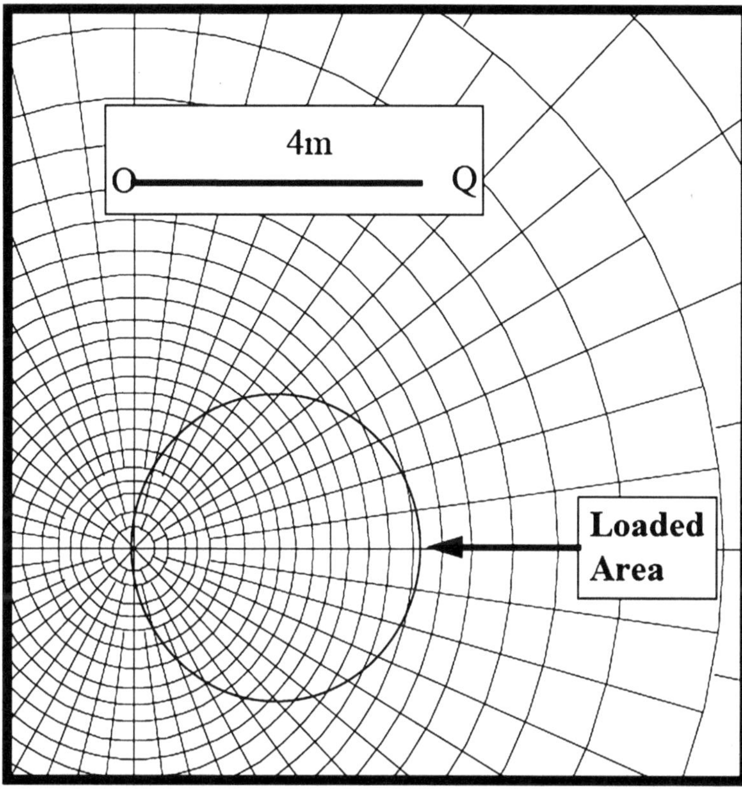

Fig. 5.17 Newmark's chart

Thus, the increase in stress

$$\Delta\sigma_{zz} = 12.1 - 16.7 + 25.0 + -16.4 = 3.0\ \text{kPa} \tag{5.28e}$$

Stresses due to foundation loads of ***an arbitrary shape*** *applied at the ground surface.*

Newmark's chart provides a graphical method for calculating the stress increase due to a uniformly-loaded region of an arbitrary shape resting on a deep homogeneous isotropic elastic region.

Newmark's chart is generally given in Data sheets and is reproduced in part in Fig. 5.17. The procedure for its use is outlined below

1. The scale for this procedure is determined by the depth z at which the stress is to be evaluated; thus, z is equal to the distance OQ shown on the chart.
2. Draw the loaded area to scale so that the point of interest (more correctly, its vertical projection on the surface) is at the origin of the chart; the orientation of the drawing does not matter.

3. Count the number of squares (N) within the loaded area; if more than half the square is inside the area; then count the squares; otherwise neglect it.
4. The vertical stress increase $\Delta\sigma_{zz} = N \times$ [scale factor(0.001)] $\times$ [surface stress (p)]

The procedure is most easily illustrated by an example.

Example. Suppose that a uniformly-loaded circle of radius 2 m carries a uniform stress of 100 kPa. It is required to calculate the vertical stress at a depth of 4 m below the edge of the circle.

The loaded area is drawn on Newmark's chart to the appropriate scale (i.e., the length OQ is set to represent 4 m) as shown in Fig. 5.17.

It is found that the number of squares, N = 194, and thus, the stress increase is found to be

$$\Delta\sigma_{zz} = 194 \times 0.001 \times 100 = 19.4\,\text{kPa} \tag{5.29}$$

This result can also be checked using the influence charts for circular loading, and it is then found that:

$$z/a = 2,\ r/a = 1.\Delta\sigma_{zz}/p = 0.2 \text{ and so } \Delta\sigma_{zz} = 20\,\text{kPa} \tag{5.30}$$

5.7 Closed-Form Solutions for Special Loadings

5.7.1 *A Finite Set of Wells Pumping in a Semi-Infinite Domain*

Consider a semi-infinite aquifer, and consider a vertical well screened at a depth interval of $a[L]$ to $b[L]$ $(0 < b < a)$; the screen length is simply equal to $a - b$.

For a given mass extraction (or injection) $Q_w\,[M/T]$, the goals are to derive an analytical solution for transient pressure and soil subsidence induced by a single well.

In this section, we present a closed-form solution of the Biot equations using the Helmotz representation of the displacement field and the Green functions techniques.

Lin and Lu derived a closed form solution of the point source problem for the decoupled form of the Biot equations. They use the Hankel transform technique.

Figure 5.18 presents an impulsive point sink buried in a saturated porous stratum at a depth h. The aquifer is considered as a homogeneous isotropic porous medium with a vertical axis of symmetry.

The following assumptions are made for further simplifying the problem:

(a) The aquifer is isotropic at least in the horizontal plane;
(b) The extraction (or injection) rate is uniformly distributed over the screen interval;

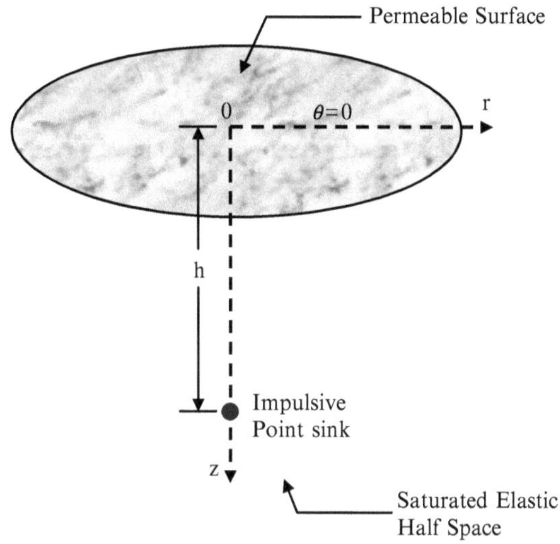

Fig. 5.18 An impulsive point sink-induced land deformation problem

(c) During the water pumping, the water pressure and the soil movement are assumed to be governed by the Biot model;
(d) The coordinate is chosen such that the origin is at the land surface and the vertical (z) axis positive downwards (Fig. 5.18); and
(e) The water pressure $p(r, z, t)$ and the soil displacement $\vec{u}\,(r, z) = (u_x(r, z, t),$ $u_y(r, z, t),\ u_z(r, z, t))$ are functions of *r*, z, and *t* only.

This last assumption implies that the displacement field is rotational and therefore, following theorem 2, the Biot equations may be separated into two separated equations. These equations use the potential function $\varphi(r, z, t)$ and the pressure head $p(r, z, t)$.

$$K_r \frac{\partial^2 p}{\partial x^2} + K_r \frac{\partial^2 p}{\partial y^2} + K_z \frac{\partial^2 p}{\partial z^2} = \frac{1}{\alpha} \frac{\partial p}{\partial t};\ t \geq 0 \tag{5.31a}$$

$$\frac{\partial^2 \varphi}{\partial x^2} + \frac{\partial^2 \varphi}{\partial y^2} + \frac{\partial^2 \varphi}{\partial z^2} = p(r,z,t) \tag{5.31b}$$

The boundary conditions and initial conditions may be summarized as:

$$p(r, z, 0) = P0;\ \varphi(r, z, 0) = 0; \tag{5.32a}$$

$$p(r, 0, t) = P0;\quad \varphi(r, 0, t) = 0; \tag{5.32b}$$

The well is set at an arbitrary location, (x_m, y_m), and the we consider an infinitely small increment $d\,z_m$ at a point z_m in the well screen $b < z_m < a$.

The point (x_m, y_m, z_m) may be treated as a point source (or sink) with strength of $Q_w\, d\, z_m \,/(b-a) = C d z_m$.

Following Carslaw and Jaeger (1959, p. 261), the solution of the water pressure $U(x, y, z, t) = p(r, z, t) - P_0$ for the point source (or sink) with an initial condition and boundary conditions presented above is:

$$\mathrm{d\,U(x,y,z,t)} = \mathrm{C}\,\frac{\mathrm{erfc}\left[\sqrt{\mathrm{r}^2 + (\mathrm{z}-\mathrm{z_w})^2}\,/\sqrt{4\,\alpha\,\mathrm{t}}\right]}{\sqrt{\mathrm{r}^2 + (\mathrm{z}-\mathrm{z_w})^2}}\,\mathrm{dz_w} \tag{5.33}$$

Here erfc(x) is the complementary error function.

$$erfc(x) = 1 - \int_x^{\infty} \exp(-t^2)dt$$

The normalized radial distance r is given by:

$$\mathrm{r} = \sqrt{\frac{\mathrm{K_z}}{\mathrm{K_r}}}\sqrt{(\mathrm{x}-\mathrm{x_w})^2 + (\mathrm{y}-\mathrm{y_w})^2} \tag{5.34}$$

However, the semi-infinite medium is bounded above by a constant pressure. The solution that satisfies the boundary conditions can be obtained by the method of images (see Mandel et al. 2003).

$$\mathrm{d\,U(x,y,z,t)} =$$

$$\mathrm{C}\left[\frac{\mathrm{erfc}\left[\sqrt{\mathrm{r}^2 + (\mathrm{z}-\mathrm{z_w})^2}\,/\sqrt{4\,\alpha\,\mathrm{t}}\right]}{\sqrt{\mathrm{r}^2 + (\mathrm{z}-\mathrm{z_w})^2}} - \frac{\mathrm{erfc}\left[\sqrt{\mathrm{r}^2 + (\mathrm{z}+\mathrm{z_w})^2}\,/\sqrt{4\,\alpha\,\mathrm{t}}\right]}{\sqrt{\mathrm{r}^2 + (\mathrm{z}+\mathrm{z_w})^2}}\right]\mathrm{dz_w} \tag{5.35a}$$

The solution for a continuous line source (or sink) is simply the integral of this last expression when z_w varies from b to a.

$$\mathrm{U(x, y, z, t)} =$$

$$\mathrm{C}\left[\int_{\mathrm{b}}^{\mathrm{a}}\frac{\mathrm{erfc}\left[\sqrt{\mathrm{r}^2 + (\mathrm{z}-\mathrm{z_w})^2}/\sqrt{4\,\alpha\,\mathrm{t}}\right]}{\sqrt{\mathrm{r}^2 + (\mathrm{z}-\mathrm{z_w})^2}} - \frac{\mathrm{erfc}\left[\sqrt{\mathrm{r}^2 + (\mathrm{z}+\mathrm{z_w})^2}\,/\sqrt{4\,\alpha\,\mathrm{t}}\right]}{\sqrt{\mathrm{r}^2 + (\mathrm{z}+\mathrm{z_w})^2}}\right]\mathrm{dz_w} \tag{5.35b}$$

The solution of the displacement potential $\phi(x, y, z, t)$ may be formally written as:

$$\varphi(\mathrm{x,y,z,t}) = \mathrm{C}\int_{-\infty}^{\infty}\int_{-\infty}^{\infty}\int_{0}^{\infty}\left[\frac{[\mathrm{P_0} + \mathrm{U}(\eta, \mu, \xi, \mathrm{t})]}{\sqrt{(\mathrm{x}-\eta)^2 + (\mathrm{y}-\mu)^2 + (\mathrm{z}-\xi)^2}} - \frac{[\mathrm{P_0} + \mathrm{U}(\eta, \mu, \xi, \mathrm{t})]}{\sqrt{(\mathrm{x}-\eta)^2 + (\mathrm{y}-\mu)^2 + (\mathrm{z}+\xi)^2}}\right] \mathrm{d}\eta\, \mathrm{d}\mu\, \mathrm{d}\xi \tag{5.35c}$$

This last equation may be simplified using cylindrical coordinates.

$$\varphi(x, y, z, t) = C\int_{0}^{\infty}\int_{0}^{2\pi}\int_{0}^{\infty}\left[\frac{[P_0 + U(r, \theta, \xi, t)]}{\sqrt{(x - r\cos\theta)^2 + (y - r\sin\theta)^2 + (z-\xi)^2}} - \frac{[P_0 + U(r, \theta, \xi, t)]}{\sqrt{(x - r\cos\theta)^2 + (y - r\sin\theta)^2 + (z+\xi)^2}}\right] r dr\, d\theta\, d\xi \tag{5.36}$$

5.7.2 *A Finite Set of Wells Pumping in a Finite Strip Layer Domain*

Let us now extend the solution to multi-well cases by means of the principle of superposition.

Consider an aquifer of depth $H\,[L]$ and consider a vertical well screened at a depth interval of $a[L]$ to $b[L]$ $(0 < b < a < H)$, the screen length is simply equal to $a - b$.

For a given mass extraction (or injection) $Q_w\,[M/T]$, the goals are to derive an analytical solution for transient pressure and soil subsidence induced by a single well.

Also later, we will extend the solution to multi-well cases by means of the principle of superposition.

The following assumptions are made for further simplifying the problem:

(a) The aquifer is isotropic at least in the horizontal plane;
(b) The extraction (or injection) rate is uniformly distributed over the screen interval;
(c) During the water pumping, the water pressure and the soil movement are assumed to be governed by the Biot model;

(d) The coordinate is chosen such that the origin is at the land surface and the vertical (z) axis positive downwards (Fig. 5.18) and
(e) The water pressure $p(r,z,t)$ and the soil displacement $\vec{u}(r,z) = (u_x(r,z,t), u_y(r,z,t), u_z(r,z,t))$ are functions of *r, z,* and *t* only.

This last assumption implies that the displacement field is rotational and therefore, following Theorem 1 and Equations 5.3a and 5.9a, the Biot equations may be separated into two separated equations. These equations use the potential function $\phi(r,z,t)$ and the pressure head $p(r,z,t)$.

$$K_r\frac{\partial^2 p}{\partial x^2} + K_r\frac{\partial^2 p}{\partial y^2} + K_z\frac{\partial^2 p}{\partial z^2} = \frac{1}{\alpha}\frac{\partial p}{\partial t}; \ t \geq 0$$

$$\frac{\partial^2 \phi}{\partial x^2} + \frac{\partial^2 \phi}{\partial y^2} + \frac{\partial^2 \phi}{\partial z^2} = p(r,z,t)$$

Boundary conditions and initial conditions may be summarized as:

$$p(r,z,0) = P0; \ \varphi(r,z,0) = 0;$$

$$p(r,0,t) = P0; \ \varphi(r,0,t) = 0;$$

$$\left.\frac{\partial p}{\partial z}\right|_{z=H} = \left.\frac{\partial \varphi}{\partial z}\right|_{z=H} = 0$$

The well is set at an arbitrary location, (x_m, y_m), and the we consider an infinitely small increment $d\,z_m$ at a point z_m in the well screen $b < z_m < a$. The point (x_m, y_m, z_m) may be treated as a point source (or sink) with a strength of $Q_w\,d\,z_m\,/(b-a) = C d z_m$.

Following Carslaw and Jaeger (1959, p. 261), the solution of the water pressure $U(x,y,z,t) = p(r,z,t) - P_0$ for the point source (or sink) with an initial condition and boundary conditions presented above is:

$$d\,U(x,y,z,t) = C\,\frac{erfc\left[\sqrt{r^2 + (z - z_w)^2}\,/\sqrt{4\,\alpha\,t}\right]}{\sqrt{r^2 + (z - z_w)^2}}\,d\,z_w$$

Here, erfc(x) is the complementary error function.

$$erfc(x) = 1 - \int_x^\infty \exp(-t^2)dt$$

The normalized radial distance r is given by:

$$r = \sqrt{\frac{K_z}{K_r}}\sqrt{(x - x_w)^2 + (y - y_w)^2}$$

However, the semi-infinite medium is bounded above by a constant pressure. The solution that satisfies the boundary conditions can be obtained by the method of images (see Shan et al. 1992).

$$d\,U(x,y,z,t) = C\,d z_w \sum_{n=0}^{\infty} \left[\frac{erfc\left[\sqrt{r^2 + (z - z^+{}_n)^2} \,/\, \sqrt{4\,\alpha\,t} \right]}{\sqrt{r^2 + (z - z^+{}_n)^2}} \right] - \left[\frac{erfc\left[\sqrt{r^2 + (z + z^-{}_n)^2} \,/\, \sqrt{4\,\alpha\,t} \right]}{\sqrt{r^2 + (z + z^-{}_n)^2}} \right] \tag{5.37}$$

Where the locations of the images are calculated by a series of images points in periodic strips of depths H:

$$z^+{}_n = (-1)^n\, z_w \pm 2nH \qquad (n = 0, 1, 2, \ldots..)$$

$$z^-{}_n = (-1)^{n+1}\, z_w \pm 2nH \qquad (n = 0, 1, 2, \ldots..)$$

The solution for a continuous line source (or sink) is simply the integral of this last expression when z_w varies from b to a.

$$U(x,y,z,t) = C \left[\int_b^a \sum_{n=0}^{\infty} \left[\frac{erfc\left[\sqrt{r^2 + (z - z^+{}_n)^2} \,/\, \sqrt{4\,\alpha\,t} \right]}{\sqrt{r^2 + (z - z^+{}_n)^2}} \right] - \left[\frac{erfc\left[\sqrt{r^2 + (z + z^-{}_n)^2} \,/\, \sqrt{4\,\alpha\,t} \right]}{\sqrt{r^2 + (z + z^-{}_n)^2}} \right] \right] d z_w \tag{5.38}$$

The solution of the displacement potential $\varphi(x,y,z,t)$ may be formally written as:

$$\varphi(x,y,z,t) = C \int_{-\infty}^{\infty} \int_{-\infty}^{\infty} \int_{0}^{\infty} \left[\frac{[P_0 + U(\eta,\mu,\xi,t)]}{\sqrt{(x-\eta)^2 + (y-\mu)^2 + (z-\xi)^2}} - \frac{[P_0 + U(\eta,\mu,\xi,t)]}{\sqrt{(x-\eta)^2 + (y-\mu)^2 + (z+\xi)^2}} \right] d\eta\, d\mu\, d\xi \tag{5.39}$$

This last equation may be simplified using cylindrical coordinates.

$$\varphi(x,y,z,t) = C\int_0^{\infty}\int_0^{2\pi}\int_0^{\infty}\left[\frac{[\mathrm{P}_0 + U(r,\theta,\xi,t)]}{\sqrt{(x-r\cos\theta)^2+(y-r\sin\theta)^2+(z-\xi)^2}}\right.$$
$$\left.-\frac{[P_0 + U(r,\theta,\xi,t)]}{\sqrt{(x-r\cos\theta)^2+(y-r\sin\theta)^2+(z+\xi)^2}}\right] rdr\,d\theta\,d\xi \tag{5.40}$$

5.8 Comparison Between the Coupled and Decoupled Solutions

Consider the Biot system of equation in cylindrical coordinates.

$$\mathrm{G}\nabla^2\mathrm{u_r} + \frac{\mathrm{G}}{1-2\nu}\frac{\partial\varepsilon}{\partial r} - \mathrm{G}\frac{\mathrm{u_r}}{\mathrm{r}^2} = \frac{\partial \mathrm{p}}{\partial \mathrm{r}} + \mathrm{b_r}, \tag{5.41a}$$

$$\mathrm{G}\nabla^2\mathrm{u_z} + \frac{\mathrm{G}}{1-2\nu}\frac{\partial\varepsilon}{\partial z} = \frac{\partial \mathrm{p}}{\partial \mathrm{z}} + \mathrm{b_z}, \tag{5.41b}$$

$$-\frac{k}{\gamma_w}\nabla^2 p = n\beta\frac{\partial p}{\partial t} + \alpha\frac{\partial\varepsilon}{\partial t} + \frac{Q_0}{2\pi r}\delta(r)\delta\,(z-h)\,\delta(t), \tag{5.41c}$$

Using the Helmholtz representation, that the system of Eqs. (5.41a) through and (5.41c) may be solved using a decomposition of the displacement field.

$$\mathrm{u_r} = \mathrm{u^{(1)}}_\mathrm{r} + \mathrm{u^{(2)}}_\mathrm{r}; \tag{5.42a}$$

$$\mathrm{u_z} = \mathrm{u^{(1)}}_\mathrm{z} + \mathrm{u^{(2)}}_\mathrm{z} \tag{5.42b}$$

Where $\mathrm{u^{(1)}}_\mathrm{r}$ and $\mathrm{u^{(1)}}_\mathrm{z}$ are solutions of the equations:

$$\mathrm{G}\nabla^2\mathrm{u^{(1)}}_\mathrm{r} = \frac{1-2\upsilon}{2(1-\upsilon)}\frac{\partial \mathrm{p}}{\partial \mathrm{r}} \tag{5.43a}$$

$$\mathrm{G}\nabla^2\mathrm{u^{(1)}}_\mathrm{z} = \frac{1-2\upsilon}{2(1-\upsilon)}\frac{\partial \mathrm{p}}{\partial \mathrm{z}} \tag{5.43b}$$

And $\mathrm{u^{(2)}}_\mathrm{r}$ and $\mathrm{u^{(2)}}_\mathrm{z}$ are solutions of the equations:

$$\mathrm{G}\nabla^2\mathrm{u^{(2)}}_\mathrm{r} = \mathrm{b_r} \tag{5.43c}$$

$$\mathrm{G}\nabla^2\mathrm{u^{(2)}}_\mathrm{z} = \mathrm{b_z} \tag{5.43d}$$

Also the water pressure is the solution of the equations:

$$-\frac{k}{\gamma_w}\nabla^2 p = \chi\frac{\partial p}{\partial t} - \frac{Q_0}{2\pi r}\delta(r)\delta\left(z-h\right)\delta(t)$$

$$\chi = n\beta + \alpha\frac{1-2\upsilon}{2G(1-\upsilon)}; \quad Q_0 \geq 0 \tag{5.43e}$$

Boundary Conditions:

Consider the half space surface, $z = 0$, as a traction-free boundary for all time $t \geq 0$. From Eqs. (1c) and (1d), the mechanical boundary conditions at $z = 0$ are expressed in terms of $u^{(1)}{}_r$; $u^{(2)}{}_r$ and $u^{(1)}{}_z$; $u^{(2)}{}_z$ by

$$\frac{2G\nu}{1-2\nu}\left[\frac{\partial u^{(i)}{}_r\left(r,0,t\right)}{\partial r} + \frac{u^{(i)}{}_r\left(r,0,t\right)}{r}\right] + \frac{2G\left(1-\nu\right)}{1-2\nu}\frac{\partial u^{(i)}{}_z\left(r,0,t\right)}{\partial z} = 0, \tag{5.44a}$$

$$G\left[\frac{\partial u^{(i)}{}_r\left(r,0,t\right)}{\partial z} + \frac{\partial u^{(i)}{}_z\left(r,0,t\right)}{\partial r}\right] = 0. \tag{5.44b}$$

$$i = 1,2$$

The far boundary $z \to \infty$ due to the effect of an impulsive point sink must vanish at any time. Hence

$$\lim_{z\to\infty} u^{(i)}{}_r\left(r,z,t\right) = 0 \tag{5.45a}$$

$$\lim_{z\to\infty} u^{(i)}{}_z\left(r,z,t\right) = 0; \quad i = 1,2 \tag{5.45b}$$

$$\lim_{z\to\infty} p\left(r,z,t\right) = 0 \tag{5.45c}$$

Initial Conditions:

Assuming no initial changes in displacements and seepage of the aquifer, then the initial conditions at time $t = 0$ of the mathematical model is:

$$u^{(i)}{}_r\left(r,z,0\right) = 0; \quad i = 1,2,3 \tag{5.46a}$$

$$u^{(i)}{}_z\left(r,z,0\right) = 0; \quad i = 1,2,3, \tag{5.46b}$$

$$p\left(r,z,0\right) = 0; \quad i = 1,2,3. \tag{5.46c}$$

5.8.1 Solutions of Coupled Versus Decoupled

The generalized Helmholtz decomposition of the displacement field of the Biot equations, presented in Sect. 5.2, is valid for both the coupled and the decoupled problems. The only difference in the resulting system of equations is the value of the parameter χ (see Eq. 5.43c).

A solution of the governing partial differential Eqs. (5.41a), through (5.41c) has been solved for the decoupled problem $\alpha = 0$ (Lu and Lin 2007).

The system of equations can be reduced to a set of ordinary differential equations by performing appropriate Laplace and Hankel transforms (Sneddon 1951) with respect to the time variable t and the radial coordinate r:

The transient horizontal displacement $u_r(r, 0, t)$ and vertical displacement $u_z(r, 0, t)$ of the pervious ground surface due to an impulsive point sink have been derived as below:

$$u_r(r,0,t) = \frac{Q_0\gamma_w}{2(2\eta-1)\pi Gk}\left\{-\frac{cr}{(h^2+r^2)^{3/2}} + \int_0^{ct}\frac{chr}{16\tau^3}\exp\left(-\frac{r^2+2h^2}{8\tau}\right) \cdot \left[I_0\left(\frac{r^2}{8\tau}\right) - I_1\left(\frac{r^2}{8\tau}\right)\right]d\tau\right\} \tag{5.47a}$$

$$u_z(r,0,t) = \frac{Q_0\gamma_w}{2(2\eta-1)\pi Gk}\left\{\frac{ch}{(h^2+r^2)^{3/2}}\operatorname{erf}\left(\frac{\sqrt{h^2+r^2}}{2\sqrt{ct}}\right) - \frac{ch}{h^2+r^2}\frac{1}{\sqrt{\pi ct}}\exp\left(-\frac{h^2+r^2}{4ct}\right)\right\} \tag{5.47b}$$

erf(x) denotes error function; and $I_\alpha(x)$ is known as the modified Bessel function of the first kind of order α.

The instantaneous ground surface horizontal and vertical displacements of the pervious half space at $t \to 0^+$ are obtained as following:

$$u_r\left(r,0,0^+\right) = -\frac{cQ_0\gamma_w}{2(2\eta-1)\pi Gk}\frac{r}{(h^2+r^2)^{3/2}} \tag{5.48a}$$

$$u_z\left(r,0,0^+\right) = \frac{cQ_0\gamma_w}{2(2\eta-1)\pi Gk}\frac{h}{(h^2+r^2)^{3/2}}. \tag{5.48b}$$

The parameters η and c are defined as $\eta = (1-\nu)/(1-2\nu)$; $\mathrm{c} = \mathrm{k}/\chi\gamma_\mathrm{w}$

The maximum ground surface horizontal displacement $u_{r\ \max}$ of the half space due to an impulsive point sink is derived from Eq. (5.48a) using $r = h\Big/\sqrt{2} \approx 0.707\,h$. After doing so, we have

Table 5.1 Typical values of the elastic saturated aquifer of medium dense sand

Parameter	Symbol	Value	Units
Shear modulus[a]	G	20.10^6	N/m^2
Porosity[b]	n	0.3	–
Poisson's ratio[a]	ν	0.3	–
Permeability[a]	k	1.10^{-5}	m/s
Compressibility of water[c]	$1/\beta$	$2.14.10^9$	N/m^2
Unit weight of water[a]	γ_ω	9.81	N/m^3

[a]Booker and Carter (1986a)
[b]Slough et al. (1999)
[c]Vardoulakis and Beskos (1986)

$$u_{r\ \max} = u_r\left(h\left/\sqrt{2}\right.,0,0^+\right) = -\frac{\sqrt{3}cQ_0\gamma_w}{9\,(2\eta-1)\,\pi Gkh^2} \tag{5.49}$$

in which the value $r = h/\sqrt{2}$ is derived when $du_r\left(r,0,0^+\right)/dr$ is equal to zero:

$$\frac{du_r\left(r,0,0^+\right)}{dr} = -\frac{Q_0\gamma_w}{2\,(2\eta-1)\,\pi Gk}\frac{h^2-2r^2}{\left(h^2+r^2\right)^{5/2}} = 0. \tag{5.50}$$

This leads to solutions of $r = \pm h/\sqrt{2}$, but only $r = h/\sqrt{2}$ is reasonable for $r \in [0,\infty)$.

The maximum ground surface vertical displacement $u_{z\ \max}$ of the pervious half space due to an impulsive point sink is derived from Eq. (5.48b) with $r = 0$. Hence

$$u_{z\ \max} = u_z\left(0,0,0^+\right) = \frac{cQ_0\gamma_w}{2\,(2\eta-1)\,\pi Gkh^2}. \tag{5.51}$$

The absolute value of the ratio of $u_{z\ \max\,(coupled)}\big/u_{z\ \max\,(uncoupled)}$ can be derived from Eq. (5.51) as below:

$$\left|\frac{u_{z\ \max\,(coupled)}}{u_{z\ \max(uncoupled)}}\right| = \left|\frac{\chi_{coupled}}{\chi_{uncoupled}}\right| = \frac{1}{1+\frac{1-2\nu}{2nG\beta(1-\upsilon)}}. \tag{5.52}$$

Also from Eq. (5.52) where $\chi_{coupled}$ is defined in Eq. (5.43e) one can easily see that

$$\left|\frac{u_{z\ (coupled)}(r,0,0^+)}{u_{z\ (uncoupled)}(r,0,0^+)}\right| = \left|\frac{u_{z\ \max\,(coupled)}}{u_{z\ \max(uncoupled)}}\right| = \left|\frac{\chi_{coupled}}{\chi_{uncoupled}}\right| = \frac{1}{1+\frac{1-2\nu}{2nG\beta(1-\upsilon)}} \tag{5.53}$$

For medium dense sand, typical values of the parameters are summarized in Table 5.1.

If the pumping depth h is 30 m and the instantaneous amount of groundwater is pumped from the aquifer at $Q_0 = 1\ \mathrm{m}^3$ initially, then it can have a maximum horizontal displacement and vertical deformation at the pervious ground surface of 0.97 cm and 2.52 cm, respectively.

An estimate of the ration $u_{z\ \max(coupled)} / u_{z\ \max(uncoupled)}$ gives:

$$\left| \frac{u_{z\ \max(coupled)}}{u_{z\ \max(uncoupled)}} \right| = \frac{1}{1 + \frac{1-2\upsilon}{2nG\beta(1-\upsilon)}} = 167.67 \tag{5.54}$$

The above result shows the maximum vertical ground surface horizontal displacement due to an impulsive point sink computed with the full coupling of the Biot equations is around 168 times larger than the same displacement computed with a non coupled system. This results was computed for typical parameter values for a dense sand soil. For gravel soils the shear modulus is higher and the ratio is around 1.5.

5.8.2 Numerical Results

The particular interest is the vertical displacement of stratum at each stage of the consolidation process, then the average consolidation ratio U is defined as:

$$U = \frac{u_z(r, 0, t)}{u_{z\ \max}}, \tag{5.55}$$

where $u_z(r, 0, t)$ and $u_{z\ \max}$ follows the definitions in Eqs. (5.48b) and (5.55), respectively. For pervious half space, U can be derived for both coupled and decoupled problem as below:

$$U_{uncoupled}(r, 0, t) = \frac{h^3}{(h^2 + r^2)^{3/2}} \operatorname{erf}\left(\frac{\sqrt{h^2 + r^2}}{2\sqrt{c_1 t}} \right) - \frac{h^3}{h^2 + r^2} \frac{1}{\sqrt{\pi c_1 t}} \exp\left(-\frac{h^2 + r^2}{4c_1 t} \right); \ c_1 = k / n\beta\gamma_w. \tag{5.56a}$$

$$U_{coupled}(r, 0, t) = \frac{h^3}{(h^2 + r^2)^{3/2}} \operatorname{erf}\left(\frac{\sqrt{h^2 + r^2}}{2\sqrt{c_2 t}} \right) - \frac{h^3}{h^2 + r^2} \frac{1}{\sqrt{\pi c_2 t}} \exp\left(-\frac{h^2 + r^2}{4c_2 t} \right); \ c_2 = c_1 \left(1 + \frac{1 - 2\upsilon}{2nG\beta(1 - \upsilon)} \right) \tag{5.56b}$$

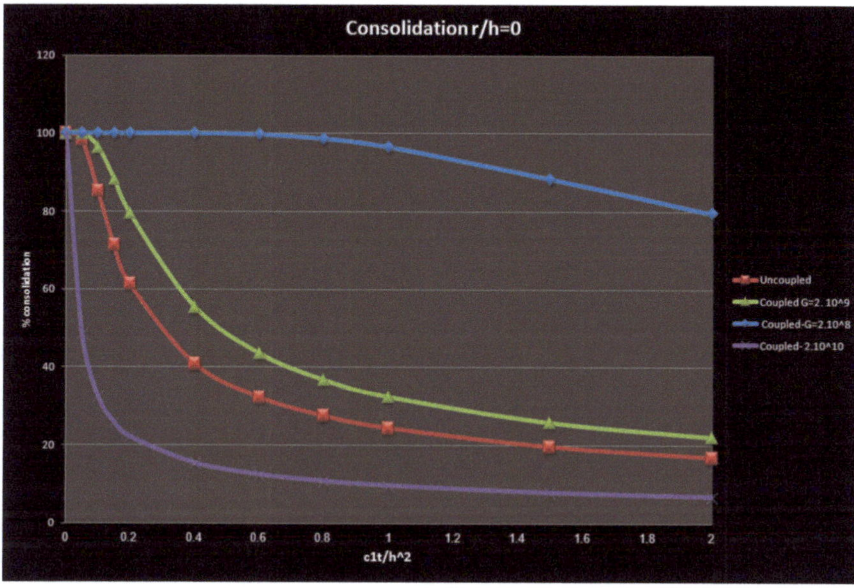

Fig. 5.19 Comparison of the Average consolidation ratio U for coupled and decoupled models at $r/h = 0$

These last expressions of the average consolidation ratio U for both coupled and non coupled formulations may be written in terms of the ratio $u = r/h$, the time ratio $T = c_1 t/h^2$ and

$$\kappa = 1 + \frac{1 - 2\nu}{2nG\beta(1 - \upsilon)} \tag{5.57a}$$

$$U_{uncoupled}(r, 0, t) = \frac{1}{(1 + u^2)^{3/2}} \operatorname{erf}\left(\frac{\sqrt{1 + u^2}}{2T}\right) - \frac{1}{1 + u^2} \frac{1}{\sqrt{\pi T}} \exp\left(-\frac{1 + u^2}{4T}\right); \tag{5.57b}$$

$$U_{coupled}(r, 0, t) = U_{uncoupled}(r, 0, t) = \frac{1}{(1 + u^2)^{3/2}} \operatorname{erf}\left(\frac{\sqrt{1 + u^2}}{2\kappa T}\right) - \frac{1}{1 + u^2} \frac{1}{\sqrt{\pi \kappa T}} \exp\left(-\frac{1 + u^2}{4\kappa T}\right); \tag{5.57c}$$

Figures 5.19 and 5.20 shows the average consolidation ratio U at $r/h = 0$ and for the impulsive pumping. Note that U initially decreases rapidly, and then the rate of vertical displacement at the ground surface reduces gradually. Each final value of U will disappear for the saturated aquifer is treated as linear elastic porous medium.

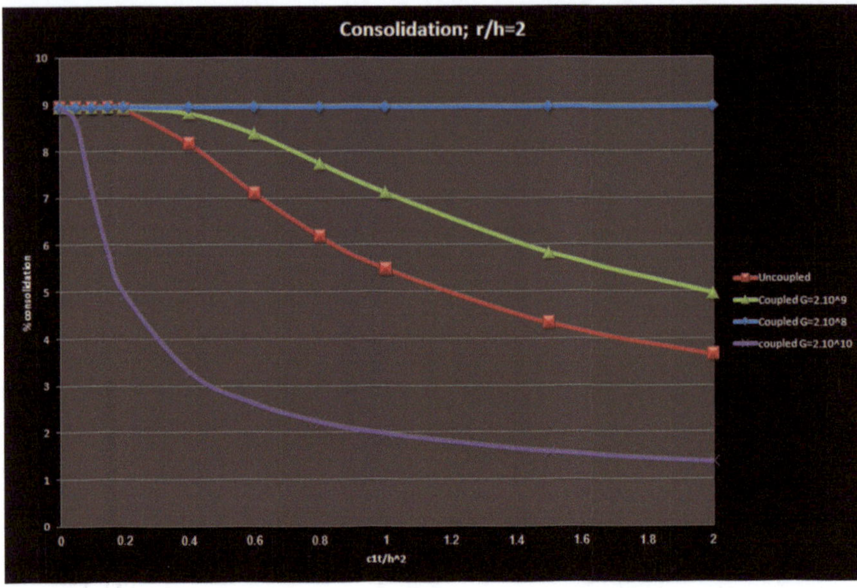

Fig. 5.20 Comparison of the Average consolidation ratio U for coupled and decoupled models at $r/h = 2$

For large values of the shear stress, the decrease of the curve is sharper than the result curve for decoupled problem. However for small values of the elastic shear modulus the consolidation takes places even far from the pumping well. The ground deformations due to pumping will fully recover after the excess pore water pressure disappears in this model.

5.9 Appendix A. The Helmholtz Theorem

In this appendix, the proof of existence and uniqueness of the Helmholtz theorem is presented for an infinite three dimensional space. Similar proofs may be found in classical treatise of mathematical methods (See Morse and Feshbach (1953). Recently, Zhou (2006) presented a generalized Helmholtz decomposition theorem for closed volume separated by a regular boundary. They show that the proposed decomposition may be unique providing that the vector field is assigned on the external surface for both the normal and the tangent directions. A generalization of the generalized Helmholtz decomposition theorem for multi connected domain was presented by Gui and Dou (2007) and Gui and Li (2011).

In this appendix the three dimensional vector is written in bold instead $(\mathbf{F} = \vec{\mathrm{F}};\ \nabla = \vec{\nabla})$.

If a vector field **F** whose divergence $\nabla \bullet \mathbf{F} = b(\mathbf{r})$ and curl $\nabla \times \mathbf{F} = \mathbf{c}(\mathbf{r})$ are known and entirely contained within a finite region of space, then **F** is uniquely determined and has the value

$$\mathbf{F} = -\nabla\phi + \nabla \times \mathbf{A} \tag{5.A.1}$$

where

$$\phi(\mathbf{r}) \equiv \frac{1}{4\pi}\int_V \frac{\mathrm{b}(\mathbf{r}')}{\mathrm{R}}\mathrm{dv}' \tag{5.A.2a}$$

$$\mathbf{A}(\mathbf{r}) \equiv \frac{1}{4\pi}\int_V \frac{\mathrm{c}(\mathbf{r}')}{R}\mathrm{dv}' \tag{5.A.2b}$$

The variables in Eq. (5.A.2a) and (5.A.2b) are defined as

$$\mathbf{r} \equiv x\mathbf{e}_x + y\mathbf{e}_y + z\mathbf{e}_z$$

$$\mathbf{r}' \equiv x'\mathbf{e}_x + y'\mathbf{e}_y + z'\mathbf{e}_z$$

$$R \equiv |\mathbf{r} - \mathbf{r}'|$$

The prime coordinates represent the source coordinates whereas the unprimed coordinates represent the field coordinates.

Proof. It can be shown that

$$\nabla^2\left(\frac{1}{\mathrm{R}}\right) = -4\pi\delta(\mathrm{R}) \tag{5.A.3}$$

$\delta(R)$ is the Dirac delta function defined by the following properties

$$\int \delta(\mathbf{r} - \mathbf{r}')\mathrm{dv}' = 1 \tag{5.A.4a}$$

$$\int \mathbf{F}(\mathbf{r}')\delta(\mathbf{r} - \mathbf{r}')\mathrm{dv}' = \mathbf{F}(\mathbf{r}) \tag{5.A.4b}$$

Using Eqs. (5.A.3) and (5.A.4b) **F** can be expressed as

$$\mathbf{F}(\mathbf{r}) = \int_V \mathbf{F}(\mathbf{r}')\left[-\frac{1}{4\pi}\nabla^2\left(\frac{1}{\mathrm{R}}\right)\right]\mathrm{dv}' = -\int_V \mathbf{F}(\mathbf{r}')\left[\frac{1}{4\pi}\nabla^2\left(\frac{1}{\mathrm{R}}\right)\right]\mathrm{dv}' \tag{5.A.5}$$

It is important to remember that ∇^2 is defined in terms of the unprimed coordinates. As such it can be brought through the integral to the outside to yield

$$\mathbf{F}(\mathbf{r}) = -\nabla^2\left(\int_V \frac{\mathbf{F}(\mathbf{r}')}{4\pi R}dv'\right) = -\nabla^2\mathbf{C} \tag{5.A.6a}$$

$$\mathbf{C} \equiv \int_V \frac{\mathbf{F}(\mathbf{r}')}{4\pi R}dv' \tag{5.A.6b}$$

Using the identity $-\nabla^2\mathbf{C} = \nabla(\nabla\bullet\mathbf{C}) - \nabla\times(\nabla\times\mathbf{C})$ Eq. (5.A.6a) can be written as

$$\mathbf{F} = -\nabla^2\mathbf{C} = -\nabla(\nabla\bullet\mathbf{C}) + \nabla\times(\nabla\times\mathbf{C}) \tag{5.A.7}$$

Define φ and $\mathbf{A}$ as follows

$$\varphi \equiv \nabla\bullet\mathbf{C} \tag{5.A.8a}$$

$$\mathbf{A} \equiv \nabla\times\mathbf{C} \tag{5.A.8b}$$

In terms of ϕ and $\mathbf{A}$ we can write Eq. (5.A.7) as

$$\mathbf{F} = -\nabla\phi + \nabla\times\mathbf{A} \tag{5.A.9}$$

∇ and ∇' are defined and related by

$$\nabla \equiv \mathbf{e}_x\frac{\partial}{\partial x} + \mathbf{e}_y\frac{\partial}{\partial y} + \mathbf{e}_z\frac{\partial}{\partial z} \tag{5.A.10a}$$

$$\nabla' \equiv \mathbf{e}_x\frac{\partial}{\partial x'} + \mathbf{e}_y\frac{\partial}{\partial y'} + \mathbf{e}_z\frac{\partial}{\partial z'} \tag{5.A.10b}$$

$$\nabla = -\nabla' \tag{5.A.10c}$$

It can be show that

$$\nabla\left(\frac{1}{R}\right) = -\nabla'\left(\frac{1}{R}\right) \tag{5.A.11}$$

To find ϕ in terms of $b(\mathbf{r}')$ we substitute Eq. (5.A.6b) into the definition of ϕ (Eq. 5.A.8a) to obtain

$$\phi = \nabla\bullet\mathbf{C} = \nabla\bullet\left(\int_V \frac{\mathbf{F}(\mathbf{r}')}{4\pi R}dv'\right) = \int_V \nabla\bullet\left(\frac{\mathbf{F}(\mathbf{r}')}{4\pi R}\right)dv' \tag{5.A.12}$$

$\mathbf{F}$ can be brought outside the divergence by using the identity

$$\nabla \bullet \left(\frac{\mathbf{F}(\mathbf{r}')}{\mathrm{R}}\right) = \mathbf{F}(\mathbf{r}') \bullet \nabla\left(\frac{1}{\mathrm{R}}\right) + \frac{1}{\mathrm{R}}\nabla \bullet \mathbf{F}(\mathbf{r}') = \mathbf{F}(\mathbf{r}') \bullet \nabla\left(\frac{1}{\mathrm{R}}\right) \tag{5.A.13}$$

The relation The $\nabla \bullet \mathbf{F}(\mathbf{r}') = 0$ has been employed in Eq. (5.A.13). This follows from the fact that $\mathbf{F}$ is a function only of the primed coordinates whereas the derivatives in the divergence are with respect to the unprimed coordinates. Equation (5.A.13) is substituted into Eq. (5.A.12) to obtain

$$\phi = \int_V \mathbf{F}(\mathbf{r}') \bullet \nabla\left(\frac{1}{4\pi \mathrm{R}}\right) \mathrm{d}v' = -\int_V \mathbf{F}(\mathbf{r}') \bullet \nabla'\left(\frac{1}{4\pi \mathrm{R}}\right) \mathrm{d}v' \tag{5.A.14}$$

Substitute the identity in Eq. (5.A.13), replacing ∇ with ∇' to obtain

$$\begin{aligned} \phi &= -\frac{1}{4\pi}\int_V \left\{\nabla' \bullet \left(\frac{\mathbf{F}(\mathbf{r}')}{\mathrm{R}}\right) - \frac{\nabla' \bullet \mathbf{F}(\mathbf{r}')}{\mathrm{R}}\right\} \mathrm{d}v' \\ &= -\frac{1}{4\pi}\int_V \left\{\nabla' \bullet \left(\frac{\mathbf{F}(\mathbf{r}')}{\mathrm{R}}\right)\right\} \mathrm{d}v' + \frac{1}{4\pi}\int_V \frac{\nabla' \bullet \mathbf{F}(\mathbf{r}')}{\mathrm{R}} \mathrm{d}v' \end{aligned} \tag{5.A.15}$$

Utilizing the divergence theorem, and substituting $\nabla' \bullet \mathbf{F}\ (\mathbf{r}') = b(\mathbf{r}')$ into Eq. (5.A.15), we get

$$\phi = -\frac{1}{4\pi}\oint_S \frac{\mathbf{F}(\mathbf{r}') \bullet \mathrm{d}\mathbf{S}'}{\mathrm{R}} + \frac{1}{4\pi}\int_V \frac{\mathrm{b}(\mathbf{r}')}{\mathrm{R}} \mathrm{d}v' \tag{5.A.16}$$

Since, by assumption, the source, b, is restricted to a finite region and thus, can be contained entirely contained within a surface S, then S and V can be chosen to be a sphere of radius r which entirely contains the source within it. The last integral on the right side of Eq. (5.A.16) remains finite. Regarding the first integral on the right; as the radius of the surface increases as r, then the area of the surface increases as r^2. However the integrand decreases as r^3. Therefore, as we let the radius go to infinity, we see that the first integral vanishes. Eq. (5.A.18) then becomes

$$\phi = \frac{1}{4\pi}\int_V \frac{\mathrm{b}(\mathbf{r}')}{\mathrm{R}} \mathrm{d}v' \tag{5.A.17}$$

This is just the relation in Eq. (5.A.2a) that we wished to arrive at.

To find $\mathbf{A}$ in terms of $\mathbf{c}(\mathbf{r}')$, we follow a very similar procedure to what we have done above. Substitute the definition of $\mathbf{C}$ in Eq. (5.A.6b) into the definition of $\mathbf{A}$ into Eq. (5.A.8b) to obtain

$$\phi = \nabla \times \mathbf{C} = \nabla \times \left(\int\limits_V \frac{\mathbf{F}(\mathbf{r}')}{4\pi R} dv' \right) = \frac{1}{4\pi} \int\limits_V \nabla \times \left(\frac{\mathbf{F}(\mathbf{r}')}{R} \right) dv' \tag{5.A.18}$$

F can be brought outside the curl by using the identity

$$\nabla \times \left(\frac{\mathbf{F}(\mathbf{r}')}{R} \right) = \nabla \left(\frac{1}{R} \right) \times \mathbf{F}(\mathbf{r}') + \frac{1}{R} \nabla \times \mathbf{F}(\mathbf{r}') \tag{5.A.19}$$

The last term on the right side of Eq. (5.A.19) vanishes, i.e., $\nabla \times \mathbf{F}(\mathbf{r}') = 0$. This follows from the fact that **F** is entirely a function of the primed coordinates whereas the derivatives in the curl are with respect to the unprimed coordinates. Substitute Eq. (5.A.19) into Eq. (5.A.18), and then employ the identity in Eq. (5.A.11) to obtain

$$\begin{aligned} \mathbf{A} &= \frac{1}{4\pi} \int\limits_V \nabla \left(\frac{1}{4\pi R} \right) \times \mathbf{F}(\mathbf{r}') dv' = -\frac{1}{4\pi} \int\limits_V \nabla' \left(\frac{1}{R} \right) \times \mathbf{F}(\mathbf{r}') dv' \\ &= \frac{1}{4\pi} \int\limits_V \mathbf{F}(\mathbf{r}') \times \nabla' \left(\frac{1}{R} \right) dv' \end{aligned} \tag{5.A.20}$$

Utilizing the identity in Eq. (5.A.19), replacing ∇ with ∇', Eq. (5.A.20) becomes

$$\begin{aligned} \mathbf{A} &= \frac{1}{4\pi} \int\limits_V \left\{ \frac{\nabla \times \mathbf{F}(\mathbf{r}')}{R} - \nabla' \times \left(\frac{\mathbf{F}(\mathbf{r}')}{R} \right) \right\} dv' \\ &= \frac{1}{4\pi} \int\limits_V \frac{\mathbf{c}(\mathbf{r}')}{R} dv' - \frac{1}{4\pi} \int\limits_V \nabla' \times \left(\frac{\mathbf{F}(\mathbf{r}')}{R} \right) dv' \end{aligned} \tag{5.A.21}$$

Applying a modified version of Gauss's theorem, Eq. (5.A.21) becomes

$$\mathbf{A} = \frac{1}{4\pi} \int\limits_V \frac{\mathbf{c}(\mathbf{r}')}{R} dv' + \frac{1}{4\pi} \int\limits_S \frac{\mathbf{F}(\mathbf{r}') \times d\mathbf{S}}{R} \tag{5.A.22}$$

For reasons similar to those applied to Eq. (5.A.16), the first integral vanishes yielding the final result

$$\mathbf{A} = \frac{1}{4\pi} \int\limits_V \frac{\mathbf{c}(\mathbf{r}')}{R} dv' \tag{5.A.23}$$

This is just the relation in Eq. (5.A.2b) that we wished to arrive at. This completes the proof of the Helmholtz theorem.

The uniqueness theorem The solution **F** which satisfies $\nabla \bullet \mathbf{F} = b(\mathbf{r})$ and $\nabla \times \mathbf{F} = \mathbf{c}(\mathbf{r})$ is unique if the sources are bounded in space and the fields vanish at sufficiently large distances from the sources.

Proof. Suppose there are two functions, $\mathbf{F}_1$ and $\mathbf{F}_2$, which satisfies$\nabla \bullet \mathbf{F} = b(\mathbf{r})$ and $\nabla \times \mathbf{F} = \mathbf{c}(\mathbf{r})$, i.e.

$$\nabla \times \mathbf{F}_1 = \mathbf{c}(\mathbf{r}), \quad \nabla \times \mathbf{F}_2 = \mathbf{c}(\mathbf{r}) \tag{5.A.24a}$$

$$\nabla \bullet \mathbf{F}_1 = \mathrm{b}(\mathbf{r}), \quad \nabla \bullet \mathbf{F}_2 = \mathrm{b}(\mathbf{r}) \tag{5.A.24b}$$

Define the vector $\mathbf{W} \equiv \mathbf{F}_1 - \mathbf{F}_2$. It follows from Eq. (5.A.24a, b) that

$$\nabla \times \mathbf{W} = 0 \tag{5.A.25a}$$

$$\nabla \bullet \mathbf{W} = 0 \tag{5.A.25b}$$

Since the curl of **W** is zero, it can be expressed as the gradient of a scalar function, i.e.,

$$\mathbf{W} = \nabla \Psi \tag{5.A.26}$$

It follows from Eq. (5.A.26) and Eq. (5.A.25b) that

$$\nabla^2 \Psi = 0 \tag{5.A.27}$$

Apply the divergence theorem to the vector $\mathbf{D} \equiv \Psi \nabla \Psi$ gives

$$\int_S \mathbf{D} \bullet d\mathbf{S} = \int_V \nabla \bullet \mathbf{D} dv \rightarrow \int_S (\Psi \nabla \Psi) \bullet d\mathbf{S} = \int_V \nabla \bullet (\Psi \nabla \Psi)\, dv$$

$$\int_S (\Psi \nabla \Psi) \bullet d\mathbf{S} = \int_V \left[\Psi \nabla^2 \Psi + (\nabla \Psi)^2 \right] dv \tag{5.A.28}$$

Inserting Eq. (5.A.27) reduces Eq. (5.A.28) to give

$$\int_S (\Psi \nabla \Psi) \bullet d\mathbf{S} = \int_V (\nabla \Psi)^2 dv \tag{5.A.29}$$

Let S be any spherical surface of radius r that encloses all sources. As r becomes large compared to the dimensions of the sources, Ψ goes to zero as $1/r$, and $\nabla \Psi$

goes to $1/r^2$. Therefore, $\Psi\nabla\Psi$ goes to zero as $1/r^3$. Since the surface increases as r^2, it follows that the surface integral on the left-hand side goes to zero. Since the integrand on the right side is always positive, it follows that $\nabla\ \Psi = 0$ and thus, $\mathbf{W} = 0$, i.e.,

$$\mathbf{F}_1 = \mathbf{F}_2 \tag{5.A.30}$$

Therefore, **F** as defined in Eq. (5.A.1) is unique.

It should be noted that the gradient of an arbitrary continuous function Ψ can be added to **A** and leave **F** unchanged. To see this, make the substitutions $\mathbf{A} \to \mathbf{A} + \nabla\ \Psi$

$$\text{(A)}\ \mathbf{F} = -\nabla\varphi + \nabla \times (\mathbf{A} + \nabla\Psi) = -\nabla\varphi + \nabla \times \mathbf{A}$$

References

Bear J (1972) Dynamics of fluids in porous media. Dover, New York

Belotserkovets A, Prevost JH (2011) Thermoporoelastic response of a fluid-saturated porous sphere: an analytical solution. Int J Eng Sci 49:1415–1423

Biot MA (1941) General theory of three-dimensional consolidation. J Appl Phys 12:155–164

Biot MA (1955) Theory of elasticity and consolidation for a porous anisotropic solid. J Appl Phys 26:182–185

Booker JR, Carter JP (1986a) Analysis of a point sink embedded in a porous elastic half space. Int J Numer Anal Method Geomech 10:137–150

Booker JR, Carter JP (1987a) Elastic consolidation around a point sink embedded in a half-space with anisotropic permeability. Int J Numer Anal Method Geomech 11:61–77

Booker JR, Carter JP (1986b) Long term subsidence due to fluid extraction from a saturated, anisotropic, elastic soil mass. Q J Mech Appl Math 39:85–98

Booker JR, Carter JP (1987b) Withdrawal of a compressible pore fluid from a point sink in an isotropic elastic half space with anisotropic permeability. Int J Solids Struct 23:369–385

Carslaw HS, Jaeger JC (1959) Conduction of heat in solids, 2nd edn. Oxford University Press, London

Castelletto N, Ferronato M. Gambolati G, Janna C, Teatini P (2010a) Thermo-poro-elastic effects in the anthropogenic uplift of Venice by deep seawater injection. In: Proceedings of the EISOLS 2010, Queretaro, Mexico, 17–22 Oct 2010. IAHS Publication 339

Castelletto N, Ferronato M, Gambolati G, Teatini P (2010b) An analytical solution of plane-strain consolidation due to a point sink within a fluid-saturated porous elastic media. In: Proceedings of the EISOLS 2010, Queretaro, Mexico, 17–22 Oct 2010. IAHS Publication 339

Chen GJ (2002) Analysis of pumping in multilayered and poroelastic half space. Comput Geotech 30:1–26

Chen GJ (2005) Steady-state solutions of multilayered and cross-anisotropic poroelastic half-space due to a point sink. Int J Geomech 5:45–57

Cours de mecanique des sols (1980) – enseignement specialise – Ecole des Ponts et Chaussees – Les fondations

Detournay E, Cheng AH-D (1993) Fundamentals of poroelasticity, Chapter 5. In: Fairhurst C (ed) Comprehensive rock engineering: principle, practice and projects, vol II, Analysis and design method. Pergamon Press, Oxford, pp 113–171

Ferronato M, Pini G, Gambolati G (2009) The role of preconditioning in the solution to FE coupled consolidation equations by Krylov subspace methods. Int J Numer Anal Method Geomech 33:405–423

Geertsma J (1973) Land subsidence above compacting oil and gas reservoirs. J Petrol Technol 25:734–744

Gui YF, Dou WB (2007) A rigorous and completed statement on Helmhotz theorem. In: Progress in electromagnetics research, PIER 69, pp 287–307

Gui YF, Li P (2011) On the uniqueness theorem of time-harmonic electromagnetic fields. J Electromagn Anal Appl 3:13–21

Harbaugh AW (2005) MODFLOW-2005, the US Geological Survey modular ground-water model – the ground-water flow process techniques and methods 6-A16. US Geological Survey, Reston

Harbaugh AW, Banta ER, Hill MC, McDonald MG (2000) MODFLOW-2000, the US Geological Survey modular ground-water model – user guide to modularization concepts and the ground-water flow process. US Geological Survey Open-File-Report 00–92

Hsieh PA (1996) Deformation-induced changes in hydraulic head during ground-water withdrawal. Ground Water 34:1082–1089

Jian-Fei Lu F, Sheng-Leng D (2006) A semi-analytical solution of a circular tunnel surrounded by a poroelastic medium and subjected to a moving load. Research report no R857. University of Civil Engineering- Sydney NSW 2006 Australia

Kaynia AM, Banerjee P (1993) Fundamental solutions of Biot's equations of dynamic poroelasticity. Int J Eng Sci 31:817–830

Lu J C-C, Lin F-T (2005) Analysis of transient ground surface displacements due to a point sink in a porous elastic half-space. In: Proceedings of the 10th conference of Advanced Technology Council in Mathematics, Cheong-Ju, Korea, 12–16 December, pp 135–144

Lu J C-C, Lin F-T (2006) The transient ground surface displacements due to a point sink/heat source in an elastic half-space. In: ASCE, Proceedings of GeoShanghai international conference, Shanghai, China, Geotechnical special publication no. 148, pp 210–218

Lu J C-C, Lin F-T (2007) Analysis of the transient ground surface displacements subject to a point sink in a poroelastic half space. Chung Hua J Sci Eng 5(1):77–86

Magnan J-P (2001) Cours de mécanique des sols et des roches. Ecole Nationale des Ponts et Chaussées, 2 vols

Mandel AH, Zeitoun DG, Dagan G (2003) Salinity sources of Kfar Uria wells in the Judea group aquifer of Israel. Quantitative identification model. J Hydol 270:39–48

Morse PM, Feshbach H (1953) Methods of theoretical physics, vol II. Mc Graw Hill Company, New York

Poland JF, Davis GH (1969) Land subsidence due to withdrawal of fluids. In Varnes DJ, Kiersch G (eds) Reviews in engineering geology II. Geological Society of America Inc., Boulder

Poland GF (1976) Land subsidence stopped by artesian- head recovery- San Joaquim Valley. In: Proceedings of the second international symposium land subsidence held at Anaheim, California, December 1976

Poulos HG, Davis EH (1974) Elastic solutions for soil and rock mechanics. John Wiley & sons Inc., New York

Phienwej N, Giao PH, Nutalaya P (2006) Land subsidence in Bangkok, Thailand. Eng Geol 82:187–201

Shan C, Falta RW, Javandel I (1992) Analytical solutions for steady state gas flow to a soil vapor extraction well. Water Resour Res 28:1105–1120

Slough KJ, Sudicky EA, Forsyth PA (1999) Numerical simulation of multiphase flow and phase partitioning in discretely fractured geologic media. J Contam Hydrol 40(2):107–136

Sneddon IN (1951) Fourier transforms. McGraw-Hill, New York, pp 48–70

Tarn J-Q, Lu C-C (1991) Analysis of subsidence due to a point sink in an anisotropic porous elastic half space. Int J Numer Anal Method Geomech 15:573–592

Vardoulakis I (2009) Lecture notes on Cosserat continuum mechanics with application to the mechanics of granular media. N.T.U. Athens, Greece, P.O. box 144, Paiania Gr-19002, http://geolab.mechan.ntua.gr/

Wang HF (2000) Theory of linear poroelasticity. Princeton University press, Princeton

Wilson AM, Gorelick S (1996) The effects of pulsed pumping on land subsidence in the Santa Clara valley, California. J Hydrol 174:375–396

Zhang C, Guo H (2006) Time-space coupled predicting model of land subsidence caused by groundwater exploitation based on stochastic medium theory. International conference on safety science and technology, Shanghai

Zhang H, Baray DA, Hocking GC (1999) Analysis of continuous and pulsed pumping of a phreatic aquifer. Adv Water Res 22:623–632

Zhou XL (2006) On uniqueness theorem of a vector function. In: Progress in Electromagnetics research, PIER 65, pp 93–102

Chapter 6
The Numerical Solution of the Biot Equations

Abstract In this chapter, first a review of the iterative methods proposed for the numerical solution of the coupled Biot equations is presented. Then, the numerical scheme based on the decoupling method of the solution of the Biot model is presented. The numerical method, called the compartmental model is similar to the finite-volume method. We describe the treatment of the different types of boundary conditions.

6.1 A Review of the Computer Codes for Solving the Biot System of Equations

The equations to be solved are the following:

The Biot model has been presented in the previous chapters. It may be summarized as:

$$\begin{cases} (\lambda + G)\dfrac{\partial \varepsilon}{\partial x} + G\nabla^2 u = \chi\dfrac{\partial p}{\partial x} + g_x(x,y,z,t) \\ (\lambda + G)\dfrac{\partial \varepsilon}{\partial y} + G\nabla^2 v = \chi\dfrac{\partial p}{\partial y} + g_y(x,y,z,t) \\ (\lambda + G)\dfrac{\partial \varepsilon}{\partial z} + G\nabla^2 w = \chi\dfrac{\partial p}{\partial z} + g_z(x,y,z,t) \end{cases}$$

$$(x,y,z) \in \Omega;\ t > 0$$

$$\vec{\nabla}.\left(K\vec{\nabla}p\right) = \frac{1}{S}\frac{\partial p}{\partial t} + \frac{\partial \varepsilon}{\partial t} + Q :\ ;(x,y,z) \in \Omega;\quad t > 0 \tag{6.1}$$

where the components of the displacement vector $\vec{d}$ are u, v, w and the pore pressure is p, $\vec{g} = (g_x, g_y, g_z)$ is the loading vector.

Ω is the aquifer domain. The solution $\vec{d}$ belongs to the Sobolev space:

D.G. Zeitoun and E. Wakshal, *Land Subsidence Analysis in Urban Areas*, Springer Environmental Science and Engineering, DOI 10.1007/978-94-007-5506-2_6,

$$\vec{d} \in \{\vec{v} \in H^1(\Omega) \times H^1(\Omega) \times H^1(\Omega);\ \vec{v} = \vec{u}\, on\ \partial\Omega\} \times [0.T] = [H_0^1(\Omega, T)]^3$$

The boundary conditions:

For the water pressure p:

$$p(x,t) = g_1(x,t) \qquad (x \in \partial\Omega^1,\ t > 0);$$
$$-\mathbf{K}\nabla p.\vec{n}_{\partial\Omega} = g_2(x,t) \qquad (x \in \partial\Omega^2,\ t > 0); \tag{6.2}$$

where Ω is a bounded, polygonal open set of R^2, $\partial\Omega^1$ and $\partial\Omega^2$ are partitions of the boundary $\partial\Omega$ of Ω corresponding to Dirichlet and Neumann boundary conditions and $\vec{n}_{\partial\Omega}$ the unit outward normal to the boundary $\partial\Omega$.

The solution p belongs to the Sobolev space:

$$p \in \{ f \in H^1(\Omega);\ f(x,t) = g_1(x,t)$$
$$(x \in \partial\Omega^1,\ t > 0); -\mathbf{K}\nabla f.\vec{n}_{\partial\Omega} = g_2(x,t)$$
$$(x \in \partial\Omega^2,\ t > 0);\} \times [0,T] = H(div,\Omega) \times [0,T]$$

For the displacement field:

$$u(x,t) = u_1(x,t) \quad (x \in \partial\Omega,\ t > 0);$$
$$v(x,t) = v_1(x,t) \quad (x \in \partial\Omega,\ t > 0);$$
$$w(x,t) = w_1(x,t) \quad (x \in \partial\Omega,\ t > 0); \tag{6.3}$$

- **Elastic parameters of the Biot equations**

  The classical Biot model takes into consideration an elastic behavior of the matrix. However, in this work, we assume an elastic model with a different elastic modulus similar to hysteresis behavior, frequently used in soil mechanics, in the consolidation theory.

  In our approach, we are using the equivalent elastic modulus computed from the compression modulus Cs, Cr, and Cc. Thus, the reduction in the voids ratio is included as an irreversible effect in the Biot theory. These parameters are obtained from an oedometric test. Typical results of the Bangkok clay are shown in Fig. 6.1.
- **A solution of the Biot equations**

  Computational solution of the Biot system of equations in its dynamic form and/or in its static form has been intensively studied (see for a review, Carcione et al. 2010). In this section, only the static Biot equations are discussed.

  Land subsidence caused by fluid extraction from productive reservoirs has been investigated since the beginning of the nineteenth century. But it was only later that Biot (1941) developed the mathematical framework of the physical processes governing the poroelastic problem. Several researchers have

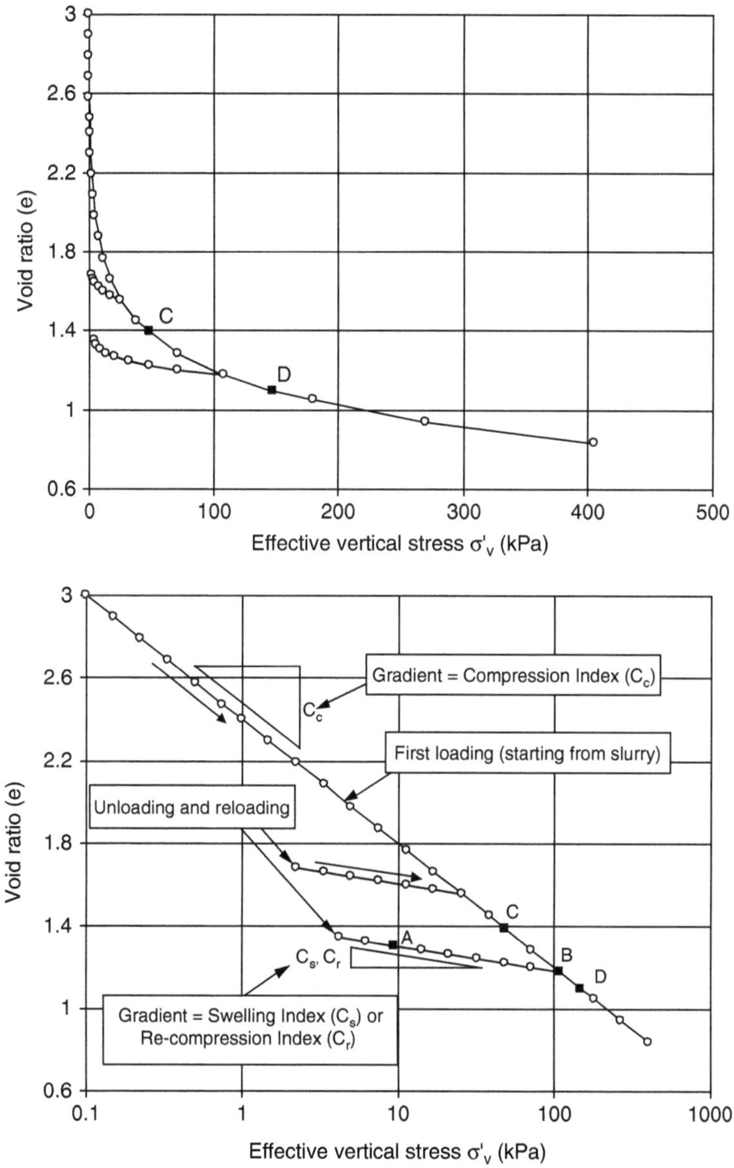

Fig. 6.1 Definitions of Cr, Cs, and Cc

contributed to this subject, since a subsidence prediction over depleted gas, oil, and water fields is of great environmental and economical relevance to many countries. A major debate involving the scientific community in the last decades concerns whether to treat the prediction of land subsidence induced by reservoir exploitation with coupled (Hsieh 1996) or uncoupled (Geertsma 1973) models.

The former approach supplies the time evolution of both fluid pressures and the soil displacements by solving simultaneously the equations governing the mechanical equilibrium and the groundwater flow. The uncoupled approach, instead, first calculates the time evolution of the pressure field using the groundwater flow equation and then solves separately the equilibrium equation for the soil displacements at the times of interest. In the previous chapters, we show that the Terzaghi theory of consolidation may be understood as a way of decoupling the Biot system of equations. From a physical and mathematical point of view, the coupled model represents the more rigorous approach to land subsidence prediction, but it is very computationally expensive, and, in addition, its numerical solutions are often affected by numerical instability and convergence problems (Hsieh 1996; Ferronato et al. 2010). A tremendous effort on developing an accurate and stable numerical scheme for the coupled system of equations has been performed (Lippmann et al. 1976; Hicks et al. 1996; Narasimhan and Witherspoon 1976; Aguilar et al. 2008; Barbeiro and Wheeler 2008; Hörlin 2010). Frequently finite element approximations were used in the numerical solution of the Biot system (Ohnishi et al. 1987; Shimizu 1991; Szostak-Chrzanowski and Chrzanowski 1991; Dassargues and Li 1991; Guvanasen and Chan 1995; Ferronato et al. 2010). A comparison between an iterative solution of symmetric and non-symmetric forms of Biot's FEM equations has been presented in Toh and Phoon (2007).

On the other hand, the approximate uncoupled model, although theoretically less accurate, requires less computational resources allowing a finer description of the medium properties. In addition, since the uncoupled model does not suffer all the numerical problems rising with the coupled model, non-specialized users can easily employ it, as long as the coupling is weak. In this respect, an investigation concerning the importance of coupling in the sedimentary basin of the Po River plain, Italy, has been performed by Gambolati et al. (2000). For both models, a standard Galerkin finite element formulation for an axis-symmetric geological setting is developed with an unconditionally stable implicit time integration scheme. Although the coupled and uncoupled models are based on a similar numerical formulation, the computational resources needed for the uncoupled simulator are significantly less than for the coupled model, both in memory and CPU requirements. For these reasons, it is of great interest to investigate under which conditions and configurations the uncoupled model supplies results with acceptable errors in practical applications.

The method we propose for estimating land subsidence, due to the withdrawal of groundwater and/or overloading due to urbanization, calculates the groundwater seepage in a 3-D-condition and calculates the land subsidence one-dimensionally. The governing equation on groundwater seepage is based on the three-dimensional mass conservation law and the principle of effective stress. The land subsidence calculation method is derived based on the following assumptions: (1) displacements occur only in the vertical direction; and (2) in a vertical direction, the total stresses do not change. The governing equation is solved by a numerical method, i.e., a finite element method (FEM) in spatial discretization and a finite difference method (FDM) in a time series discretization. In FEM, the Galerkin method is adopted, and in FDM, the lumped matrix method is employed.

In the present model, we use the SUB-T package of the MODFLOW model developed by the Department of Groundwater Management at the USGS (Hoffmann et al. 2003) to compute the first estimation of ε_0 and p_0. The additional iterations were computed using a compartmental model described in the next section.

6.2 Numerical Methods for the Solution of Groundwater Problems

In this chapter, we use the formulation of the general solution of the Biot system of equations presented in Chap. 5 for the derivation of a numerical solution of the Biot system of equations.

The decoupled system of equations permits the getting of a non-iterative numerical scheme.

As shown in the previous chapter, the general solution of the Biot equations may be computed by solving three partial differential systems:

(a) **The solution for the water pressure p:**

$$\left[\frac{\chi(1-2\nu)}{2G(1-\nu)}+\frac{1}{S}\right]\frac{\partial p}{\partial t}+Q-\vec{\nabla}.(K\vec{\nabla}p)=0 \tag{6.4}$$

(b) **The solution of $\vec{d}_1$ via Eq.** (6.2)

$$\begin{aligned}
\vec{d}_1 &= \vec{\nabla}\phi; \\
\nabla^2\phi &= \frac{\chi(1-2\nu)}{2G(1-\nu)}p(x,y,z,t)+\varepsilon_0 \quad (x,y,z)\in\Omega;\ t>0 \\
u_1(x,t) &= \frac{\partial\phi}{\partial x}=0 \quad (x\in\partial\Omega,\, t>0); \\
v_1(x,t) &= \frac{\partial\phi}{\partial y}=0 \quad (x\in\partial\Omega,\, t>0); \\
w_1(x,t) &= \frac{\partial\phi}{\partial z}=0 \quad (x\in\partial\Omega,\, t>0);
\end{aligned} \tag{6.5}$$

(c) **The solution of $\vec{d}_2$ via Eq.** (6.3)

$$
\begin{aligned}
G\vec{\nabla}.\vec{\nabla}\vec{d}_2 &= \vec{g} \\
d_{2x}(x,t) &= u_1(x,t) \qquad (x \in \partial\Omega,\ t > 0); \\
d_{2y} &= v_1(x,t) \qquad (x \in \partial\Omega,\ t > 0); \\
d_{2z} &= w_1(x,t) \qquad (x \in \partial\Omega,\ t > 0);
\end{aligned} \tag{6.6}
$$

The solution of the Eq. (6.1) is given by $\vec{d} = \vec{d}_1 + \vec{d}_2$.

6.2.1 *The General Formulation for the Numerical Solution of the Flow Problem*

Numerous mathematical models are based on conservation principles and constitutive laws, which are formulated by

$$S\frac{\partial h(x,t)}{\partial t} + \nabla.\vec{q} = f \quad x \in \Omega \tag{6.7}$$

$$\vec{q} = -K\nabla h \tag{6.8}$$

Where S is a storage coefficient, K is the flux related parameter, and $\vec{q}$ is the flux of the associated state variable h. Equation (6.7) is derived from the mass conservation principle, and Eq. (6.8) corresponds to the constitutive law such as Darcy's law *(h* is the water or hydraulic level), or Fick's law *(h* is the concentration of a solute), Ohm's law *(h* is the electric potential), or Fourier's law *(h* is the temperature).

The associated initial and boundary conditions are of the Dirichlet or Neumann type:

The initial conditions:

$$h(x,0) = h_0(x) \qquad x \in \Omega; \tag{6.9}$$

The boundary conditions:

$$
\begin{aligned}
h(x,t) &= g_1(x,t) \quad (x \in \partial\Omega^1,\ t > 0); \\
-\mathbf{K}\nabla h\,.\vec{n}_{\partial\Omega} &= g_2(x,t) \quad (x \in \partial\Omega^2,\ t > 0);
\end{aligned} \tag{6.10}
$$

Where Ω is a bounded, polygonal open set of R^2, $\partial\Omega^1$ and $\partial\Omega^2$ are partitions of the boundary $\partial\Omega$ of Ω corresponding to Dirichlet and Neumann boundary conditions, and $\vec{n}_{\partial\Omega}$ the unit outward normal to the boundary $\partial\Omega$.

This system is generally solved by finite volume (FV) or finite element (FE) methods of the lower order (Ern and Guermond 2004, among others). FV ensures an exact mass balance over each element and continuous fluxes across common element boundaries. FE ensures an exact mass balance on a dual mesh but leads to discontinuous fluxes at common elements edges. However, the finite element is more flexible because of its capacity of dealing with complex geometry.

A mixed finite element formulation (MFE) where both hydraulic heads and fluxes are unknown variables has been developed. The MFE preserves the advantages of both methods: accurate mass balance at the element level, continuity of the flux from one element to its neighbor, and mesh flexibility.

The method solves simultaneously the state variable and its gradient with the same order of accuracy (Babuska et al. 1977; Brezzi and Fortin 1991; Girault and Raviart 1986; Raviart and Thomas 1977).

Some numerical experiments have shown the superiority of the mixed finite element method relative to the other classic methods (Darlow et al. 1984; Durlofsky 1994; Kaasschieter and Huijben 1992; Mose et al. 1994).

However, its implementation leads to a system matrix with significantly more unknowns than the FV and FE methods.

When the lowest-order Raviart-Thomas space (Girault and Raviart 1986; Raviart and Thomas 1977) is used, the resolution of the system (6.1), (6.2) and (6.3), (6.4) leads to a system with one scalar unknown per edge for the hybrid formulation of the MFE (Brezzi and Fortin 1991; Chavent and Roberts 1991).

Attempts to reduce the number of unknowns have been investigated by various researchers. For rectangular meshes, mixed finite elements of the lowest order reduce to the standard cell-centered finite volume method (Chavent and Roberts 1991), provided that numerical integration is used. Cordes and Kinzelbach (1996) provided similar results for triangles and showed the equivalence between the mixed finite element and finite volumes without any numerical integration. However, such equivalence is restricted to a steady state and without sink/source terms inside the domain. Moreover, the mixed finite element method does not require a Delaunay triangulation (Carroll et al. 2008), unlike a finite volume scheme (Eymard et al. 2006; Younes et al. 2004).

6.2.2 The Difficulty of Calibration

Efficient management of an aquifer depends primarily on adequate knowledge of its hydro-geological parameters. The problem of estimating aquifer parameters with the aid of a numerical model is often referred to as inverse modeling. The fundamental benefit of inverse modeling is its ability to auto -calibrate parameter values that produce the best fit between observed and simulated hydraulic heads.

Beside the numerical considerations presented in the previous section, the choice of the mixed formulation reduces the number of parameters to calibrate. When a mixed formulation is used, only the storage coefficient has to be calibrated. Therefore, a mixed formulation is appropriate when no head data are available in given region. On the other hand, when head data, the classical formulation using the water head as an unknown, leads to more accurate results (Mose et al. 1994).

6.3 The Multiple-Cell Models

The present section deals with the mathematical modeling of large-scale aquifers by using an advanced cell model (ACM), where the cells are defined according to the hydrogeology of the aquifer, and the state variables are both the water level and the pumping rate in each cell. Such a model is a generalization of a compartmental model developed by several others (Bachmat 1995; Campana et al. 2001).

The cell model proposed is a finite volume method (FVM) (see Barth and Ohlberger 2004) combined with the general Darcy law. As mentioned above, the FVM ensures an exact mass balance over each element and continuous fluxes across common element boundaries.

The classical formulation of the FVM solves the cell equations with the water head as the unknown. This formulation requires the knowledge of the conductivity parameters in each cell.

The second innovation of the present work consists in using a mixed formulation where the unknown at a given cell may be whether the flux entering/leaving the cell or the head or both. Also, we use two separate formulations of the groundwater equations:

1. A classical formulation

   In this classical formulation, the state variable is the hydraulic head computed for each cell. The formulation of the cell model results in a linear system of equations where the vector of the head is unknown; the matrix contains known parameters and/or parameters to be calibrated, and the right-hand side of the matrix equation contains known quantities such as pumping rates, or spring discharges and/or quantities to be calibrated.

6.4 A Cell Model for Large-Scale Aquifers

The cell model proposed is a water balance model derived from the general flow equations that describe transient flow in groundwater.

6.4.1 Basic Assumptions

The hydrologic model presented in this chapter is based on the following assumptions:

(a) The lateral boundaries of the basin are separated into two types of connection to external aquifers:
(b) The hydraulic boundary conditions imposed on these external layers are of a head (Type 1) varying with time. The head is constant at each time step.
(c) The basin can be subdivided laterally into a set of interconnected straight prismatic cells. The same area may be occupied by two cells or more: for example, a cell for the upper aquifer and an underlying cell for the lower aquifer. The basin acts as a system of distinct, yet hydraulically interconnected, aquifers: for example, an upper aquifer and a lower one.
(d) The hydraulic characteristics of a cell are relatively uniform throughout the cell, yet may vary from one cell to another. All hydraulic characteristics are fixed in time.
(e) A cell may be phreatic, confined, partly confined, and partly phreatic, or leaky. In this work, we do not consider the problem of the free surface; the conductivity is integrated on the vertical from the bottom of the layer to the phreatic surface leading to a constant transmissivity value.
(f) The discharge across a pervious boundary between adjacent cells is proportional to the difference of their piezometric heads and directed towards the cell with the lower head.
(g) A phreatic cell is replenished by infiltration of rainwater and, to some extent, by return flow from water consumption and disposal systems in the overlying area. Cells may also be artificially recharged through wells.
(h) Leakage, into a cell or out of it, may also occur at its bottom.
(i) The volume of natural replenishment ${N^t}_i$ at the cell $Cell_i$ in a given time t is proportional to the difference between annual depth of rainfall and evapotranspiration. ${N^t}_i = \alpha_i(Precipitation_i - evapotranspiration_i) + P_0$:
(j) The salinity of the groundwater has a negligible effect on its velocity.
(k) We assume a constant time step increment Δt.

6.4.2 A General Finite Volume Formulation

Consider now the entire three-dimensional domain V of the aquifer region that we want to model and each of the boundaries of the area V; $\Gamma_j; j = 1, m_r \ldots$ (m_r: number of adjacent aquifers or boundaries). The general flow equations through the aquifer may be written in a vector form as (Bear 1979, Chap. 10):

$$-\vec{\nabla}.\,\vec{q} + \sum_{i=1}^{i=N_{well}} Q_i + R = S\frac{\partial h}{\partial t}; x \in V \qquad (6.11)$$

$$\vec{q} = -K\overrightarrow{\nabla h} \tag{6.12}$$

$$h(x) = h_j; x \in \Gamma_j;\ j = 1, m_r \tag{6.13}$$

where:

$\overrightarrow{\nabla}$ is the gradient operator $\left(\overrightarrow{\nabla} = \left(\frac{\partial}{\partial x}, \frac{\partial}{\partial y}, \frac{\partial}{\partial z}\right)\right)$
S is the storativity of the porous media
$\vec{q}$ is the flux term
$h(x,t)$ is the hydraulic head at a time t and point x.
K is the hydraulic conductivity tensor.
$\sum_{i=1}^{i=N_{well}} Q_i + R$ is the general input and output fluxes entering the aquifer. N_{well} is the total number of wells. Q_i is the input flow rate of the well i and R is the general output fluxes.

The aquifer area V is separated into N_c regions of uniform hydrological behavior. Each one of these regions is called a cell.

In order to derive the mass balance equation, we integrate the basic equations over V and use the Green Stockes theorem:

$$\int_V \left(\sum_{i=1}^{i=N_{well}} Q_i + R - S\frac{\partial h}{\partial t}\right) dx = \int_V \overrightarrow{\nabla}.\,\vec{q}\,dx \tag{6.14}$$

then:

$$\sum_{i=1}^{i=N_{well}} Q_i + \int_V R\,dx - \int_V S\frac{\partial h}{\partial t}\,dx = \sum_{j=1}^{j=m_r} \int_{\Gamma_j} \vec{q}.\overrightarrow{n_j}\,dx \tag{6.15}$$

Equation (6.15) represents the global mass balance of the aquifer.

In order to derive the mass balance for each cell $Cell_i$, we define the following sets of indices:

$C_i = \{1, 2, \cdots, N_i\}$ Where N_i is the number of cells connected with $Cell_i$ including external cells (cells representing the external aquifers).

$$F_i = \{j \in C_i, \text{Cell}_j \text{ is connected with } \text{Cell}_i\};$$

$$E_i = \{j \in C_i, \text{Cell}_i \text{ is connected with External aquifers}\}$$

$$\overrightarrow{n_j} : \text{is the external normal to the part of the boundary } \Gamma_j$$

In order to derive the mass balance equation (Eqs. 6.7 and 6.8) for cell $Cell_i$, we integrate the basic equations over $Cell_i$ and use the Green Stockes theorem:

$$Q_{Cell_i} + \int_{Cell_i} R_i dx - \int_{Cell_i} S_i \frac{\partial h_i}{\partial t} dx = \sum_{j=1}^{j=N_i} \int_{\Gamma^{Cell_i}{}_j} \vec{q}.\vec{n_j} dx \tag{6.16}$$

In Eq. (6.16), each term has an index i that represents the value of the specific variable at cell $Cell_i$. Equation (6.16) may also be derived from Eq. (6.15) and represents the water mass balance of cell $Cell_i$. $\Gamma^{Cell_i}{}_j$ represents the boundary between $Cell_i$ and $Cell_j$. Q_{Cell_i} represents the total pumping from $Cell_i$.

The water balance for a given cell $Cell_i$ (Eq. 6.16) can be stated for a balanced period between t and $t + \Delta t$ and discretized using an implicit scheme:

$$Q_i^{t+\Delta t} + R^{t+\Delta t}{}_i \int_{Cell_i} dx - \left(\frac{h_i^{t+\Delta t} - h_i^t}{\Delta t} \right) \int_{Cell_i} S_i dx = \sum_{j=1}^{j=N_i} \int_{\Gamma^{Cell_i}{}_j} \vec{q}_j.\vec{n_j} dx|_{t+\Delta t} \tag{6.17}$$

$$\vec{q}_j = -K_j^{extc} \overrightarrow{\nabla h} \tag{6.18}$$

6.4.3 The Classical Formulation

When inserting Eq. (6.2) into Eq. (6.1), Eq. (6.1) leads to the following set of cell equations:

For $i = 1, \ldots, N_c$

$$S_i(h^{t+\Delta t}{}_i - h^t{}_i) - \sum_{j \in S_i} \Delta t K_{ij}(h^{t+\Delta t}{}_j - h^{t+\Delta t}{}_i) - \sum_{l \in E_i} \Delta t K_{il}(h_l{}^{t+\Delta t} - h^{t+\Delta t}{}_i)$$
$$= \Delta t[L^{t+\Delta t}{}_i + D^{t+\Delta t}{}_i - P^{t+\Delta t}{}_i - B_i{}^{t+\Delta t}] \tag{6.19}$$

- In Eq. (6.6), the cell volume and each of the common areas between cell $Cell_i$ and the adjacent cells have been incorporated into the integrated storativity and the integrated conductivity parameters.
- Also the general term $Q_i^{t+\Delta t} + R_i{}^{t+\Delta t} \int_{Cell_i} dx$ has been separated into $[L^{t+\Delta t}{}_i + D^{t+\Delta t}{}_i - P^{t+\Delta t}{}_i - B_i{}^{t+\Delta t}]$

  Where:

  S_i: Volume storativity of the cell $Cell_i$ (dimension L^2)
  $h^t{}_i$: Average head of the cell $Cell_i$ at time t (L)
  K_{ij}: Integrated conductivity of water between $Cell_i$ and $Cell_j$ (L^2/T).
  K_{lj}: Integrated conductivity of water between $Cell_i$ and external aquifers (L^2/T)
  Δt: time step (T)

$L^t{}_i = \alpha_i\ Precipitation_i + P_0$: Replenishment in cell $Cell_i$ at time t as a function of the precipitation. (L^3/T)
$D^t{}_i$: Recharge at cell $Cell_i$ at time t (L^3/T)
$P^t{}_i$: Pumping rate at cell $Cell_i$ at time t (L^3/T)
$B^t{}_i$: Spring flow at cell $Cell_i$ at time t (L^3/T)

Notice that the term K_{ij} is not the classical hydraulic conductivity. It includes the sectional area and the distances ($\Delta x_{ij}, \Delta y_{ij}, \Delta z_{ij}$) between cells i and j. When K_i is isotropic and the conductivity between $Cell_i$ and $Cell_j$ is represented by K^*_{ij}, K_{ij} is derived as $AK^*_{ij}\left(\frac{n_{xj}}{\Delta x_{ij}} + \frac{n_{yj}}{\Delta y_{ij}} + \frac{n_{zj}}{\Delta z_{ij}}\right)$ from the definition.

Let us define the vector of the unknown:

$$\overrightarrow{H}^{(t+\Delta t)} = \left[h_1^{t+\Delta t}, h_2^{t+\Delta t}, \cdots, h_{N_c}^{t+\Delta t}\right] \tag{6.20}$$

It is formed by the unknown head at each cell.

Equation (6.6) may be written as a determined linear system to solve at each time step:

$$E\overrightarrow{H}^{(t+\Delta t)} = \Delta t\,\overrightarrow{F}^{(t+\Delta t)} + ST\overrightarrow{H}^{(t)} \tag{6.21}$$

where the $N_c \times N_c$ matrix E is the sum of the storativity ST matrix and the conductivity matrix C;

$$E = ST - \Delta t C$$

The components of the different matrices and vectors are:

$$\begin{aligned} ST_{ij} &= S_i\,\delta_{ij}\,; \\ C_{ij} &= K_{ij} + \delta_{il} K_{lj}; \\ F_i &= [L^{t+\Delta t}{}_i + D^{t+\Delta t}{}_i - P^{t+\Delta t}{}_i - B_i{}^{t+\Delta t}]; \\ \delta_{ij} &= 1\ if \quad i = j;\ \delta_{ij} = 0 \quad if \quad i \neq j \end{aligned}$$

The system of Eq. (6.21) is the classical formulation. In terms of parameters to calibrate, it requires the knowledge of all the hydraulic conductivity and all the storativity parameters. This may be a difficult task for large-scale aquifers where the head data are scarce and/or known on only a small part of the aquifer.

Moreover, the stability of the linear system (6.21) with respect to a variation of conductivity parameters depends on the conditioning of the matrix E. For large-scale cell systems, the external conductivity parameters cause the matrix C to be badly conditioned (see Cordes and Kinzelbach 1996). Therefore, we propose a mixed formulation that permits the reduction of the number of parameters to be calibrated.

6.4.4 The Mixed Formulation

In this formulation, the unknown vectors are both the head at each cell and the flux entering/leaving the cell.

$$\begin{aligned}\overrightarrow{H}^{(t+\Delta t)} &= \left[h_1^{(t+\Delta t)}, h_2^{(t+\Delta t)}, \cdots, h_{N_c}^{(t+\Delta t)}\right]\\ \overrightarrow{q}^{(t+\Delta t)} &= \left[q_1^{(t+\Delta t)}, q_2^{(t+\Delta t)}, \cdots, q_{N_q}^{(t+\Delta t)}\right]\end{aligned} \tag{6.22}$$

$\overrightarrow{q}^{(t+\Delta t)}$ is the vector of fluxes and N_q is the number of unknown fluxes. Notice that for a given cell, fluxes coming from different adjacent cells are present. We define the index $N_q(i)$ as the number of fluxes for the cell i.

Equation (6.21) leads to the linear system of equations to solve at each time step:

$$ST\overrightarrow{H}^{(t+\Delta t)} + \Delta t B\vec{q}^{(t+\Delta t)} = \Delta t\overrightarrow{F}^{(t+\Delta t)} + ST\overrightarrow{H}^{(t)} \tag{6.23}$$

The size of matrix B is $N_c \times N_q(N_c$ lines times N_q columns). Also, for a number $N_K \leq N_c$ of cells, we may use the Darcy law. For these special cells $Cell_l$, the fluxes between $Cell_l$ and let us say $Cell_j$; ${q^{(t+\Delta t)}}_{lj}$ may be expressed as:

${q^{(t+\Delta t)}}_{lj} = -K_{lj}({h^{(t+\Delta t)}}_j - {h^{(t+\Delta t)}}_l)$ The Darcy law results in an additional matrix equation. This matrix equation in terms of the unknown vectors: $\overrightarrow{H}^{(t+\Delta t)}$; $\vec{q}^{(t+\Delta t)}$ may be expressed as:

$$I_d\,\vec{q}^{(t+\Delta t)} = -R\overrightarrow{H}^{(t+\Delta t)} \tag{6.24a}$$

The index matrix I_d has N_K lines and N_q columns.

The matrix R contains the conductivity parameters and has N_K lines and N_c columns.

In order to solve Eqs. (6.22) and (6.23), we use the least-squares formulation:

(a) The least-squares formulation

The constraint least-squares formulation of Eqs. (6.22) and (6.23) may be formulated in terms of the residual vector of size N_c: $\overrightarrow{\mathrm{Re}\,s}^{(t+\Delta t)}$ and the residual vector of size N_K: $\overrightarrow{Cs}^{(t+\Delta t)}$

$$\overrightarrow{\mathrm{Re}\,s}^{(t+\Delta t)}(\vec{q}^{(t+\Delta t)}, \overrightarrow{H}^{(t+\Delta t)}) = ST\overrightarrow{H}^{(t+\Delta t)} + \Delta t B\vec{q}^{(t+\Delta t)} - \Delta t\overrightarrow{F}^{(t+\Delta t)} - ST\overrightarrow{H}^{(t)} \tag{6.24b}$$

$$\overrightarrow{Cs}^{(t+\Delta t)}(\vec{q}^{(t+\Delta t)}, \overrightarrow{H}^{(t+\Delta t)}) = I_d\vec{q}^{(t+\Delta t)} + R\overrightarrow{H}^{(t+\Delta t)} \tag{6.24c}$$

The least-squares formulation used in this work is the minimization of the sum of the squares of each equation of Eq. (6.22) subject to Eq. (6.23).

$$\min_{(\vec{q}^{(t+\Delta t)},\overrightarrow{H}^{(t+\Delta t)})} \sum_{i=1}^{i=N_c} [\mathrm{Re}\, s_i^{(t+\Delta t)}]^2$$

$$\text{Subject to } \overrightarrow{Cs}^{(t+\Delta t)}(\vec{q}^{(t+\Delta t)}, \overrightarrow{H}^{(t+\Delta t)}) = \vec{0} \tag{6.25}$$

This minimization problem is solved using an augmented Lagrangian approach where the minimization problem is replaced by the following minimization problem:

$$\min_{(\vec{q}^{(t+\Delta t)},\overrightarrow{H}^{(t+\Delta t)})} \sum_{i=1}^{i=N_c} [\mathrm{Re}\, s_i^{(t+\Delta t)}]^2 + c \sum_{j=1}^{j=N_K} [C s_j^{(t+\Delta t)}]^2 \tag{6.26}$$

The first-order condition for a minimum leads to the following system of equations:

$$[(ST)^T ST + c\, R^T R]\overrightarrow{H}^{(t+\Delta t)} + [\Delta t (ST)^T B + c\ R^T I_d]\vec{q}^{(t+\Delta t)}$$
$$= \Delta t (ST)^T \overrightarrow{F}^{(t+\Delta t)} + (ST)^T ST \overrightarrow{H}^{(t)} \tag{6.27a}$$

$$[(B)^T ST + c\, {I_d}^T R]\overrightarrow{H}^{(t+\Delta t)} + [\Delta t (B)^T B + c\ {I_d}^T I_d]\vec{q}^{(t+\Delta t)}$$
$$= \Delta t (B)^T \overrightarrow{F}^{(t+\Delta t)} + (B)^T ST \overrightarrow{H}^{(t)} \tag{6.27b}$$

Several algorithms for the solution of system (6.27a) and (6.27b) have been proposed in the literature (an augmented Lagrangian method combined with a least-squares, preconditioned conjugate gradient (see Girault and Raviart 1986; Raviart and Thomas 1977, Zeitoun 2012).

(b) The algorithm

The iterative algorithm used for the solution of Eqs. (6.26, 6.27a) and (6.27b) is as follows:

Step1 Initial guess vector of size N_K: $\overrightarrow{\lambda_0}$

Step2 Solution of $\overrightarrow{H_0}$ and $\overrightarrow{q_0}$

$$(ST)^T ST\, \overrightarrow{H_0} + \Delta t (ST)^T B\, \overrightarrow{q_0} + c\, R^T \overrightarrow{\lambda}_0$$
$$= \Delta t (ST)^T \overrightarrow{F}^{(t+\Delta t)} + (ST)^T ST \overrightarrow{H}^{(t)} \tag{6.28a}$$

$$(B)^T ST\, \overrightarrow{H_0} + \Delta t (B)^T B\, \overrightarrow{q_0} + c\, {I_d}^T \overrightarrow{\lambda}_0$$

$$= \Delta t (ST)^T \overrightarrow{F}^{(t+\Delta t)} + (ST)^T ST \overrightarrow{H}^{(t)} \tag{6.28b}$$

Step3 Computation of the vector $\overrightarrow{\lambda_1} : \overrightarrow{\lambda_1} = \overrightarrow{\lambda_0} + \rho(\overrightarrow{\lambda_0} - I_d\, \vec{q}_0 - R\overrightarrow{H}_0)$

Step n $\overrightarrow{H_n}$ and $\overrightarrow{q_n}$

$$(ST)^T ST\, \overrightarrow{H_n} + \Delta t (ST)^T B\, \overrightarrow{q_n} + c\, R^T \overrightarrow{\lambda}_n$$

$$= \Delta t (ST)^T \overrightarrow{F}^{(t+\Delta t)} + (ST)^T ST \overrightarrow{H}^{(t)} \tag{6.28c}$$

$$(B)^T ST\, \overrightarrow{H_n} + \Delta t (B)^T B\, \overrightarrow{q_n} + c\, {I_d}^T \overrightarrow{\lambda}_n$$

$$= \Delta t (ST)^T \overrightarrow{F}^{(t+\Delta t)} + (ST)^T ST \overrightarrow{H}^{(t)} \tag{6.28d}$$

$$\overrightarrow{\lambda_{n+1}} = \overrightarrow{\lambda_n} + \rho(\overrightarrow{\lambda_n} - I_d\, \vec{q}_n - R\overrightarrow{H}_n)$$

Convergence of the algorithm to the solution of Eqs. (6.27a) and (6.27b) may be found in Zeitoun et al. (1997, 1995).

The advantages of the present formulation are the reduction of the number of parameters to calibrate and the good conditioning of the inverse matrix. On the other hand, the fluxes vectors has to be found.

Also, we may reduce the number of fluxes using geo-hydrological considerations.

6.4.5 *The Components of the Mass Balances*

From the above equations, one can derive each of the components of the mass balances.

1. *The cell mass balance for a period of time* $t + \Delta t$:

    $S_i({h_i}^{t+\Delta t} - {h_i}^t)$: A change in storativity

    $-\sum_{l \in E_i} K_{lj}({h_l}^{t+\Delta t} - {h_i}^{t+\Delta t})$: inflow or outflow between $Cell_i$ and external cells of the aquifer. E_i is the set of external cells connected to $Cell_i$

    $-\sum_{j \in S_i} K_{ij}({h_j}^{t+\Delta t} - {h_i}^t)$: inflow or outflow between $Cell_i$ and internal cells of the aquifer

    $\Delta t[{P_i}^{t+\Delta t} - {D_i}^{t+\Delta t}]$: Recharges and Pumping

    $\Delta t\, {L_i}^{t+\Delta t}$: Replenishment from rain and evapo-transpiration.

    $-\Delta t\, {B_i}^{t+\Delta t}$: Volume flow at the outlets of the springs.

A groundwater simulation occurs over time for the different cells. A time loop, boundary-line loop, and cell-polygon loop are created to account for all of the dimensions. Initially (time step zero), the total head is known for each cell. The sum of all the flows across a cell's boundary, including flux, recharge, and pumping, are then used to determine new values for the total head. The time step is then incremented, and calculations over the lines and polygons are repeated.

When warranted, other terms expressing inflows or outflows (Chap. 3) may be added to the water balance of a single-cell aquifer model, as follows:

1. evapo-transpiration, (which may be a function of h when the water table is sufficiently close to the surface);
2. infiltration from influent streams passing through the region and drainage into effluent streams;
3. drainage;
4. leakage from a water supply system;
5. a return flow from irrigation (which may be a function of the volume of water pumped from the aquifer), from septic tanks, etc.; and
6. leakage into or out of the specific aquifer through overlying and/or underlying semipervious layers.

Obviously, in a confined aquifer, we do not have those components of the balance which involve percolation of water to the aquifer through the unsaturated zone. In general, some of the cells may be phreatic and the others confined.

As more components are included in the water balance of a cell, the magnitude of some of them may be a priori unknown and/or must be determined as part of the calibration procedure.

The unknown aquifer parameters which must be determined during the calibration phase are usually S and L. The calibration is based on known values of h, R, and P for a number of periods in the past.

6.4.6 The Calibration of the Model

The calibration, or identification, of a model is the process in which the various model parameters (and those may also include its geometry, inputs, etc.) are determined, if no previous knowledge of them is available, or verified, if such information is available. The calibration is based on data obtained from observations of the behavior of the aquifer in the past. Such data usually includes water levels, pumping and recharge rates and volumes, water quality, interface positions, spring discharges, etc. The calibration, or identification, procedure is often referred to as the inverse problem (Bear 1979).

The calibration involves taking a past period for which data are available on the behavior of the real aquifer system (e.g., water levels) and running a simulation of the aquifer model. When the model is run in accordance with recorded input data for that period (e.g., pumpage and natural replenishment), its response (e.g., in the form of water levels) is compared with the recorded past response observed in the aquifer. The model is said to be calibrated when the difference between these two responses is less than some value specified by the planner. At the end of the calibration phase, we have a well-defined model of the aquifer system under consideration. All its parameters are well defined, and it can now be used with confidence for forecasting the future response of the aquifer to intended operations. The mathematical scheme for the calibration algorithm is presented in Section 7.

The least-squares solution of Eq. (6.28a, 6.28b, 6.28c, 6.28d) may be obtained by various algorithms such as multi-regression analysis, SVD (singular value decomposition), and others. The use of the MATLAB software enables the use of a large library of algorithms and determination of the most efficient one.

6.5 Discretization of the Displacement Equations

At each time step, the potential function ϕ has to be computed. For each displacement components, the differential equations to solve given by:

Solution of $\vec{d}_1$ via Eq. (6.2)

$$\begin{aligned}
\vec{d}_1 &= \vec{\nabla}\phi; \\
\nabla^2\phi &= \frac{\chi(1-2\nu)}{2G(1-\nu)} p(x,y,z,t) + \varepsilon_0 \quad (x,y,z) \in \Omega;\ t > 0 \\
d_{x1}(x,t) &= 0 \quad (x \in \partial\Omega, t > 0); \\
d_{y1}(x,t) &= 0 \quad (x \in \partial\Omega, t > 0); \\
d_{z1}(x,t) &= 0 \quad (x \in \partial\Omega, t > 0);
\end{aligned}$$

Solution of $\vec{d}_2$ via Eq. (6.3)

$$\begin{aligned}
G\ \nabla^2 d_{2x} &= g_1; \\
G\ \nabla^2 d_{2y} &= g_2 \\
G\ \nabla^2 d_{2z} &= g_3 \\
d_{2x}(x,t) &= u_1(x,t) \quad (x \in \partial\Omega, t > 0); \\
d_{2y} &= v_1(x,t) \quad (x \in \partial\Omega, t > 0); \\
d_{2z} &= w_1(x,t) \quad (x \in \partial\Omega, t > 0);
\end{aligned}$$

The solution $\vec{d} = \vec{d}_1 + \vec{d}_2$ is the solution of the displacement field appearing in the Biot system of equations.

Consider now the entire volume V of the aquifer region that we want to model and the various boundaries $\Gamma_j; j = 1, n_r$ (n_r: the number of different adjacent aquifers or boundaries). The uncoupled formulation of the general flow and the consolidation equations through the aquifer may be written in a vector form (see Sect. 6.2):

The aquifer is separated into N_c regions of uniform hydrological behavior. Each one of this region is a cell.

In order to derive the mass balance equation, let us integrate the basic equations over any volume V and use the Green Stockes theorem:

$$\int_V \left(\vec{\nabla}.(\vec{C}.\vec{\nabla} w)\right) dV = \int_V g_z(x,y,t)\, dV \tag{6.29}$$

then

$$\sum_{j=1}^{j=n_r} \int_{\Gamma_j} C\overrightarrow{\nabla w}.\vec{n_j}\, dS = \int_V g_z(x,y,t)\, dV \tag{6.30}$$

Equation (6.30) represents the global mass balance of the aquifer. The three times three tensor C is defined as:

$$C = \begin{pmatrix} G & 0 & 0 \\ 0 & G & 0 \\ 0 & 0 & \lambda + 2G \end{pmatrix} \tag{6.31}$$

In order to derive the mass balance for each cell $Cell_i$, let us define the following sets of indices:

$C_i = \{1, 2, \cdots, N_i\}$ Where N_i is the number of cells connected with the $Cell_i$ including external cells (the cells representing the external aquifers.)

$$S_i = \{j \in C, Cell_j \text{ is connected with } Cell_i\}$$

In order to derive the mass balance equation for the cell $Cell_i$, let us integrate the basic equations over $Cell_i$ and use the Green Stockes theorem:

$$\sum_{j=1}^{j=n_r} \int_{\Gamma_{Cell_j}} C^{(j)}\overrightarrow{\nabla w}.\vec{n_j}\, dS = \int_{Cell_i} g_z(x,y,t) dV \tag{6.32}$$

In Eq. (6.32), each term has an index i that represents the value of the specific variable at the cell $Cell_i$.

This last equation leads to the following set of cell equations:

For $i = 1, \ldots, N_c$

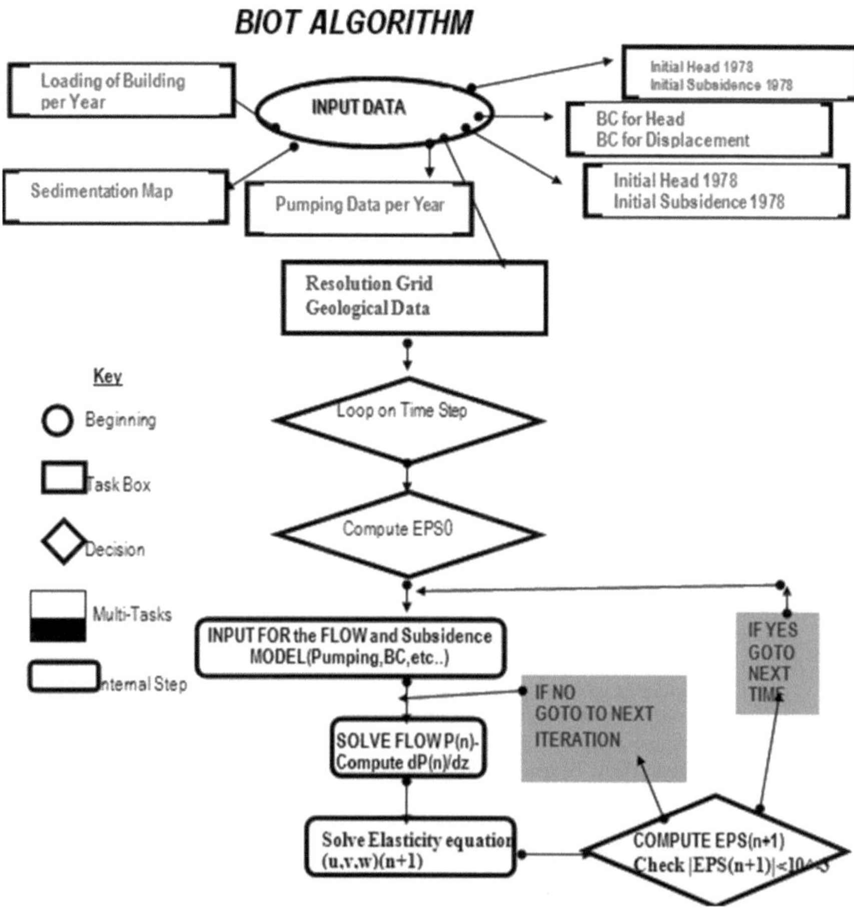

Fig. 6.2 A flowchart of the Biot algorithm

$$-\sum_{j\in S_i}\frac{C_{ij}}{\chi_i}({w^{t+\Delta t}}_j - {w^{t+\Delta t}}_i) - \sum_{l\in E_i}\frac{C_{lj}}{\chi_i}({w_l}^{t+\Delta t} - {w^{t+\Delta t}}_i) = {g_i}^{t+\Delta t}/\chi_i \quad (6.33)$$

For the consolidation equation, the different terms are:
$U_{ij} = \frac{C_{ij}}{\chi_i}$: Relative compressibility parameter between $Cell_i$ and $Cell_j$ (F/L^3)
${g_i}^{t+\Delta t}/\chi$: The loading pressure due to building (F/L^2)

The solution of the Biot equations

The unknown functions are the water pressure $p(x, y, z, t)$ and the vertical displacement field $(u(x, y, z, t), v(x, y, z, yt), w(x, y, z, t))$.

The procedure consists of assuming an initial state of deformation ε_0 computed from a Boussinesq solution.

Then for each time step, the pressure field is computed using the procedure described in the previous section (Eqs. 6.26, 6.27a and 6.27b).

Then, using Eq. (6.33), compute the displacement $\vec{d}_1$ and $\vec{d}_2$.

Then, we compute the displacement $\vec{d} = \vec{d}_1 + \vec{d}_2$ and the corresponding new deformation state ε_1. We iterate this process until the differences between two successive deformation states are small enough ($|\varepsilon_{n+1} - \varepsilon_n| < 10^{(-8)}$).

The following flowchart (Fig. 6.2) describes the explained algorithm:

References

Aguilar G, Gaspar F, Lisbona F, Rodrigo C (2008) Numerical stabilization of Biot's consolidation model by a perturbation on the flow equation. Int J Numer Method Eng 75:1282–1300

Babuska I, Oden JT, Lee JL (1977) Mixed-hybrid finite element method approximations of second-order elliptic boundary value problems. Comput Method Appl M 1:175–206

Bachmat Y (1995) A compartmental model for the Yarkon Taninim aquifer. Report of the Hydrological Service of Israel - Israel

Barbeiro S, Wheeler MF (2008) A priori error estimates for the numerical solution of a coupled geomechanics and reservoir flow model with stress-dependent permeability. Comput Geosci 14:755–766

Barth T, Ohlberger M (2004) Finite volume methods: foundation and analysis. In: Stein E, de Borst R, Hughes TJR (eds) Encyclopedia of computational mechanics. Wiley, New York

Bear J (1979) Hydraulics of groundwater. McGraw-Hill, New York

Brezzi F, Fortin M (1991) Mixed and hybrid finite element methods, vol 15, Springer series in computational mathematics. Springer, Berlin

Campana ME, Harrington GA, Tezcan L (2001) Compartmental model approaches to groundwater flow simulation. In: Mook W (ed) Environmental isotopes in the hydrological cycle: principles and applications, vol vi. UNESCO, Paris, pp 37–73

Carcione JM, Morency C, Santos JE (2010) Computational poroelasticity- a review. Geophysics 75:229–243

Carroll RWH, Pohll G, Earman S, Hershey RL (2008) A comparison of groundwater fluxes computed with MODFLOW and a mixing model using deuterium: application to the eastern Nevada test site and vicinity. J Hydrol 361:371–385

Chavent G, Roberts JE (1991) A unified physical presentation of mixed, mixed-hybrid finite elements and standard finite difference approximations for the determination of velocities in waterflow problems. Adv Water Resour 14:329–348

Cordes C, Kinzelbach W (1996) Comment on: Mose et al. Application of the mixed hybrid finite approximation in a groundwater flow model: luxury or necessity? Water Resour Res 32: 1905–1909

Darlow BL, Ewing RE, Wheeler MF (1984) Mixed finite elements methods for miscible displacement problems in porous media. Soc Pet Eng J 24:391–398

Dassargues A, Li XL (1991) Computing the land subsidence of Shanghai by a finite element method. In: Proceedings of the fourth international symposium on land subsidence, May 1991. IAHS Publication – 200

Durlofsky LJ (1994) Accuracy of mixed and control volume finite element approximations to Darcy velocity and related quantities. Water Resour Res 30:965–973

Ern A, Guermond JL (2004) Theory and practice of finite elements. Springer, New York

Eymard R, Gallouet T, Herbin R (2006) Finite volume methods, update of the preprint no 97-19. In: Ciarlet PG, Lions JL (eds) Handbook of numerical analysis, vol 7, pp 713–1020

Ferronato M, Castelletto N, Gambolati G (2010) A fully coupled 3-D mixed finite element model of Biot consolidation. J Comput Phys 229:4813–4830

Gambolati G, Teatini P, Bau D, Ferronato M (2000) The importance of poro-elastic coupling in dynamically active aquifers of the Po River basin, Italy. Water Resour Res 36:2443–2459
Geertsma J (1973) Land subsidence above compacting oil and gas reservoirs. J Pet Technol 25:734–744
Girault V, Raviart PA (1986) Finite element methods for Navier–Stokes equations: theory and algorithms, vol 5, Springer series in computational mathematics. Springer, Berlin
Guvanasen V, Chan T (1995) A new three-dimensional finite-element analysis of hysteresis thermo-hydromechanical deformation of fractured rock mass with dilatance in fractures. The second conference on mechanics of jointed and faulted rocks, Vienna, Austria
Hicks TW, Pine RJ et al (1996) A hydro-thermo-mechanical numerical model for HDR geothermal reservoir evaluation. Int J Rock Mech Min 33:499–511
Hörlin NE (2010) A symmetric weak form of Biot's equations based on redundant variables representing the fluid, using a Helmholtz decomposition of the fluid displacement vector field. Int J Numer Method Eng 84:1613–1637
Hoffmann J, Leake SA, Galloway DL, Wilson AM (2003) MODFLOW-2000 ground-water model – user guide to the subsidence and aquifer-system compaction (SUB) package. US Geological Survey Open-File Report 03–233
Hsieh PA (1996) Deformation-induced changes in hydraulic head during ground-water withdrawal. Ground Water 34:1082–1089
Kaasschieter EF, Huijben AJM (1992) Mixed-hybrid finite elements and streamline computation for the potential flow problem. Numer Method Partial D E 8:221–266
Lippmann MJ, Narasimhan TN, Witherspoon PA (1976) Numerical simulation of reservoir compaction in liquid dominated geothermal systems. In: Proceedings of the second international symposium on land subsidence, Anaheim, CA, December 1976
Mose R, Siegel P, Ackerer P, Chavent G (1994) Application of the mixed hybrid finite element approximation in a groundwater flow model: luxury or necessity? Water Resour Res 30: 3001–3012
Narasimhan TN, Witherspoon PA (1976) Numerical model for land subsidence in shallow groundwater system. In: Proceedings of the second international symposium on land subsidence, Anaheim, CA, December 1976
Ohnishi Y, Shibata H et al (1987) Development of finite element code for the analysis of coupled thermo-hydro-mechanical behavior of a saturated-unsaturated medium. Coupled processes associated with nuclear waste repositories. Academic, Orlando, pp 679–696
Raviart PA, Thomas JM (1977) A mixed finite element method for second order elliptic problems. In: Mathematical aspects of finite element method. Lecture notes in mathematics, vol 606, pp 292–315. Springe, New York
Shimizu M (1991) Finite element analysis of land subsidence due to the variation of groundwater level. In: Proceedings of the fourth international symposium on land subsidence, IAHS Publication – 200, May 1991
Szostak-Chrzanowski A, Chrzanowski A (1991) Modeling and prediction of ground subsidence using an iterative finite element method. In: Proceedings of the fourth international symposium on land subsidence, May 1991. IAHS Publication – 200
Toh KC, Phoon KK (2007) Comparison between iterative solution of symmetric and non-symmetric forms of Biot's FEM equations using the generalized Jacobi preconditioner. Int J Numer Anal Method Geomech 32:1131–1146
Younes A, Ackerer PH, Chavent G (2004) Int J Numer Method Eng 59:365–388. doi: 10.1002/nme.874
Zeitoun DG (2012) A compartmental model for stable time- dependent modeling of surface water and groundwater. Environ Model Assess 17:673–697
Zeitoun DG, Laible JP, Pinder GF (1995) A weighted least squares method for first-order hyperbolic systems. Int J Numer Method Fl 20:191–212
Zeitoun DG, Laible JP, Pinder GF (1997) An iterative penalty method for the least squares solution of boundary value problems. Numer Method Partial D E 13:257–281

Chapter 7
General Software

Abstract In this chapter, the interface of the software is described in general terms, and we have detailed the computation of the different types of loadings and the input data.

7.1 A General Flow-Chart of the Model

The following flow-chart describes the general algorithm, as follows:

1. Initialization of the model
   - Computation of the building load for each cell in the year 1990;
   - Computation of the average water pressure for the 32 cells; and
   - Initial displacement for each cell assumed to be zero.
2. Reading data
   - Read the soil parameters data;
   - Calibrate the groundwater model; and
   - Read the displacement data for each cell.
3. Loop on time: new time step
4. Loop on the different cells:
   - Compute the stress in the cell;
   - Compute the variation of loading;
   - Compute the water pressure variation;
   - Determine the consolidation time; and
   - Compute the displacement.
5. End of cells loop
6. Output the water pressure and the settlement of the layer
7. End of time step.

D.G. Zeitoun and E. Wakshal, *Land Subsidence Analysis in Urban Areas*,
Springer Environmental Science and Engineering, DOI 10.1007/978-94-007-5506-2_7,

7.2 The MATLAB Interface

The development was done on a PC windows system. The data is a collection of arrays that the user may build.

We used the MATLAB software, because of the large library of mathematical algorithms given with this package. A graphic user interface was developed and connected with MATLAB, making it very user-friendly.

The Interface was developed using the graphic user interface tools offered by the MATLAB 8 software. The interface GUI consists of a main window where the different menus are given (see Fig. 7.1):

Four main menus corresponding to four different tasks were built:

1. Reading of the data;
2. Calibration of the parameters;
3. Validation of the model; and
4. A prediction tool.

The main windows also contain:

(a) A tool for loading the location map of the aquifer with the cells;
(b) Four plot windows; one for the water levels; one for the pumping rate, one for the rain data, and one for spring data; and
(c) A connection between these plot windows which allows all the plots created by the software to be forged into an array.

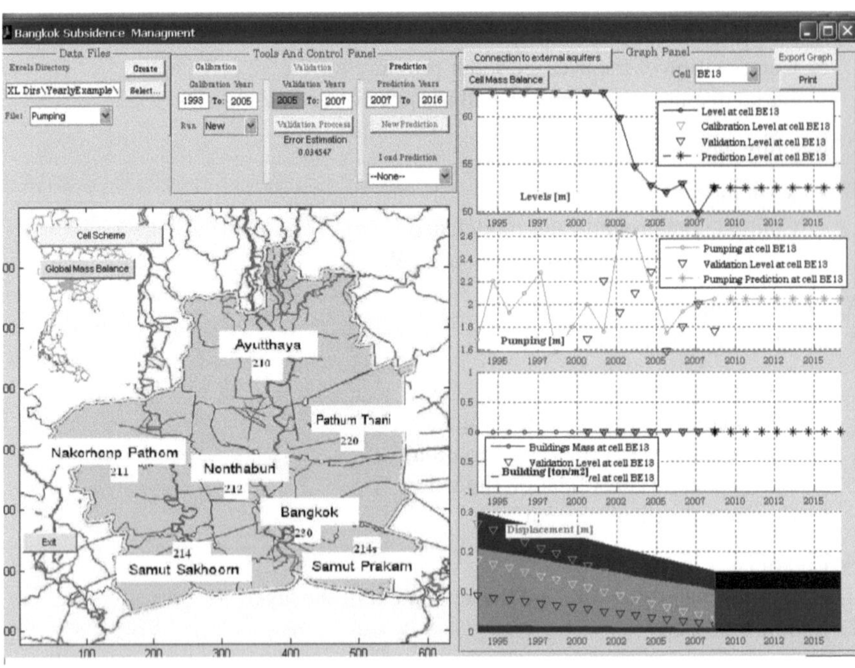

Fig. 7.1 The MATLAB interface main windows

7.2.1 *Reading the Data*

The developed software permits the reading of model data such as the list of cells, the annual water level, the annual rates of pumping, the rain data array and the spring discharges array.

The *reading data menu* contains also the connection tool where connections between the different cells are defined

7.2.2 *The Calibration Tool*

The software allows two types of calibrations:

(a) Calibration of the hydrological parameters with respect to the head data; and
(b) Calibration of the outlets of the aquifer such as springs.

In any process of calibration, we need first to define the period of the calibration (see Fig. 7.1).

7.2.3 *The Hydrological Parameters Calibration*

The software enables the building of the hydrological connections between cells in an easy way. An array called parameters is built with a specific structure:

Every line relates to a specific cell of the model, and the array columns relate to the calibration parameters, such as storativity and transmissivity between the different cells.

The hydro-geological model is introduced using a simple convention, as follows:

1. When the hydro-geological connection between cells does not exist, the user writes a zero at the corresponding line and column of the "parameter" array;
2. When the hydro-geological connection between two different cells exists, the user assigns a blank for each parameter to be calibrated; and
3. When the hydro-geological connection between two different cells exists, the user assigns a value for each known parameter.

7.2.4 *The Validation Tool*

The validation process consists of checking the difference between the model results and the measured head data for a given period.

First, the period of time is chosen as a number of time steps (generally a season or year) for the validation process. The outputs of this process are the computed head solution for the selected period and the computation of the error between measured and computed heads.

The software presents the validation results in the following three forms:

1. Calculation of the global error between the calculated results and the measured data;
2. Creation of an array containing the results of the model; and
3. A visualization tool showing the graphs of the computed head vs the measured head.

7.2.5 The Prediction Tool

First, the period of time is chosen as a number of time steps (generally season or year) for the prediction process. The first year of the period is always the last year of the data.

7.2.6 The Prediction Window

The prediction window contains two columns of data, one for the hydraulic head and one for the pumping rate for each cell and thus, for each time in the prediction period.

The user writes the known data (water level and/or pumping rates) for each year leaving a blank when the value is unknown. These results are stored in an array.

The prediction process consists of computing the unknown data using the mathematical approach described above.

The software presents the prediction results in the following three forms:

1. Creation of an array containing the results of the model;
2. A visualization tool showing the graphs of the prediction head in the prediction period; and
3. A visualization tool showing the graphs of the prediction pumping rate in the prediction period.

7.3 The Calibration of the Models

The calibration, or identification, of a model is the process in which the various model parameters (and that may also include its geometry, inputs, etc.) are determined, if no previous knowledge of them is available, or verified (if such information is available). The calibration is based on data obtained from observations of the behavior of the aquifer in the past. Such data usually includes water levels, pumping and recharge rates and volumes, water quality, interface positions, spring discharges, etc. The calibration, or identification, procedure is often referred to as the inverse problem (see Bear 1972).

The calibration involves taking a past period for which data are available on the behavior of the real aquifer system (e.g., water levels) and running a simulation of the aquifer model. When the model is calibrated in accordance with recorded input data for that period (e.g., pumpage and natural replenishment), its response (e.g., in the form of water levels) is compared with the recorded past response observed in the aquifer. The model is said to be calibrated when the difference between these two responses is less than some value specified by the planner. At the end of the calibration phase, we have a well-defined model of the aquifer system under consideration. All its parameters are well defined, and it can now be used with confidence for forecasting the response of the aquifer in the future to the planned operations.

7.3.1 A Mathematical Formulation for the Calibration of the Flow Equation

For each cell $Cell_i$, we have a number of ${N^i}_p$ to calibrate, so that:

$$N_p = \sum_{i=1}^{i=N_c} {N^i}_p \tag{7.1}$$

N_p is the total number of parameters to calibrate.
For each cell $Cell_i$, the set of parameters to calibrate may be: ${P^i}_1, {P^i}_2, \cdots, {P^i}_{N_p^i}$
Where

$P_l^i = S_i$ (storativity of $Cell_i$) or
$P_l^i = K_{ij}$ Hydraulic conductivity of water between $Cell_i$ and $Cell_j$
$P_l^i = K_{lj}$: Hydraulic conductivity of water between $Cell_i$ and external resources
$P_l^i = \alpha_i$: Replenishment parameter.

The steps of the calibration process for the flow equations are:

STEP 1: A definition of the calibration vector:
We order the unknowns parameters with the following order:

$$\vec{K}^T = \left[P_1^1, P_2^1, \cdots, P_{N_P^1}^1, P_1^2, P_2^2, \cdots, P_{N_P^2}^2, \cdots, P_1^{N_c}, P_2^{N_c}, \cdots, P_{N_P^{N_c}}^{N_c}\right] \tag{7.2}$$

Table 7.1 Calibration of the parameters

Parameters	Selected for calibration	Not selected for calibration
$P_l^i = S_i$	$M_{ij} = (p^{t+\Delta t}{}_i - p^t{}_i)$	$G_i^t = G_{iPrevious}^t - S_i(p^{t+\Delta t}{}_i - p^t{}_i)$
$P_l^i = K_{ij}$	$M_{ij} = (p_j^{t+\Delta t} - p_i^{t+\Delta t})$	$G_i^t = G_{iPrevious}^t - K_{ij}(p_j^{t+\Delta t} - p_i^{t+\Delta t})$
$P_l^i = K_{lj}$	$M_{ij} = (p_l^{t+\Delta t} - p_i^{t+\Delta t})$	$G_i^t = G_{iprevious}^t - K_{lj}(p_l^{t+\Delta t} - p_i^{t+\Delta t})$
$P_l^i = \alpha_i$	$M_{ij} = -Precipitation_i \times \Delta t$	$G_i^t = G_{iPrevious}^t + Precipitation_i \times \Delta t \times \alpha_i$

STEP 2: Building of the calibration matrix:

The calibration matrix equation is given by:

$$M\vec{K} = \vec{L} \tag{7.3}$$

The size of the matrix M is $N_p \times N_c \times N_y$ for the line number (N_y Number of years for the calibration period and N_p for the unknown parameters).

The size of $\vec{L}$ is $N_c \times N_y$, computation of the matrices and vectors $M; \vec{L}$.

In the present model, the recharge, the pumping, and the spring discharge are given data for the period of calibration. Therefore, the second member $\vec{L}$, for each cell $Cell_i$ at time t is defined as:

$$L_i^{t+\Delta t} = \Delta t[D^{t+\Delta t}{}_i - P^{t+\Delta t}{}_i - B_i{}^{t+\Delta t}] + G_i^t \tag{7.4}$$

G_i^t Represents the known part of the Equation (7.4) and M_{ij}^t are as described in Table 7.1:

For a given parameter, there are three possibilities, as follows:

(a) The hydrological connection does not exist, and therefore, no value is assigned to the parameter;
(b) The hydrological connection exists, and the value of the parameter is known. In this case, a new term is added to the right-hand side of the equation in a new value of G_i^t. This value is described in the last column of Table 7.1 for the different parameters, and
(c) The hydrological connection exists, and the value is unknown. Then, this value is determined by the calibration process. A new term is added to the M_{ij}^t matrix.

This value is described in the second column of Table 7.1 for the different parameters.

STEP 3: Computation of the matrix system:

$$M\vec{K} = \vec{S}$$

The matrix M is not invertible, so we used the least-squares method for the solution of this system. The difficulties are the constraints on the parameters.

The least-squares solution of the BIOT equations may be obtained by different algorithms such as multi-regression analysis, SVD (singular value decomposition), etc. The use of the MATLAB software permits the use of a large library of algorithms and the testing of the most efficient one.

A similar calibration method was applied for the consolidation equations where the calibration parameter is the parameter $P_l^i = U_{lj}$.

7.3.2 A Mathematical Formulation for the Calibration of the Consolidation Equation

(a) **The Terzaghi theory**

The relation between the data measurements of the displacement and the consolidation parameters have been calibrated, using multivariate statistics from 1978 to 2007.

(b) **The Biot theory**

For each cell $Cell_i$, we have a number of $N^i{}_p$ to calibrate, so that:

$$N_p = \sum_{i=1}^{i=N_c} N^i{}_p \tag{7.5}$$

N_p is the total number of parameters to calibrate.

For each cell $Cell_i$, the set of parameters to calibrate may be: $P^i{}_1, P^i{}_2, \cdots, P^i{}_{N_p^i}$

Where

$P_l^i = U_{ij}$: Compressibility parameter between $Cell_i$ and $Cell_j$.

$P_l^i = U_{lj}$: Compressibility parameter between $Cell_i$ and external resources.

The steps of the calibration process for the flow equations are:

STEP 1: Definition of the calibration vector:

We define the unknown parameters with the following order:

$$\vec{K^T} = \left[P_1^1, P_2^1, \cdots, P_{N_P^1}^1, P_1^2, P_2^2, \cdots, P_{N_P^2}^2, \cdots, P_1^{N_c}, P_2^{N_c}, \cdots, P_{N_P^{N_c}}^{N_c}\right]$$

STEP 2: Building of the calibration matrix:

The calibration matrix equation is given by:

$$M\vec{K} = \vec{L} \tag{7.6}$$

The size of the matrix M is $N_p \times N_c \times N_y$ for the line number (N_y Number of years for the calibration period and N_p for the unknown parameters).

Table 7.2 Constraints on the parameters

Parameters	Constraints
$P_l^i = S_i$	$S_i \geq 0$
$P_l^i = K_{ij}$	$K_{ij} \geq 0$
$P_l^i = K_{lj}$	$K_{lj} \geq 0$
$P_l^i = \alpha_i$	$\alpha_i \geq 0$

Table 7.3 The calibration of the parameters

Parameters	Selected for calibration	Not selected for calibration
$P_l^i = K_{ij}$	$M_{ij} = (w_j^{t+\Delta t} - w_i^{t+\Delta t})$	$G_i^t = G_{iPrevious}^t - K_{ij}(w_j^{t+\Delta t} - w_i^{t+\Delta t})$
$P_l^i = K_{lj}$	$M_{ij} = (w_l^{t+\Delta t} - w_i^{t+\Delta t})$	$G_i^t = G_{iprevious}^t - K_{lj}(w_l^{t+\Delta t} - w_i^{t+\Delta t})$

The size of $\vec{L}$ is $N_c \times N_y$, computation of the matrices and vectors $M; \vec{L}$.
In the present model, the recharge, the pumping, and the spring discharge are given data for the period of calibration. Therefore, the second member $\vec{L}$, for each cell $Cell_i$ at time t is defined as:

$$L^{t+\Delta t} = G_i^t + D^{t+\Delta t}{}_z p^{t+\Delta t}{}_i + g_i{}^{t+\Delta t}/\chi_i \tag{7.7}$$

G_i^t represents the known part of Equation (20) and M_{ij}^t are is described in Table 7.2:
For a given parameter, there are three possibilities, as follows:

(a) The hydrological connection does not exist, and therefore, no value is assigned to the parameter.
(b) The hydrological connection exists, and the value of the parameter is known. In this case, a new term is added to the right hand-side of the equation in a new value of G_i^t. This value is described in the last column of Table 7.2 for the different parameters.
(c) The hydrological connection exists, and the value is unknown. Then, this value is determined by the calibration process. A new term is added to the M_{ij}^t matrix.

This value is described in the second column of Table 7.2 for the different parameters (Table 7.3).

STEP 3: Computation of the matrix system:

$$M\vec{K} = \vec{S} \tag{7.8}$$

The matrix M is not invertible, so we used the least-squares method for the solution of this system. The difficulties are the constraints on the parameters. The least-squares solution may be obtained by different algorithms such as multi-regression analysis, SVD (singular value decomposition), etc. The use of the MATLAB software permits the use of a large library of algorithms and the testing of the most efficient one.

The model was validated against analytical solutions for a steady-state flow (see Zeitoun and Wakshal 2009).

7.4 The Validation of the Models

The goal of the model validation is to measure the differences between the data and the results of the model in a time period different from the calibration period.

The matrix system for the validation at a time t is given by the matrix equation:

$$A\vec{H} = \vec{R} \tag{7.9}$$

Where $\vec{H}^T = \left[h_1^{t+\Delta t}, h_2^{t+\Delta t}, \cdots, h_{N_c}^{t+\Delta t}\right]$.

7.5 The Prediction Using the Model

The general prediction problem for a cell model may be formulated by two different ways:

Assumption 1: "For any cell, the unknown may be either the pumping rate or the hydraulic head but not both of them and at least one of them."

Then define N_c as the cell number.

$$C = \{1, 2, \cdots, N_c\}$$

Define the following sets of indices

$$H = \{j \in C, h_j^{t+\Delta t} \textit{ is unknown}\}$$

$$Q = \{j \in C, Q_j \textit{ is unknown}\}$$

$$N_H = Card\, H;\; N_Q = Card\, Q;\; N_c = N_H + N_Q$$

Steps for the prediction process

STEP 1: Definition of the prediction vector:

$$\vec{U}^T = \left[u_1^{t+\Delta t}, u_2^{t+\Delta t}, \cdots, u_{N_c}^{t+\Delta t}\right] \tag{7.10}$$

The size of the matrix B is $N_c \times N_c$

The size of the vector $\vec{R}$ is N_c

The unknown vector is built in the following order

$$\begin{aligned} &if\ \ i \leq N_H;\ \ u_i^{t+\Delta t} = h_{l(i)}^{t+\Delta t};\\ &if\ i > N_H;\ u_i^{t+\Delta t} = Q_{p(i)}^{t+\Delta t} \end{aligned} \tag{7.11}$$

The index of the unknown head in the general system is $l(i) \in H$ and $p(i) \in Q$ is the index of the unknown pumping rate.

STEP 2: Building of the matrix

The linear system to solve for the prediction system for each new prediction year:

$$E\vec{U} = \vec{Z} \tag{7.12}$$

The matrix E is defined as follows:

$$\begin{aligned} &if\ i \in C;\\ &j \leq N_H;\ E_{ij} = A_{il(j)};\\ &j > N_H; E_{ij} = 0;\ if\ \ p(j) \neq i\\ &\qquad\qquad E_{ij} = \Delta t\ A_{p(j)};\ f\ \ p(j) = i \end{aligned} \tag{7.13}$$

The vector $\vec{Z}$ is defined as:

$$\begin{aligned} &\forall i \in C;\\ &if\ \ i \in H; z_i = S_i {h^t}_i + \sum_{l \in E_i} K_{lj} {h_l}^{t+\Delta t} + \Delta t \left[{N^{t+\Delta t}}_i \right.\\ &\qquad \left. + {D^{t+\Delta t}}_i - {P^{t+\Delta t}}_i - {B_i}^{t+\Delta t} \right]\\ &if\ \ i \in Q;\ for\ all\ j\ \in C z_j = S_j {h^t}_j + \sum_{l \in E_i} K_{lj} {h_l}^{t+\Delta t} + \Delta t [{N^{t+\Delta t}}_i\\ &\qquad + {D^{t+\Delta t}}_i - {B_i}^{t+\Delta t}] + K_{ij} h_i^{t+\Delta t} \end{aligned} \tag{7.14}$$

STEP 3 Solve

$$E\vec{U} = \vec{Z} \tag{7.15}$$

The use of the MATLAB software permits the solving of this matrix problem using the least-squares algorithm.

Assumption 2: "For any cell, the unknown may be either the pumping rate or the hydraulic head but also both of them or anyone of them."

Then define N_p as the total number of unknowns

$$C = \{1, 2, \cdots, N_p\}$$

Define the following sets of indices

$$H = \{j \in C, h_j^{t+\Delta t} \text{ is unknown}\}$$

$$Q = \{j \in C, Q_j \text{ is unknown}\}$$

$$N_H = Card\, H;\ N_Q = Card\, Q;\ N_c \neq N_H + N_Q$$

References

Bear J (1972) Dynamics of fluids in porous media. Dover, New York
Zeitoun DG, Wakshal E (2009) The analysis of subsidence in the Bangkok area and its vicinity. Report DGR, Bangkok, Thailand

Chapter 8
A Case Study: The Bangkok Plain

Abstract In this chapter, the implementation of the methodology is described on a modeling study of the Bangkok area. The use of historical satellite maps of the urbanization of the Bangkok area is connected with the GIS system. The building of the model data is described in detail. We analyze separately the different types of effects. The computational of the total subsidence and each component is presented. Also, the effect of pumping from a deep layer on subsidence in the upper layer is presented.

8.1 Geology and Hydrogeology

The study areas cover the Bangkok Metropolis and its vicinity, comprising wholly the following seven provinces, namely (see Fig. 8.1):

The city of BANGKOK
The NONTHABURI Province
The SAMUTPRAKARN Province
The SAMUT SAKHON Province
The PATHUM THANI Province
The PHRA NAKHON SI AYUTTHAYA Province (AYUTTHAYA)
The NAKHON PATHOM Province

The choice of the most appropriate conceptual model for a given aquifer system and for a given management problem is dictated first of all by the features of the aquifer itself (e.g., its geological properties), but also by the following criteria (see Bear 1972):

(a) It should be sufficiently simple so as to be amenable to mathematical treatment;
(b) It should not be too simple so as to exclude those features which are of interest to the investigation on hand. As the range of possible models between these two limits is still wide, we should add two more important criteria, namely, that:

D.G. Zeitoun and E. Wakshal, *Land Subsidence Analysis in Urban Areas*, Springer Environmental Science and Engineering, DOI 10.1007/978-94-007-5506-2_8,

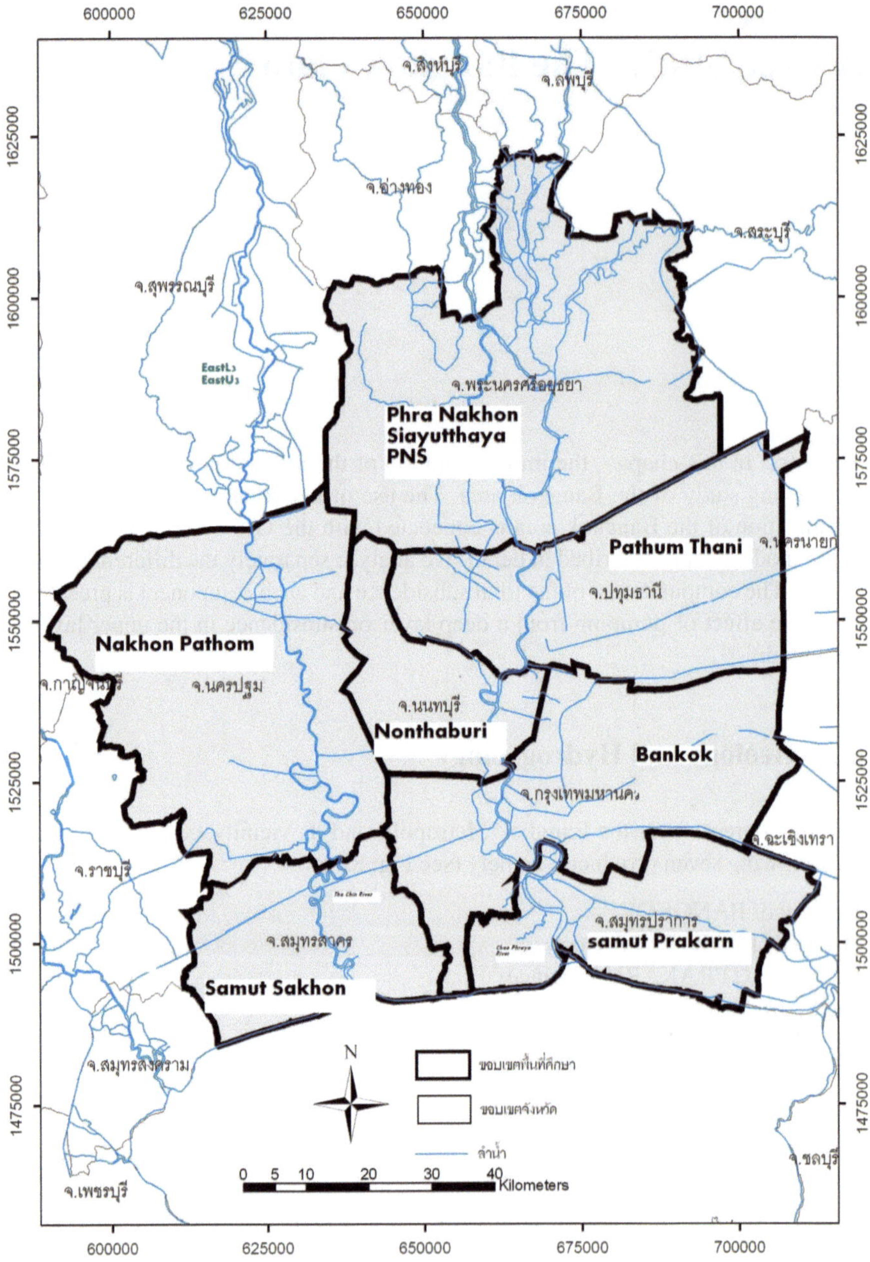

Fig. 8.1 A location map: the Bangkok region

(c) Information should be available for calibrating the model; and
(d) The model should be the most economic one for solving the problem on hand.

8.1.1 The Geology of the Bangkok Basin

1. General geology
   A simplified geological map of Thailand is shown indicating the following rock units: the Pre-Cambrian gneiss of amphibolite facies and the Lower Paleozoic sandstone and limestone are normally found side by side in the west from Mae Hong Son to Kanchanaburi, in the eastern Gulf, and in the southern Peninsula.

   The Mid-Paleozoic rocks cover large areas in both the west and the east. They are represented by a thick monotonous geosynclinal sequence mainly consisting of politic rocks and volcanic debris.

   The Upper Paleozoic rocks, on the other hand, originated from a shallower marine which consists of arenaceous and calcareous rocks. In the west, they include some Triassic rocks, either conformable or unconformable.

   The Early Mesozoic sequence in the western mountains is usually involved with the Upper Paleozoic rocks, but the Middle and Late Mesozoic rocks are found strongly unconformable with either the Early Mesozoic or the Upper Paleozoic. These later Mesozoic rocks are mainly non-marine red beds, typically underlying the Khorat Plateau with a mild and broad folded structure.

   The non-marine Cenozoic rocks are mainly found in isolated intermountain basins, and larger parts of them underlie the Chao Phraya Plain and the Gulf of Thailand.

   Lithofacies change across the country suggests that the stratigraphic successions in Thailand resulted from the successive depositional evolutions of the Paleozoic-Early Mesozoic geosynclines, post-Early Mesozoic platform basins and lately, the Cenozoic basins.

   The Gulf of Thailand was formed by rifting and a brief period of ocean spreading during the Late Cretaceous and/or Cenozoic era. The Chao Phraya depression (i.e., the Central Valley and the Central Plain) is thought to be a graben constituting the onshore extension of the Gulf. It may also be regarded as the largest in a series of Tertiary extensional basins in northern Thailand.
2. The basement structure of the Central Plain
   Based on the interpretation of the Landsat photograph, strike-slip faults in NW-SE direction are clearly identified. These faults are accompanied with secondary faults in N-S direction and divide the basement into several blocks, in which these geological blocks constitute the grabens. The depth to the basement is estimated to be more than 1,800 m, according to the oil exploratory drilling, aero magnetic and seismic data.

   In correlation with the lithologic facies in the Gulf, the sediments overlying the basement are thought to be deposits in the coastal deltaic environment from Mid-Tertiary to Quaternary. Recent studies on sedimentation in the Bangkok region may be found in Mongsrijun et al. (2008).

8.1.2 *Hydrogeology*

The Bangkok Metropolitan area is situated on the flood plain and delta of the Chao Phraya River which traverses the Lower Central Plain of Thailand. The plain, also known as the Lower Chao Phraya Basin, extends about 200 km from the north to the south and about 175 km from the east to the west. It is bounded on the east and the west by the mountain ranges and the south by the Gulf of Thailand. To the north, the plain is bordered by a series of small hills dividing it from the Upper Central Plain, with the Chao Phraya River as an interconnection.

The geological formations involved in the transmission and confinement of groundwater in the Bangkok Metropolis are a basement complex and alluvial deposits which constitute the principal aquifers in Bangkok.

The clastic sediments overlying the basement complex are unconsolidated and semi-consolidated sediments of clay, sand and gravel, ranging in age from Tertiary to Quaternary. The total thickness of the strata ranges from about 400 m in the east to more than 1,800 m to the west of Bangkok. Data from water wells drilled in the Bangkok Metropolitan area reveals that there are at least three major depositional breaks within 600 m from ground surface, as evidenced by the strata geoelectrical properties and water quality. Piancharoen and Chuamthaisong (1976) confirmed that the first depositional break is at the depth of 100–120 m, whereas the second and third breaks are at about 350–450 m and below 600 m as indicated by a well drilled to the depth of 650 m at Samutprakarn Province (after Ramnarong and Buapeng 1991, 1992).

8.2 Aquifers in the Bangkok Area

The topmost sediments, 15–30 m thick, consist of soft clay overlying stiff clay, known as the "Bangkok clay". The underlying sand, gravel, and clay constitute the principal aquifers of Bangkok. Based on data of electric logs and drilled logs, the upper 600 m of the unconsolidated deposits are subdivided into eight principal artesian aquifers, separated from each other by thick confining clay or sandy clay layers. They are as follows:

1. Bangkok Aquifer (50 m zone) BK
2. Phra Pradaeng Aquifer (100 m zone) PD
3. Nakorn Luang Aquifer (150 m zone) NL
4. Nonthaburi Aquifer (200 m zone) NB
5. Sam Khok Aquifer (300 m zone) SK
6. Phaya Thai Aquifer (350 m zone) PT
7. Thon Buri Aquifer (450 m zone) TB
8. Pak Nam Aquifer (550 m zone) PN

8.2.1 The Bangkok Aquifer (BK)

The Bangkok aquifer consists of two sub-aquifers separated from each other by clay layers. The upper Bangkok aquifer, immediately underneath the Bangkok clay, consists of yellowish brown or grayish brown, fine sand, or clayey sand on top and fine to coarse sand with gravel at bottom. The depth to the top of the upper aquifer is 16–30 m below ground surface, and its maximum thickness is 30 m. The upper aquifer is missing in some areas of eastern Bangkok and Samutprakarn Province. The lower aquifer is separated by clay and silt layers of variable thickness. It is composed of grayish brown, fine to coarse sand with gravel and gray to grayish brown clay layers. The lower aquifer directly interconnects with the top aquifer, is about 30 m below ground surface, and it may be 30–50 m thick.

The quality of water in the Bangkok aquifer is undrinkable due to high salinity except in the south – 250–500 mg/L of chloride in groundwater.

8.2.2 The Phra Pradaeng Aquifer (PD)

The Phra Pradaeng aquifer underlies a thick confining brown to grayish brown clay about 10–14 m thick. The aquifer consists of sand and gravel with clay lenses. The sand and gravel are generally of a coarse grain, poorly sorted, subrounded to rounded, and white to whitish gray. Carbonized wood, or logs, and shell fragments have been found in many places. The sediments were probably deposited under the fluviatile and shallow sea environments on an erosional surface of hard and compact clay. The aquifer is 60–80 m deep and ranges from 20 m thick in the north to 50 m in the south.

The groundwater in the Phra Pradaeng aquifer is generally of a high yield but is mostly brackish to salty. The development of groundwater in this aquifer is mainly in the area of Phra Pradaeng District and the southern Part of Samutprakarn Province.

8.2.3 The Nakorn Luang Aquifer (NL)

The Nakorn Luang aquifer underlies the Phra Pradaeng aquifer from which it is separated by a thick brown, compact clay layer 3–10 m thick. The aquifer consists of sand and gravel with clay layers. The sand and gravel are subrounded to rounded, moderately well sorted, whitish yellow to light brown and gray to grayish brown in some places. These sediments are the upper part of an older terrace. The aquifer is 100 m below the ground surface in the east to 140 m in the south and is about 50–70 m thick.

The aquifer is very permeable and gives good quality water in the area east of the Chao Phraya River. The first groundwater development for water supply in Bangkok was from the Nakorn Luang aquifer, and it has been heavily extracted.

8.2.4 The Nonthaburi Aquifer (NB)

The Nonthaburi aquifer underlies the Nakhon Luang aquifer from which it is separated by a thick clay bed. The sediments are similar to the Nakhon Luang aquifer which is composed of sand and gravel with minor clay lenses. It can be divided into at least three subaquifers separated by leaky clay beds. The aquifer is 170–200 m below the ground surface, and its thickness is 30–70 m.

The groundwater conditions of the Nonthaburi aquifer are also similar to those of the Nakhon Luang aquifer. It was not developed earlier because it suffered from overpumping and a poor quality of water.

8.2.5 The Sam Khok Aquifer (SK)

The Sam Khok aquifer consists of alternating layers of sand or gravel and clay. The clay is generally brown to yellow and moderately to highly compacted. The sand and gravel are mostly of a medium to coarse grain, fairly well sorted, and with intercalated clay lenses. The aquifer is 240–250 m below the ground surface and 40–80 m thick.

8.2.6 The Phaya Thai Aquifer (PT)

The Phaya Thai aquifer underlies the Sam Khok aquifer from which it is separated by a hard and compact clay bed 5–10 m thick. The aquifer consists of thin sand and gravel layers intercalated with clay lenses. The sand and gravel are dirty brown, angular, medium sand to gravel size, and poorly to fairly well sorted. The aquifer is 275–350 m deep and 40–60 m thick.

The water-bearing properties of the aquifer are similar to those of the Sam Khok aquifer. Good quality water is found only in the north, east, and southeast of Bangkok. The aquifer is generally not popular due to its greater depth.

8.2.7 The Thon Buri Aquifer (TB)

The Thon Buri aquifer is situated below the second depositional break. It underlies a thick clay bed up to about 50 m in some places. The strata consist of thick sand

and gravel interbedded with thin layers of clay. The sand is of a coarse grain, rounded and well sorted. The clay is generally pinkish to grayish brown, compacted, and sandy. The aquifer is 350–400 m deep and 50–100 m thick.

The test well data shows that good quality water is present, in the north, east, and southeast of Bangkok; however, the aquifer is not as productive as others due to the presence of clay in many horizons.

8.2.8 The Pak Nam Aquifer (PN)

The Pak Nam aquifer is separated from the clay. It consists of at least three thick sand and gravel beds with clay lenses. The sand and gravel are white to gray and well sorted. The clay layers are generally very compact, olive gray to dark gray, with carbonaceous matter. The aquifer is 420–500 m deep.

The Pak Nam aquifer is very permeable and yields a considerable quantity of good quality water. The best quality water is from the sand layer occurring from about 550 m downward. The water temperature is as high as 50°C. It is, however, too deep to reach by domestic wells, except in areas where there is no alternative potential aquifer, i.e., in southern and southwestern Bangkok.

It should be taken into consideration that the exploited aquifers in Bangkok Basin are composed of young sediments.

Thus, it is characterized by the chemical diagenetic changes of the cementation process, in which minerals are precipitated in the pores of sediments, forming cements that bind the clastic components.

Such cementation results in a decrease in porosity and also results in lithification, forming the diagenetic process by which soft sediment is hardened to become a rock. Another chemical diagenetic change in clastic accumulation, as in Bangkok Basin, is the chemical alteration of clay minerals fraction, originally deposited as clastic particles.

Due to all the above-mentioned reactions, the physical diagenetic change of compaction is taking place.

It results in a decrease in the volume and porosity of the aquifer's sediments, where grains are squeezed closer together by the lithostatic weight of the overlying column of sediments. In general, pure sands are fairly packed during deposition, so they do not compact much. However, newly-deposited muds are highly porous, either as a mixture with sand or pure mud layers. Such sedimentary facies could be observed at the subsurface of Bangkok Basin upon the analyses and correlation of many observation wells (see Japan International Cooperation Agency (JICA) 1995). Often, more than 60% of the sediment is water in pore space Due to it, muds compact greatly after burial, losing at least one half of their water content.

These various aspects of water-solid rock interface should play an important role in Bangkok Basin, as the static groundwater level has been decreased by tens of meters due to the well-known overpumpage.

8.2.9 The Aquifers' Characteristics

The characteristics of the aquifers are defined mainly by their hydraulic properties which are the coefficient of the transmissivity, storage, and leakage. The first constant characterizes the ability of the aquifer to transmit water, the second the ability to release water from storage, and the third the ability of the semi-confining bed to pass water vertically upward or downward. All aquifers in the Bangkok area are generally very permeable and may yield at least 100 m^3/h to as much as 300 m^3/h.

The hydraulic properties of the main producing aquifers are similar. The transmissivity of the Bangkok aquifer obtained from a pumping test at Bangpoon, Pathum Thani is 3,950 m^2/d, and its storage coefficient is about 1×10^{-4} (Ramnarong 1976). The transmissivity coefficient of the Phra Pradaeng and Nakhon Luang aquifers range from 1,200 to 4,104 m^2/d, and their storage coefficient from 1×10^{-4} to 3.4×10^{-3}. The estimated value of seepage or leakage factors ranges from 2×10^{-6}/d to 2×10^{-5}/d (see Asian Institute of Technology 1982). The transmissivities of the Nonthaburi, Sam Khok, and Phaya Thai aquifers are 3,730, 4,992 and 6,000 m^2/d, respectively. The storage coefficient is 2×10^{-3} and the seepage factor is 1.6×10^{-5}/d.

8.2.10 The Groundwater Recharge

The principal source of recharge to the Bangkok aquifer system is mainly from lateral inflow from the upper part and the periphery of the Lower Central Plain. This inflow is directly percolated from rainfall, seepage from the Chao Phraya River and its tributaries and infiltration during floods into the aquifers. Estimated from contours of piezometric levels and an environmental isotopes study, occupies the area of Ang Thong up to Chai Nat and Uthai Thani Provinces and the areas near the exposed outcrop along the eastern and western margins of the basin.

From observations of the groundwater level fluctuations in wells in the upper part of the Lower Chao Phraya Basin, the recharge seems closely related to rainfall. The water levels in wells respond to rainfall by rising during the rainy season and declining during the dry season. The rate of fluctuation is about 1–2 m/year. The results from the environmental isotopes study have shown that the recharge is a continuation from geological to present times, and the groundwater in the Lower Chao Phraya Basin is of meteoric origin.

The annual recharge for the whole basin is estimated to be 3.2% of the rainfall (see Asian Institute of Technology 1982). The recharge is highest in September when the basin receives its heaviest rainfall. Vertical inflow directly from rainfall and seepage from rivers into the aquifer systems in the Bangkok Metropolitan area are believed to provide only a very small contribution due to the very thick clay bed on the top of the formation.

8.2.11 Groundwater Movement

The regional groundwater movement in the Lower Phraya basin and the Bangkok Metropolis has been interpreted from piezometric surface maps of the aquifers. Flow directions are generally toward the south, southeast and southwest. As the cones of depression developed in the Bangkok Metropolitan area are due to pumping hydraulic gradients, this results in a reverse flow from south to north, away from the Gulf.

The groundwater flow velocity is relatively slow due to the gentle gradient. A long residence time of groundwater in the basin has been confirmed by the environmental isotopes study (see Piancharoen and Chuamthaisong 1983). The estimated groundwater movement, determined by the carbon-14 dating method, is about 4 cm/day, and it has taken more than 10,000 years for groundwater to move from the recharge areas to central Bangkok. The carbon-14 dating also indicates that the groundwater could have entered Bangkok aquifers system more than 20,000 years ago.

Piezometric surface data, from monitoring wells in the Bangkok area, indicate that groundwater moves downward from the Bangkok to Phra Pradaeng of the Nonthaburi aquifer moves upward to the Nakorn Luang aquifer. The piezometric surfaces in the deeper aquifers are generally high compared with the shallower aquifers, and the water can move upward from the underlying aquifer into the upper aquifer.

8.2.12 Hydrochemistry

A total of 237 groundwater samples from the DMR monitoring wells and 16 samples from production wells were collected from September to December 1993. The submersible pump was used to remove stagnant water from the well so as to collect reliable samples. The method of sampling is as follows (see Japan International Cooperation Agency (JICA) 1995):

(a) Remove existing monitoring unit from the well;
(b) Measure statistic groundwater level;
(c) Install a submersible pump into the well;
(d) Pump up stagnant water from the well for at least one (1) hour;
(e) Measure the water level, discharge rate, electric conductivity, pH, and temperature at an interval of 10 min during pum;.
(f) Collect a sample from pumped water at the final stage of pumping and keep it in sampling bottles;
(g) Write the necessary information on the label pasted on the bottles: sample no., well no, sampling date, electric conductivity, pH, and temperature;
(h) Transport samples to the laboratory on the same day; and
(i) Reinstall monitoring unit to the well.

8.2.13 Results of the Analysis

8.2.13.1 The Phra Pradaeng (PD) Aquifer

The elements of (Na + K) and (Ca) occupy 25–98% and 2–48% of the total cations, respectively. Most samples are chloride-rich in anion composition. However, several samples are rich in bicarbonate. Sulfate content is higher in the PD Aquifer than in the NL Aquifer or the NB Aquifer. The samples having high ion contents show chloride as dominant in the anion composition. But the cation compositions of those samples can be classified into two (2) groups: one characterized by the dominance of (Na + K) and the other is characterized by an almost equal content of (Na + K) and (Ca). The former group is in Samutprakarn and western Bangkok, while the latter group is in northern Bangkok, Pathum Thani, and Nonthaburi. The high chloride concentration lies from Samut Sakhon to Pathum Thani along the Phraya River and extends to northern Bangkok. High concentration area is also found in Samut Prakan. The concentration partly exceeds 5,000 mg/L in those areas. Major areas of high chloride concentration can be found on the map: the coastal area from Samutprakarn to Samut Sakhon showing 3,000–16,000 mg/L and along the Chao Phraya River showing 2,000–6,000 mg/L.

8.2.13.2 The Nakhon Luang (NL) Aquifer

The element of (Na + K) occupies 30–99% of the total cations. For anions, the sulfate is less than 10% except for a few samples. The samples having high ion contents are distributed along the shoreline in Samut Prakan, western Bangkok, and along the Chao Phraya River in Nonthaburi and Pathum Thani. The samples taken from the coastal area are characterized by high content of chloride and (Na + K). The saline water may have originated from sea water. However, the samples taken inland are characterized by high calcium as well as (Na + K) contents in cations. This research indicates that the source of saline water inland may be different from that in the coastal area.

8.2.13.3 The Nonthaburi (NB) Aquifer

The samples can broadly be divided into two (2) groups: The two groups differ in the anion compositions. The samples belonging to one group are chloride-rich whereas the samples of the other group are bicarbonate-rich. The samples having high ion contents are distributed in Samut Prakan, western Bangkok, Samut Sakhon, and northern Bangkok to Pathum Thani. The samples taken from the southern part of the Study Area are rich in chloride and (Na + K). However, the samples taken inland are characterized by a high calcium content in cations. From the mouth of the Chao Phraya River to Samut Sakhon shows high chloride concentration ranging from 2,400 to 13,000 mg/L. High concentration is also found in part of eastern Samut Prakan. However, a small chloride concentration below 1,000 mg/L is found in the coastal area of central Samut Prakan. An isolated high chloride concentration which ranges from 1,000 to 5,700 mg/L is found inland from Pathum Thani to northern Bangkok.

8.2.14 Saline Intrusion (After Ramnarong 1999)

Saline intrusion in the aquifers of Bangkok has been observed since 1967 when many municipal wells in the southern parts of Bangkok yielded brackish to saline water. Since then, more and more wells in these areas and on the west bank of the Chao Phraya river have been abandoned. High salinity groundwater has also been found in some wells in central Bangkok where the aquifers once yielded fresh water. It was observed that the inferior groundwater quality is due not only to an increase in chloride content but also to the increases in hardness, iron, manganese, sulphate, calcium, sodium, magnesium, and dissolved solids. The major sources of groundwater-quality degradation are believed to be the seawater and connate water entrapped under marine conditions in pore spaces of sediments at, or subsequent to, the time of their deposition.

The saline intrusion in the Bangkok Metropolitan area is limited to the shallow aquifers immediately adjacent to the coast. Seawater moves inland when water levels decline due to over-abstraction of groundwater. The groundwater flows from all directions toward the center of cones of depression, and the saltwater has been drawn northward and northeastward from its source into the fresh water of the aquifers. It has been observed that the rate of movement of saltwater differs in different parts of the Bangkok Metropolis. The rate of movement in areas of heavy abstraction is rather greater than in areas of lower abstraction. The rate of water invasion is undoubtedly accelerated in areas where water levels are most drastically depressed by over abstraction, particularly in the Phra Pradaeng, Nakhon Luang and Nonthaburi aquifers.

8.3 Implementation of the Biot Model for the Bangkok Region

In this section, we present a first implementation of the subsidence code for a series of cells closed to the different administrative provinces.

8.3.1 Cells' Description and Input Data

The basin was split into seven basic regions corresponding to regions, comprising wholly the following seven provinces, namely:

1. The city of BANGKOK;
2. The NONTHABURI Province;
3. The SAMUTPRAKARN Province;
4. The SAMUT SAKHON Province;
5. The PATHUM THANI Province;
6. The PHRA NAKHON SI AYUTTHAYA Province; and (AYUTTHAYA); and
7. The NAKHON PATHOM Province

The governmental institutions define the administrative regions, but they correspond to hydrological units having a homogenous hydrological behavior in terms of water levels.

Each region is separated into basic layers including basic aquifers, as follows:

1. Layer1: It includes the Bangkok aquifer and the Phra; Pradaeng aquifer;
2. Layer2: It includes the Nakorn Nuang aquifer;
3. Layer3: It includes the Nonthaburi aquifer;
4. Layer4: It includes the Sam Khok aquifer; and
5. Layer5: It includes the Phraya Thai aquifer, the Thon Buri aquifer, and the Pak Nam aquifer.

8.3.1.1 A Compartmental Model for the Bangkok Region

- Compartmental cells
  In the following figures, we present the spatial extension of each cell, the groundwater data, and the subsidence data associated with each cell.
  For each province, we separate into aquifers and sub-regions. The hydraulic connection exists between the same layer and between adjacent cells.
  In the following table, we summarize the 34 different cells (see Fig. 8.2, Table 8.1).
- The Bangkok area: it is separated into East and WEST with respect to the Chao Phraya River. Thus, the different cells are: BE11, BE12, BE13, BE14, and BE15 (for the first five aquifers), and for the western part, BW11, BW12, BW13, BW14, and BW15.
- Samu Phrakarn area: it is separated into the Northern and Southern parts
- Samu Sakmonit region: it is separated into the EAST and WEST.

Also, we define the geo-hydrological model of hydraulic connections between the cells.

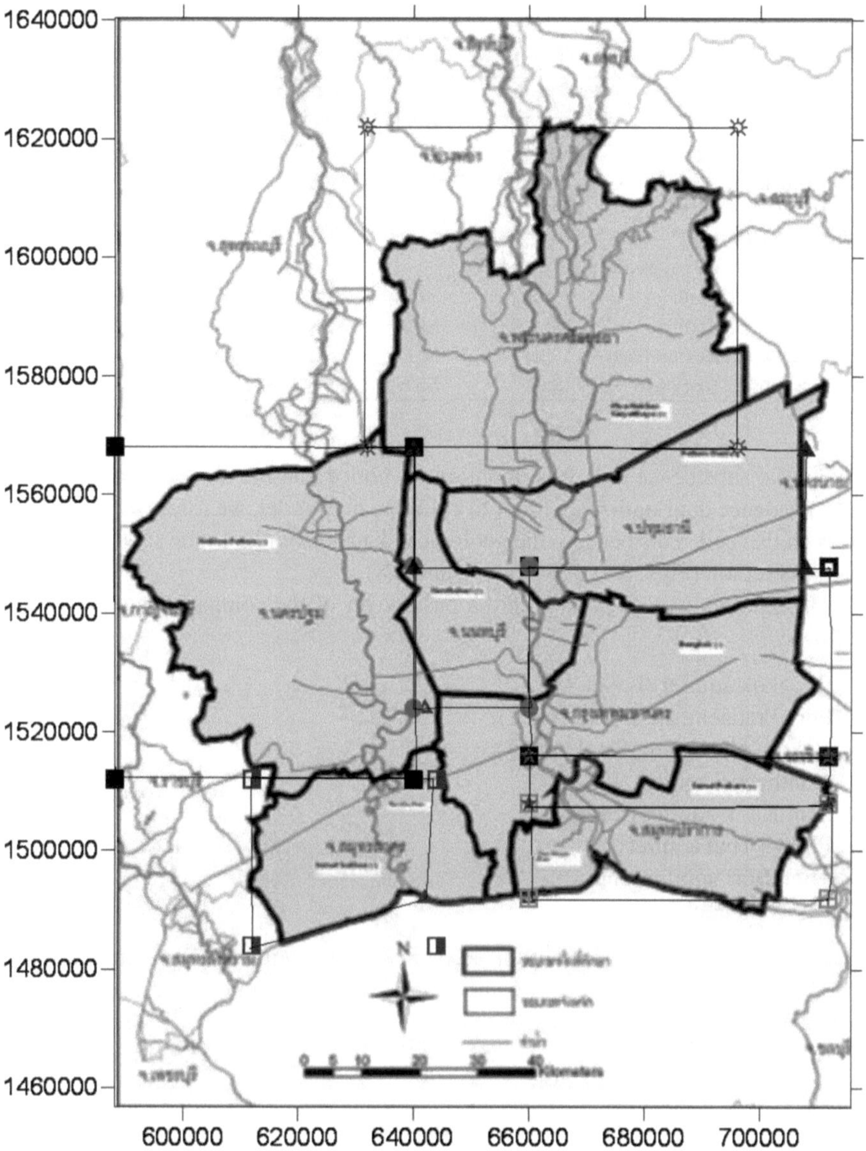

Fig. 8.2 Zoning of the various cell regions

Our assumptions are that

1. The aquifers are confined;
2. The aquifers 1, 2, and 3 are leaky; and
3. The replenishment comes from the external head BC in the northern part. For this, I need the data (Fig. 8.3).

Table 8.1 A list of layers in the simple cell model

Province	Cell Name	Layer1	Layer2	Layer3	Layer4	Layer5
Bangkok East	BE	BE11	BE12	BE13	BE14	BE15
Bangkok West	BW	BW11	BW12	BW13	BW14	BW15
Nakhon Pathom Province	NP	NP11	NP12			
NONTHABURI Region	NT	NT11	NT12			
PATHUM THANI Province	PT	PT11	PT12			
PHRA NAKHON SI AYUTTHAYA Province	PNS	PNS1				
Samu Phrakarn South	SPS	SPS1	SPS2	SPS3	SPS4	SPS5
Samu Phrakarn North	SPN	SPN1	SPN2	SPN3	SPN4	SPN5
Samu Sakmonit Region East	SSE	SSE1	SSE2	SSE3	SSE4	SSE5
Samu Sakmonit Region West	SSW	SSW1	SSW2	SSW3	SSW4	SSW5

For the subsidence data it is important to notice that the northern cell PNS has no subsidence data station. In order to calibrate our model, we associate the station 34 with this cell. This because the geological data are similar to the southern part of the PNS region (Figs. 8.4, 8.5, 8.6, 8.7, and 8.8).

The aquifers were separated into a basic layer of the compartmental model, as follows:

1. Bangkok aquifer (Layer 1);
2. Phra Pradaeng aquifer (Layer 2);
3. Nakorn Nuang aquifer (Layer 3);
4. Nonthaburi aquifer (Layer 4);
5. Sam Khok aquifer (Layer 5);
6. Phraya Thai aquifer;
7. Thon Buri aquifer; and
8. Pak Nam aquifer.

Because of the scarcity of water pressure data, the aquifers: Sam Khok aquifer, Phraya Thai aquifer, Thon Buri aquifer, and Pak Nam aquifer were considered as one layer (Layer 5).

8.3.2 The Software Used in This Study

In this work, we used the MODFLOW simulation of flow as a basic input. Thus, we developed an automatic software that translates the MODFLOW data to the input data of the Biot model.

The following flow-chart (Fig. 8.9) explains the data flow-chart.

The data transfer and file name extensions of the various data files and software are explained in Fig. 8.10.

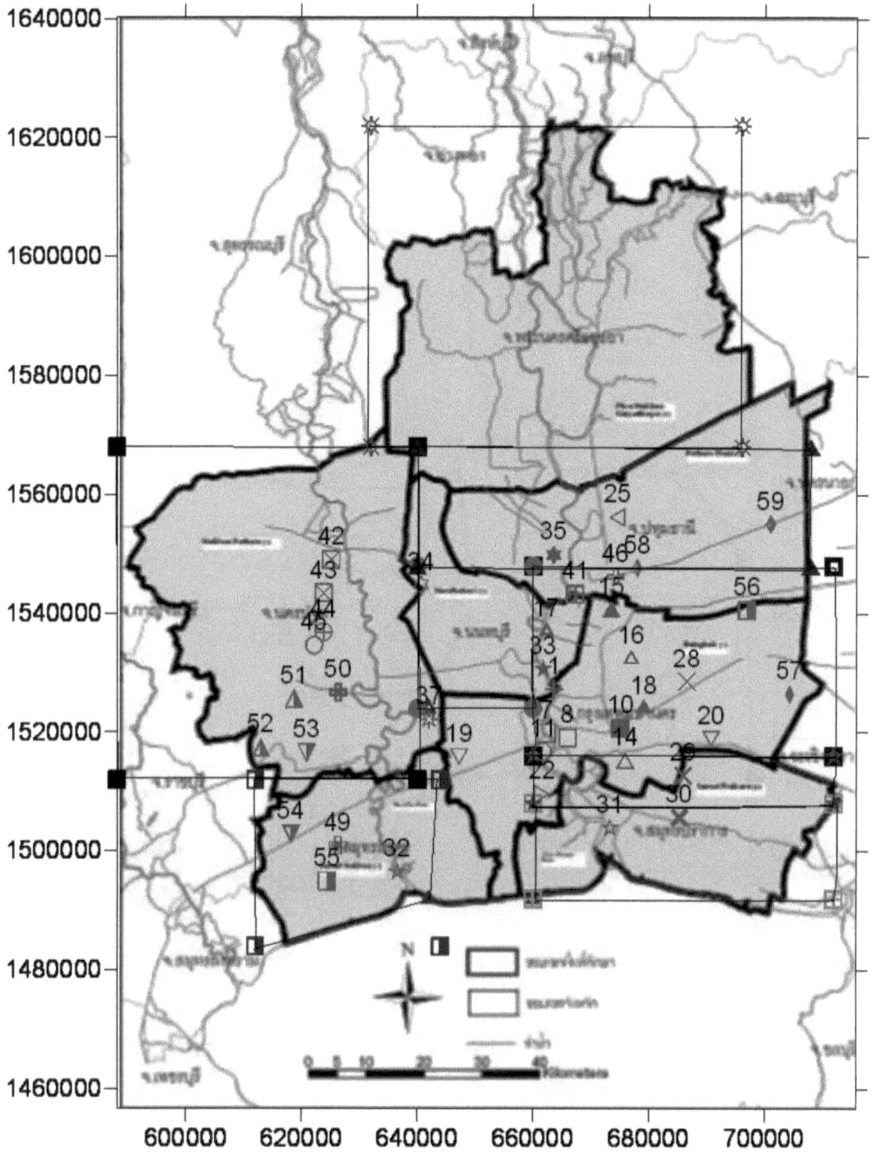

Fig. 8.3 Subsidence stations in the various zones

8.3.2.1 General Data Transfer

In the next chapter, we detail the numerical scheme that was used to solve the Biot equations.

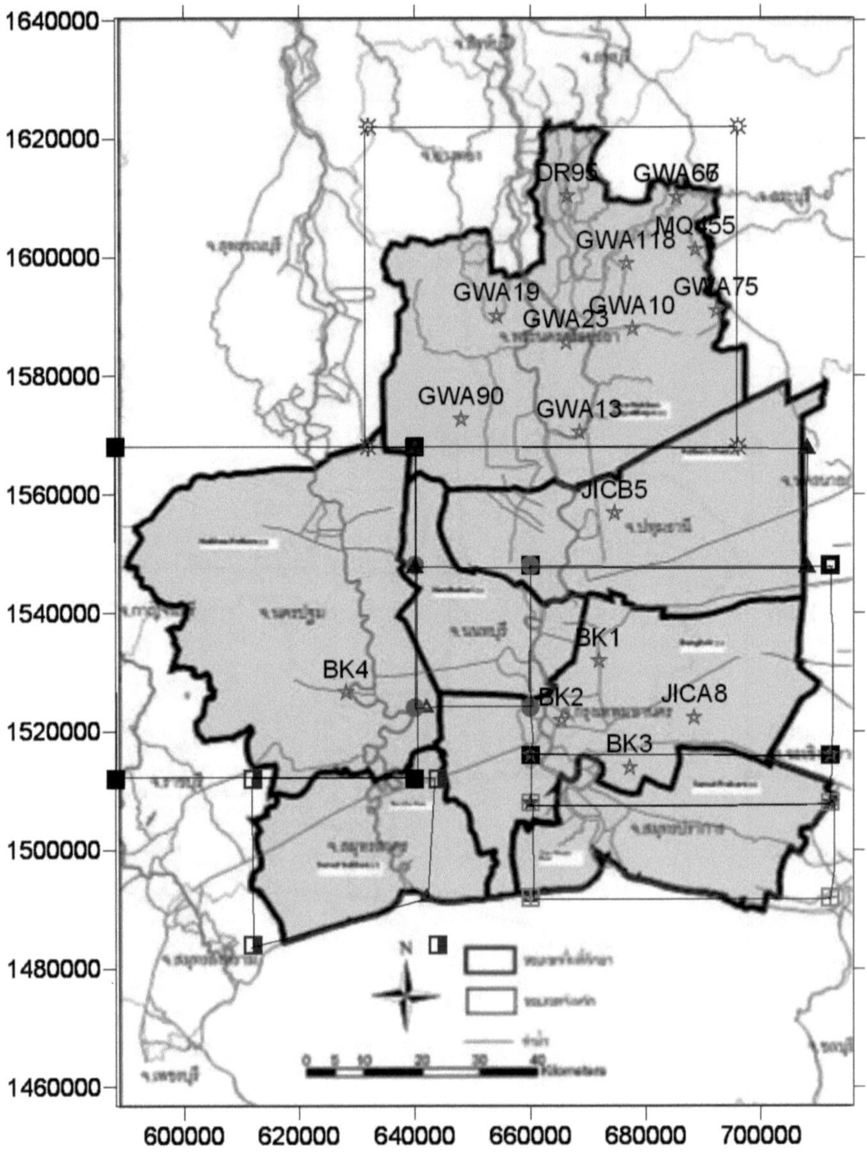

Fig. 8.4 Water pressure wells in the Bangkok aquifer: Layer1

8.3.2.2 The Hydro-Geological Model

In Figs. 8.11, 8.12, 8.13, and 8.14, we describe the assumed hydraulic connections between the different cells and the external boundary conditions.

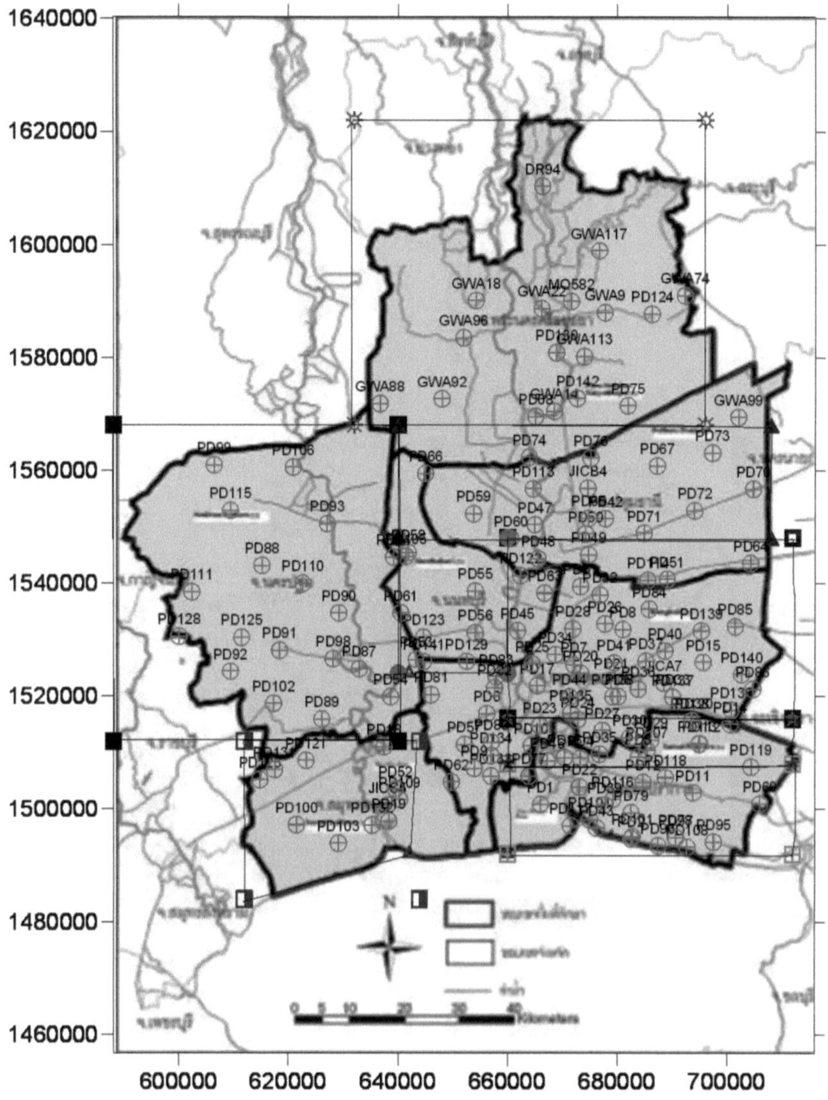

Fig. 8.5 Water pressure wells in the PD aquifer: Layer2

8.3.2.3 Boundary Conditions

We assume constant boundary conditions on the western, the northern, and the eastern part of the model. Values of the constant were taken from the MODFLOW Argus, a model developed at the Water Resources Department at the Chula University (Fig. 8.15).

The following table summarizes the boundary conditions taken in the model (Table 8.2):

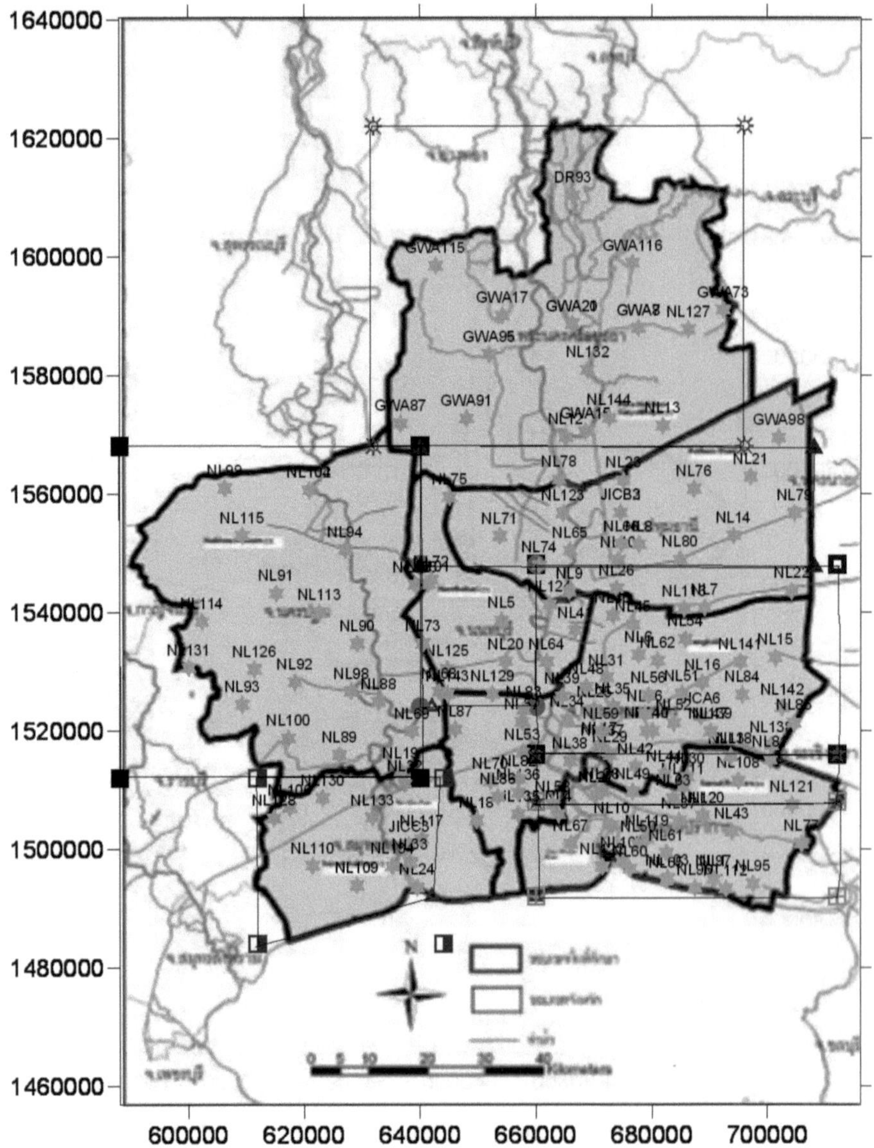

Fig. 8.6 Water pressure wells in the PD aquifer: Layer3

8.3.3 The Calculation of Stress Changes for Different Types of Piles Foundations

For piles foundations, we use the integration of the Boussinesq solution for computing the stress on the soil resulting from a load of the building (see Cours de mecanique des sols 1980).

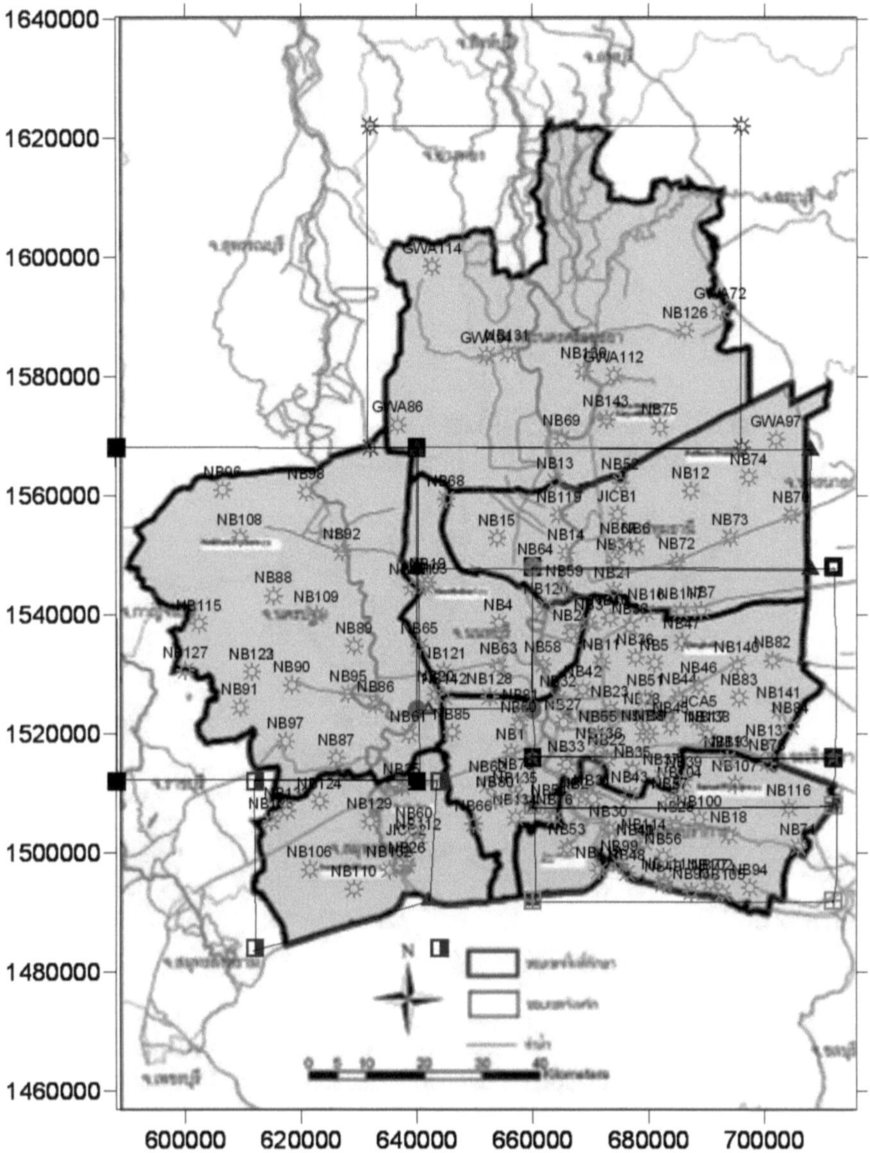

Fig. 8.7 Water pressure wells in the PD aquifer: Layer4

The following table summarizes the calculation done. Also, in the input data file of the software called CompLoading.xlsx, one can find all the various computation done for all the cells (Table 8.3).

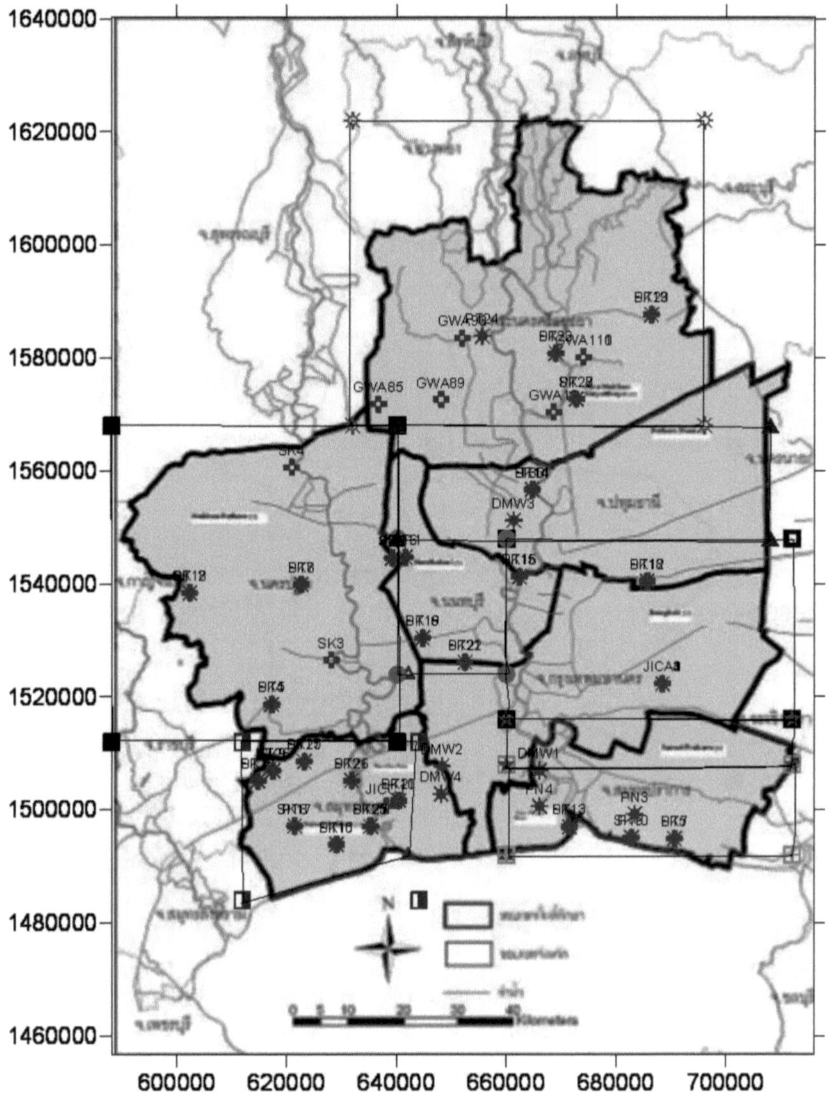

Fig. 8.8 Water pressure wells in Layer 5

8.3.4 *The Land-Use Model*

The problem that remains to be solved is the following:

For a given cell or region, how can we estimate the density of the building and the types of foundations that were in 1978?

Also, for a given cell or region, how can we estimate the density change of the buildings and the types of foundations that occur each year?

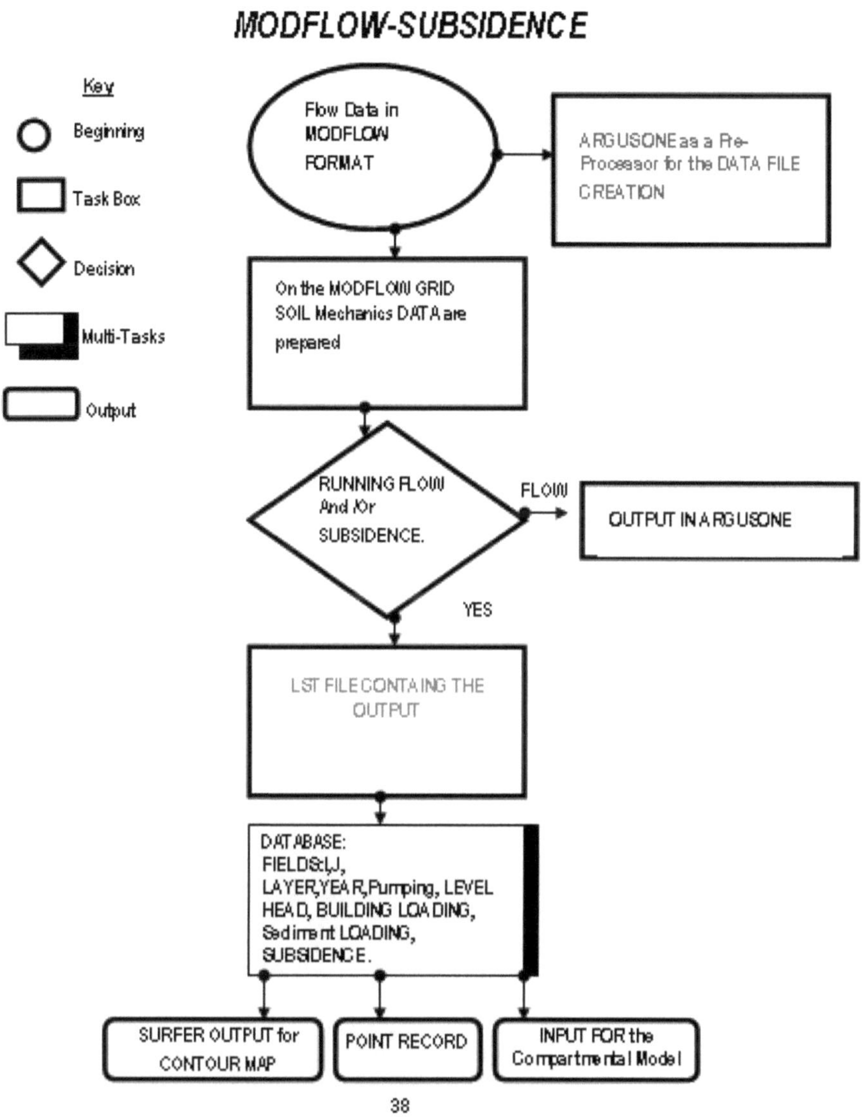

Fig. 8.9 A flow-chart of the software used for the MODFLOW-Biot model

Moreover, for a given cell or region, how can we estimate the density change of the buildings and the types of foundations that will occur in the future?

In order to answer these questions, we worked together with the Department of Land Use and the Department of the Geographic Information Systems of the Chulalongkorn University.

The method used was to digitalize data from different years and to insert it as a GIS layer.

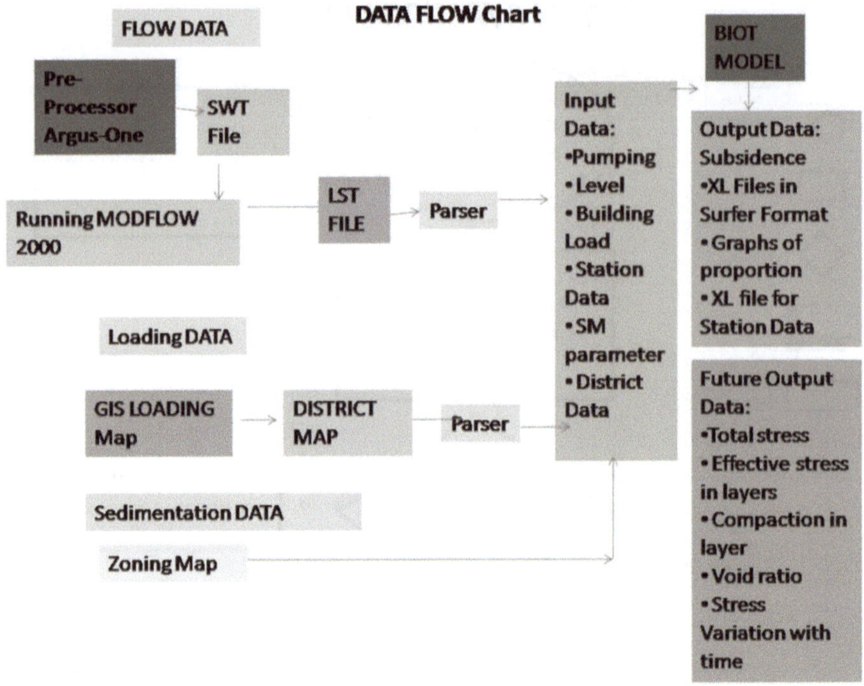

Fig. 8.10 A data flow-chart

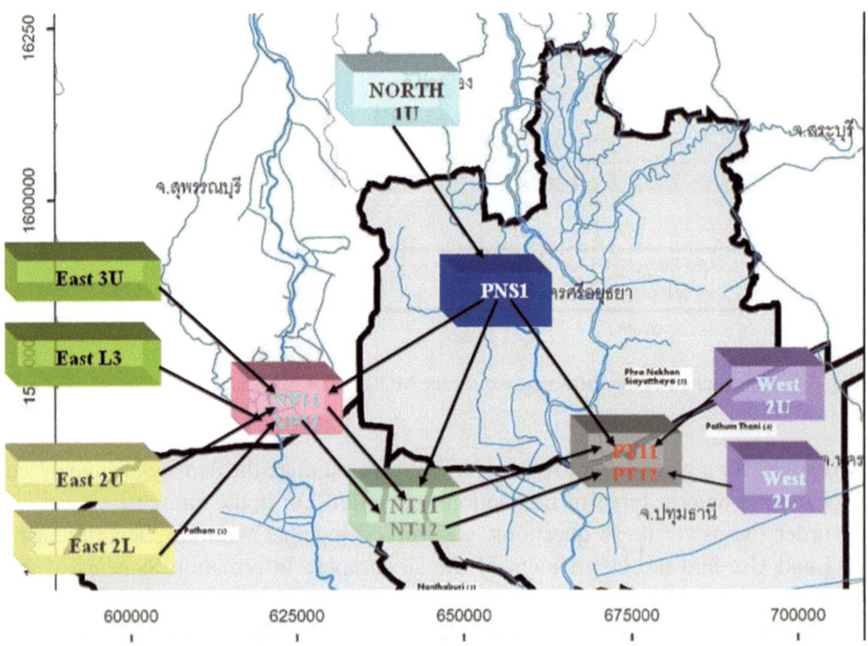

Fig. 8.11 A hydro-geological model, northern region

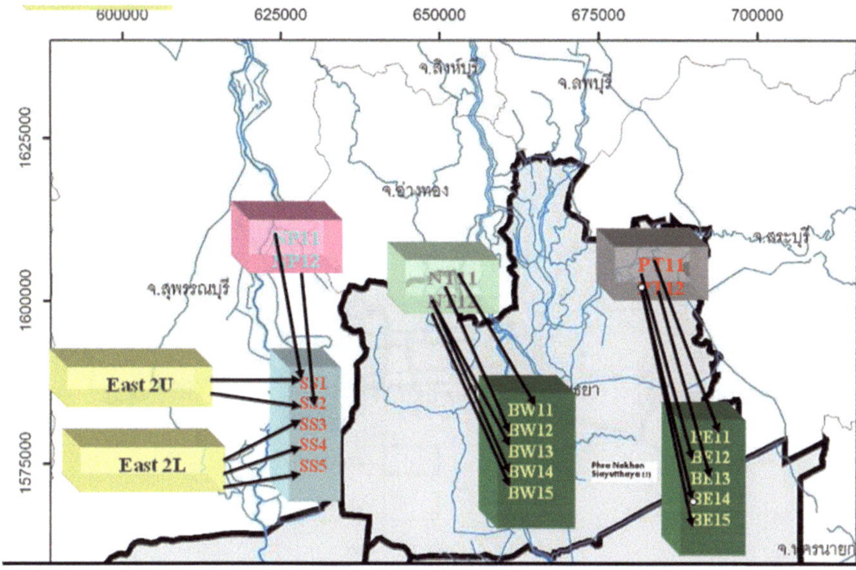

Fig. 8.12 A hydro-geological model center

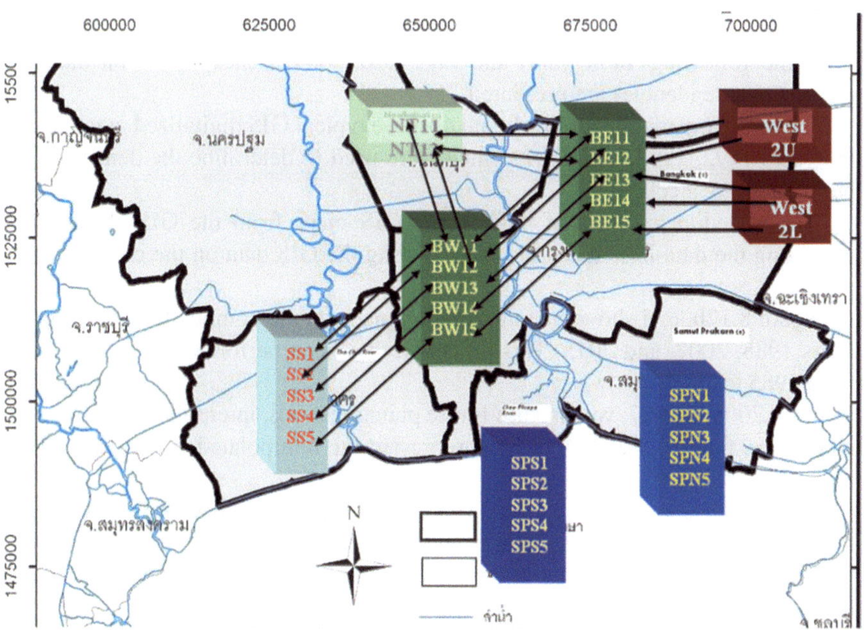

Fig. 8.13 A hydro-geological model center

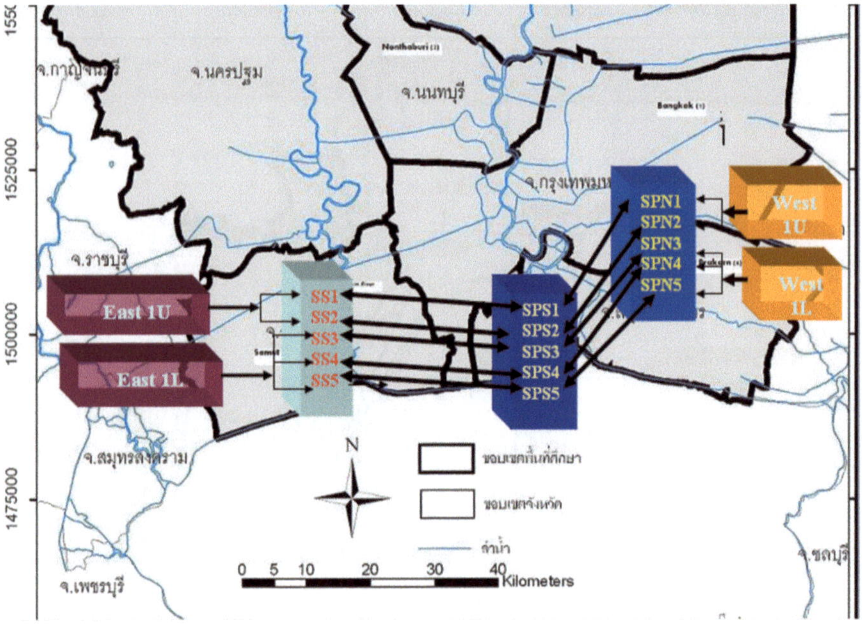

Fig. 8.14 A hydro-geological model

In the following Table (8.4) and Fig. 8.16, some basic population data and urbanization tendencies are presented.

The following maps 41.1, 42.1, and 43.1 are typical GIS digitalized map data for the years 1992, 1997, and 2007, which were used to determine the density of the building on each cell.

The procedure consists of reading the GIS maps from the GIS Department, extracting the data-base data, and interpolating the GIS data on the computational grid.

Figure 8.17b, c, d show the interpolation maps on the computational grid in the years 1985, 2007, and 2015. This procedure has been performed for all the years from 1985 to 2007.

From 2008 to 2027, we also use future planning cases, interpolated on the grid.

In these figures, the contour lines represent the interpolated values assigned in the cells.

From the data collected in 1990, we can consider three types of districts having different percentages of high, middle, and residential buildings.

The following pictures describe typical districts considered in this work, based on the work presented by the Land Use Department (Table 8.5).

Fig. 8.15 Boundary conditions for the compartmental model

Table 8.2 A table of boundary conditions for the compartmental model

	Aquifer	Formation	01/09/ 1990	01/09/ 1991	01/09/ 1992	01/09/ 1993	01/09/ 1994	01/09/ 1995	01/09/ 1996	01/09/ 1997	01/09/ 1998	01/09/ 1999	01/09/ 2000	01/09/ 2001	01/09/ 2002	01/09/ 2003	01/09/ 2004	01/09/ 2005	01/09/ 2006	01/09/ 2007
1	XX	East1U	3.1	3.1	3.1	3.1	3.1	3.1	3.1	3.1	3.1	3.1	3.1	3.1	3.1	3.1	3.1	3.1	3.1	3.1
2	XX	East1L	5	5	5	5	5	5	5	5	5	5	5	5	5	5	5	5	5	5
3	XX	East2U	5	5	5	5	5	5	5	5	5	5	5	5	5	5	5	5	5	5
4	XX	East2L	8	8	8	8	8	8	8	8	8	8	8	8	8	8	8	8	8	8
5	XX	East3U	7	7	7	7	7	7	7	7	7	7	7	7	7	7	7	7	7	7
6	XX	East3L	21	21	21	21	21	21	21	21	21	21	21	21	21	21	21	21	21	21
7	XX	North1U	25	25	25	25	25	25	25	25	25	25	25	25	25	25	25	25	25	25
8	XX	West1U	3	3	3	3	3	3	3	3	3	3	3	3	3	3	3	3	3	3
9	XX	West1L	1	1	1	1	1	1	1	1	1	1	1	1	1	1	1	1	1	1
10	XX	West2U	4.7	4.7	4.7	4.7	4.7	4.7	4.7	4.7	4.7	4.7	4.7	4.7	4.7	4.7	4.7	4.7	4.7	4.7
11	XX	West2L	5.5	5.5	5.5	5.5	5.5	5.5	5.5	5.5	5.5	5.5	5.5	5.5	5.5	5.5	5.5	5.5	5.5	5.5

Table 8.3 A pressure table for the different types of foundations

PD (%)	BKD (%)	BKK clay (%)	Pressure (T/m^2)	Loading table
0	0	100	2	1. Embankments
0	0	100	10	2. Residential buildings: up to 10 stories
0	25	75	35	3. Medium buildings:10–30 stories
15	35	50	70	4. High buildings: more than 30 stories

Table 8.4 The area and population in the Bangkok region

Province	Area	Population	%	Density
GB	7,758.10	9,785,136	100.00	1,261
BANGKOK	1,565.20	5,658,953	57.83	3,615
NAKHON PRATHOM	2,168.30	808,961	8.27	373
NONTHABURI	622.30	972,280	9.94	1,562
PATHUM THANI	1,525.90	815,402	8.33	534
SAMUT PRAKARN	1,004.10	1,077,523	11.01	1,073
SAMUT SAKHORN	872.30	452,017	4.62	518

8.3.5 The Spatial Distribution of the Loading

Let us summarize the method of computation of the loading due to building, as follows:

1. For a given cell, we associate a district type (let us say i);
2. From the GIS data, we have an estimate of the global weight acting on the surface of the cell P_i; and
3. Separating between the type of foundations

$$P_i = {\alpha_i}^1 P_i^1 + {\alpha_i}^2 P_i^2 + {\alpha_i}^3 P_i^3 + {\alpha_i}^4 P_i^4 \tag{8.1}$$

   The parameters $\alpha_i^j \;\; j = 1, \ldots, 4$ are the coefficients in the different districts tables. Each $P_i^j \;\; j = 1, \ldots, 4$ corresponds to the global loads in the cell i acting via a foundation of type j.
4. Moreover, using Fig. 8.17, for a given square meter, the following relations hold:

$$P_i^2 = \frac{10}{2} P_i^1 = 5\, P_i^1; \tag{8.2a}$$

$$P_i^3 = \frac{35}{2} P_i^1 = 17.5 P_i^1; \tag{8.2b}$$

$$P_i^4 = \frac{70}{2} P_i^1 = 35\, P_i^1 \tag{8.2c}$$

a

CHANGES OF BANGKOK AND THE VICINITY LANDUSE

LAND USE	AREA	
	SUM	%
agricultural	4,996.15	65.31
cattle	11.42	0.15
Forest	9.45	0.12
government	155.03	2.03
industrial	32.76	0.43
Other	403.16	5.27
Town and city	1,951.68	25.51
Water	90.82	1.19
TOTAL	7,650.47	100

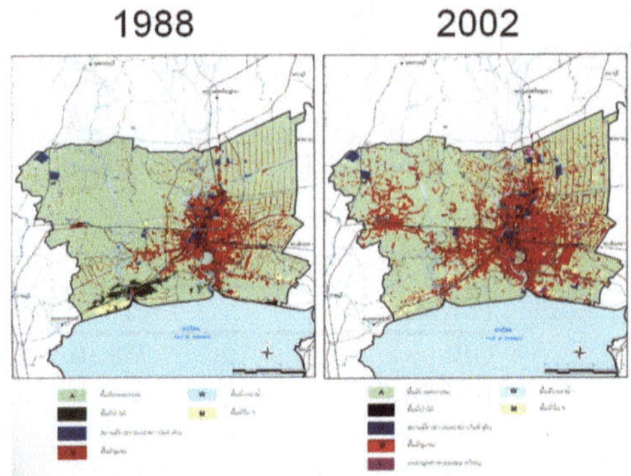

Source ;
Department of Public Work and Town and Country Planning

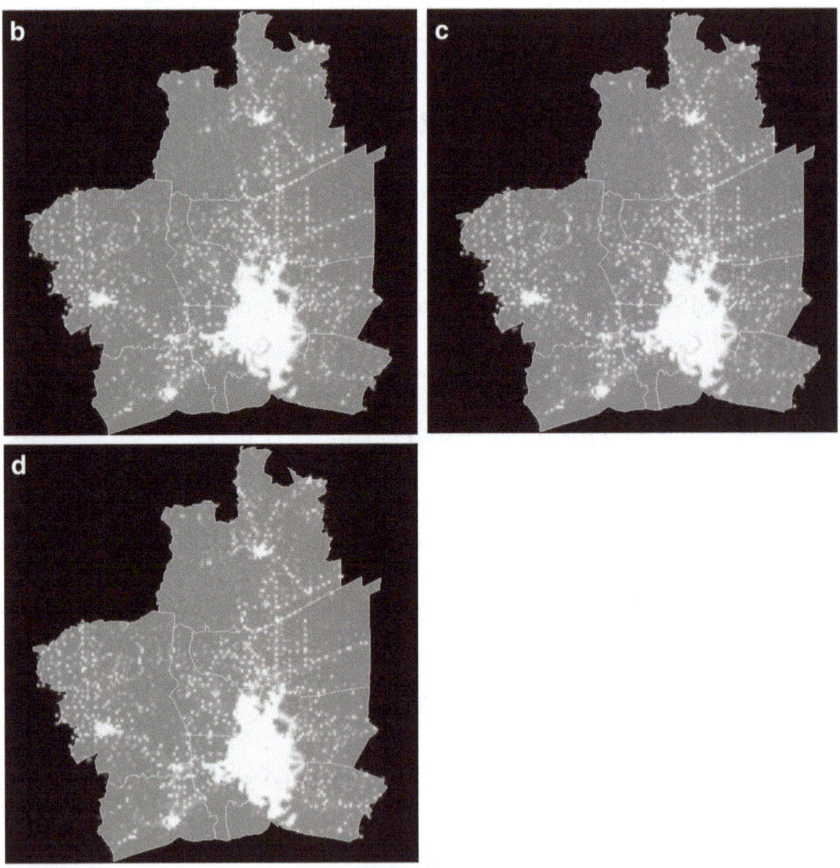

Fig. 8.16 (**a**) Land use and history (**b**) A land-use map in 1992 (**c**) A land-use map in 1997 (**d**) A land-use map in 2007

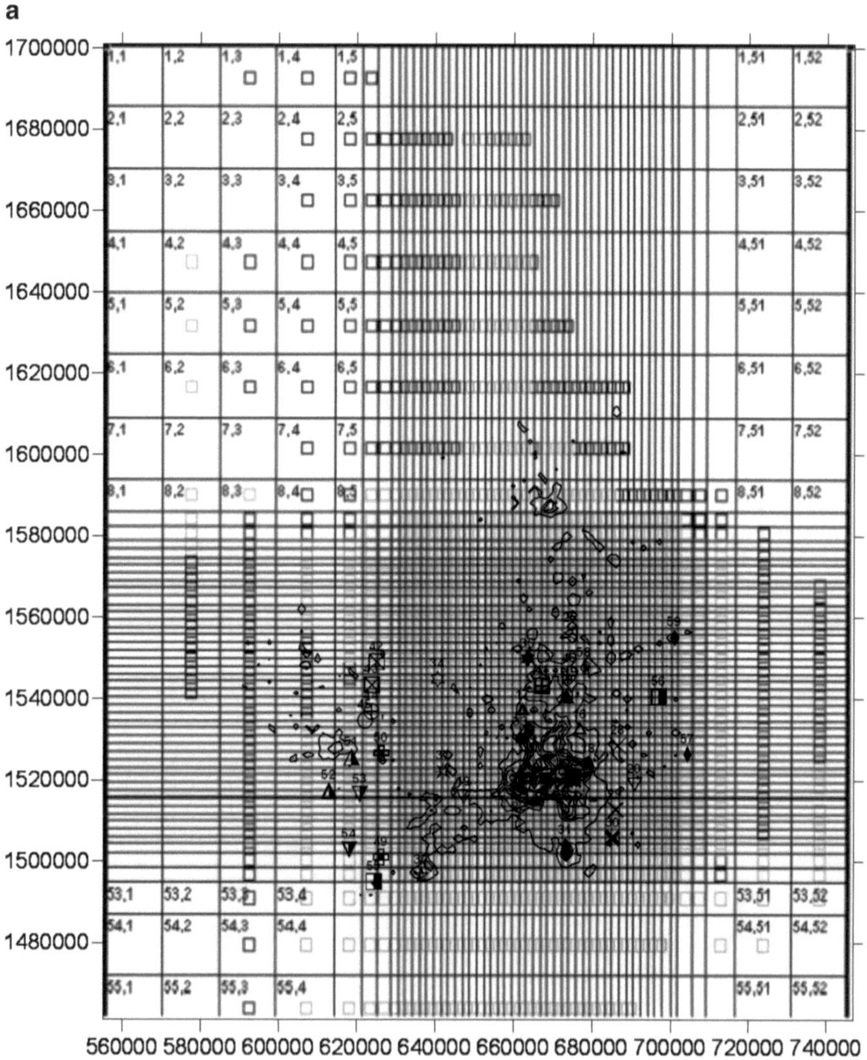

Fig. 8.17 (**a**) A building interpolation map from a land-use map in 1985 (**b**) A building interpolation map from a land-use map in 2007 (**c**) A building interpolation map from a land-use map in 2015

5. Solving for $P_i^{\,1}$

$$P_i^{\,1} = \frac{P_i}{\alpha_i^{\,1} + 5\alpha_i^{\,2} + 17.5\alpha_i^{\,3} + 35\alpha_i^{\,4}} \tag{8.3a}$$

6. Thus, finally:

$$P_i^{\,1} = \frac{P_i}{\alpha_i^{\,1} + 5\alpha_i^{\,2} + 17.5\alpha_i^{\,3} + 35\alpha_i^{\,4}}; \tag{8.3a}$$

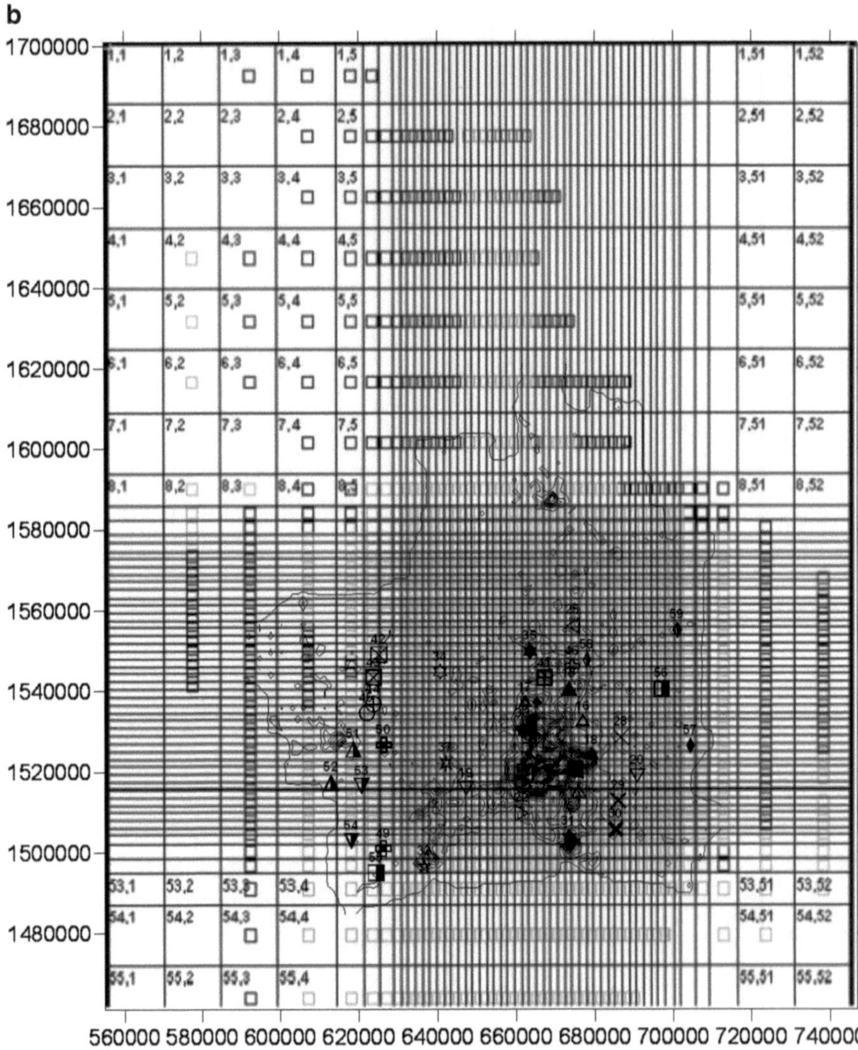

Fig. 8.17 (continued)

$$P_i{}^2 = \frac{5P_i}{\alpha_i{}^1 + 5\alpha_i{}^2 + 17.5\alpha_i{}^3 + 35\alpha_i{}^4}; \tag{8.3b}$$

$$P_i{}^3 = \frac{17.5P_i}{\alpha_i{}^1 + 5\alpha_i{}^2 + 17.5\alpha_i{}^3 + 35\alpha_i{}^4}; \tag{8.3c}$$

$$P_i{}^1 = \frac{35P_i}{\alpha_i{}^1 + 5\alpha_i{}^2 + 17.5\alpha_i{}^3 + 35\alpha_i{}^4}; \tag{8.3d}$$

In the following table, we summarize the proportion of district types for each compartmental cell (Table 8.6).

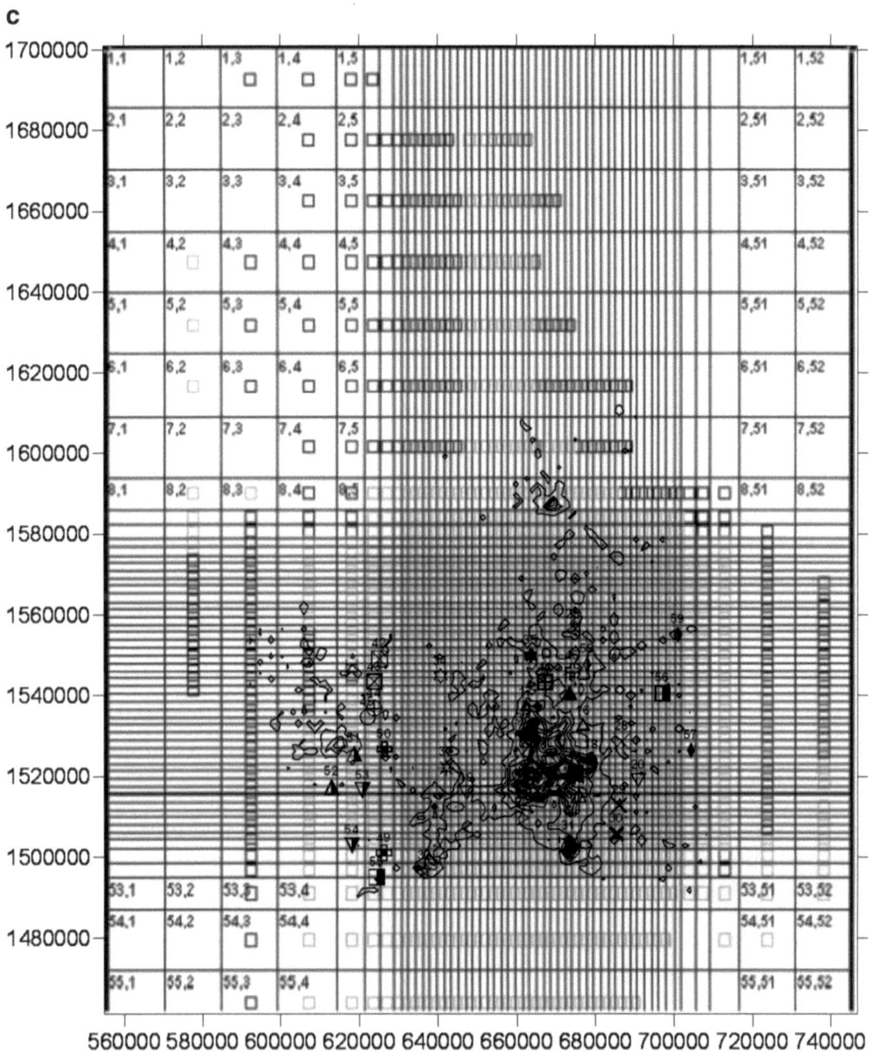

Fig. 8.17 (continued)

Table 8.5 District 3

Embankments	% Low building	% Medium building	% High building	District 3
40	40	15	5	

In summary, the computation of the loadings due to building was done, as follows:

(a) For a given compartmental cell, we first use Fig. 8.17 for computing the effective area for each type of districts;

Table 8.6 An effective area for the loading of buildings (cells to district types in 1990)

District3(%)	District2(%)	District1(%)	Area (km^2)	Model cell
0	25	75	730	**BE**
0	30	70	835	**BW**
100	0	0	1,600	**PNS**
65	30	5	622	**NT**
73	22	5	2168	**NP**
40	30	30	872	**SSS**
60	30	10	1,525	**PT**
50	30	25	504	**SPS**
35	35	30	500	**SPN**

(b) For each type of district, we use Fig. 8.18a, b, c for estimating the type of loadings;
(c) Use Fig. 8.19 to estimate the total loading and
(d) Compute the average pressure for each cell.

8.3.5.1 The Data Description

For each cell of the model, the following data were prepared for a calibration period from 1990 to 2008 with a time step of 1 year:

- The total volume of pumping rates, for each cell in each hydrologic year of the calibration period and validation period; and
- The pumping wells data.

Three categories of well data have been collected and processed within the framework of the work:

1. Fixed data which describe geometric and hydraulic characteristics of pumping wells in all cells. The fixed data set includes the following items for each of the two aquifers: Topographic elevation of pumping wells in a cell (average and standard deviation) Elevation of the bottom of pumping wells in a cell (average and standard deviation) Elevation of the top of pumping wells with the aquifer in a cell (average and standard deviation) Specific drawdown of pumping wells in a cell (average and standard deviation);
2. The total number of wells and number of pumping wells in each cell; and
3. The water level (historical) data for the calibration of the model, computed as the average value all the well measurements for each cell and measured in April of each year. An alternative method also checked is to associate a representative well for each cell, assuming that the head variation inside a given cell is negligible. Both methods gave similar data estimations.

a

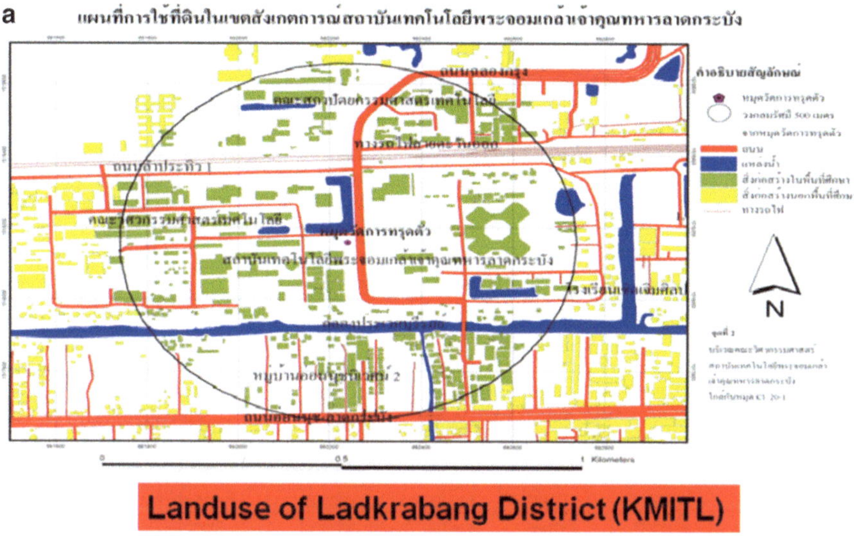

b

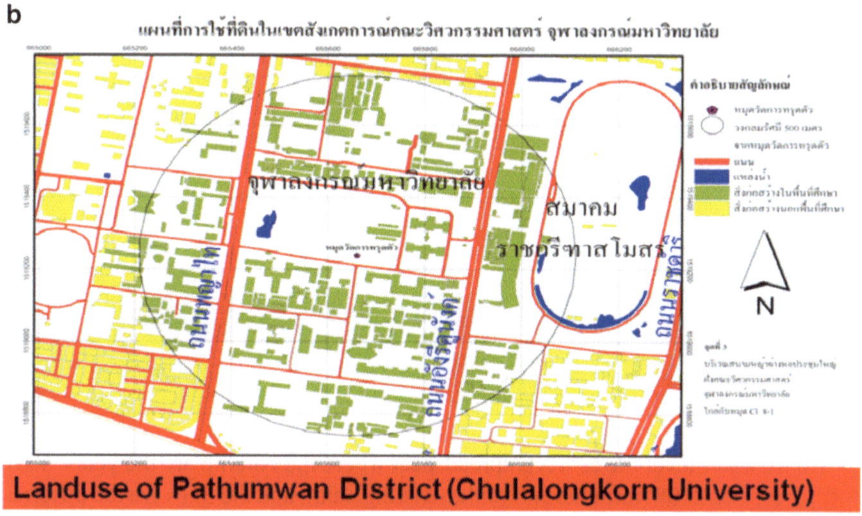

Fig. 8.18 (**a**) Land use of the Ladkrabang District (**b**) Land use of the Pathumwan District

1. In short, dynamic data include the historic series of values of the water level, the pumping rate, the recharge, the springs discharge, and the depth of rain for each cell, as described earlier.

   1.1 Water level file as a XL spreadsheet where the column are the years: 1990 …… 2008 (till the end of the existing data) and the lines are the different regions. For each region, we write the water level of a typical well in the region. (we assume here that the water levels are similar in a given region.)

	Xmin	Xmax	Ymin	YMAX	Layer	Name
BE11	642000	660000	1492000	1524000	1.00	bangkok east
BE12	642000	660000	1492000	1524000	2.00	bangkok east
BE13	642000	660000	1492000	1524000	3.00	bangkok east
BE14	642000	660000	1492000	1524000	4.00	bangkok east
BE15	642000	660000	1492000	1524000	5,6,7,8,9	bangkok east
BW11	660000	712000	1516000	1548000	1.00	Bangkok West
BW12	660000	712000	1516000	1548000	2.00	Bangkok West
BW13	660000	712000	1516000	1548000	3.00	Bangkok West
BW14	660000	712000	1516000	1548000	4.00	Bangkok West
BW15	660000	712000	1516000	1548000	5,6,7,8,9	Bangkok West
NP11	588000	640000	1512000	1568000	1.00	Nakhon Pathom
NP12	588000	640000	1512000	1568000	2.00	Nakhon Pathom
NT11	640000	660000	1524000	1548000	1.00	NanThaburi
NT12	640000	660000	1524000	1548000	2.00	NanThaburi
PT11	640000	708000	1548000	1568000	1.00	Phathum Thani
PT12	640000	708000	1548000	1568000	2.00	Phathum Thani
PNS1	632000	696000	1568000	1622000	1.00	Phra Nakhon Siayuttaya
SPS1	660000	712000	1492000	1508000	1.00	Samut Prakarn South
SPS2	660000	712000	1492000	1508000	2.00	Samut Prakarn South
SPS3	660000	712000	1492000	1508000	3.00	Samut Prakarn South
SPS4	660000	712000	1492000	1508000	4.00	Samut Prakarn South
SPS5	660000	712000	1492000	1508000	5,6,7,8,9	Samut Prakarn South
SPN1	660000	712000	1508000	1516000	1.00	Samut Prakarn North
SPN2	660000	712000	1508000	1516000	2.00	Samut Prakarn North
SPN3	660000	712000	1508000	1516000	3.00	Samut Prakarn North
SPN4	660000	712000	1508000	1516000	4.00	Samut Prakarn North
SPN5	660000	712000	1508000	1516000	5,6,7,8,9	Samut Prakarn North
SS1	612000	644000	1484000	1512000	1.00	Samut Sakhun
SS2	612000	644000	1484000	1512000	2.00	Samut Sakhun
SS3	612000	644000	1484000	1512000	3.00	Samut Sakhun
SS4	612000	644000	1484000	1512000	4.00	Samut Sakhun
SS5	612000	644000	1484000	1512000	5,6,7,8,9	Samut Sakhun

Fig. 8.19 An extension of the cell zones

1.2 A water level file as a XL spreadsheet where the column are the years: 1990 …… 2008 (until the end of the existing data), and the lines are the different regions. For each region, we write the external head at the boundary in the north.

2. A pumpage file as a spreadsheet where the column are the years: 1990 …… 2008 (until the end of the existing data), and the lines are the different regions. For each region, we write the annual pumping in the region.
3. In the zone of replenishment, we build a rain file where the columns are the years: 1990 …… 2007 (until the end of the existing data), and the lines are the different regions. For each region, we write the annual rain in the region.
4. A subsidence file as a spreadsheet where the columns are the years: 1990 …… 2007 (until the end of the existing data), and the lines are the different regions. For each region, we write the measurement of the subsidence in the area.

	01/09/1990	01/09/1991	01/09/1992	01/09/1993	01/09/1994	01/09/1995	01/09/1996	01/09/1997	01/09/1998	01/09/1999	01/09/2000	01/09/2001	01/09/2002	01/09/2003	01/09/2004	01/09/2005	01/09/2006	01/09/2007
BE11	0.00	-0.10	-1.10	-1.80	-3.00	-2.40	-3.20	-3.50	-4.20	-4.30	-4.20	-5.50	-4.60	-5.20	-5.40	-6.90	-5.70	-4.90
BE12	0.00	-0.02	-0.22	-0.3	-0.6	-0.48	-0.64	-0.7	-0.84	-0.86	-0.84	-1.1	-0.92	-1.04	-1.08	-1.38	-1.14	-0.98
BE13	0.00	-0.01	-0.11	-0.15	-0.3	-0.24	-0.32	-0.35	-0.42	-0.43	-0.42	-0.55	-0.46	-0.52	-0.54	-0.69	-0.57	-0.49
BE14	0.00	-0.0005	-0.0055	-0.0075	-0.015	-0.012	-0.016	-0.0175	-0.021	-0.0215	-0.021	-0.0275	-0.023	-0.026	-0.027	-0.0345	-0.0285	-0.0245
BE15	0.00	-0.0004	-0.0044	-0.006	-0.012	-0.0096	-0.0128	-0.014	-0.0168	-0.0172	-0.0168	-0.022	-0.0184	-0.0208	-0.0216	-0.0276	-0.0228	-0.0196
BW11	0.00	-0.1	-4.4	-7.1	-11.7	-13	-17.2	-20.8	-25.1	-28	-27	-31.9	-32.8	-35.3	-36.3	-37.7	-37.2	-39.2
BW12	0.00	-0.02	0.5	0.48	0.6	0.58	0.56	0.54	0.52	0.5	0.48	0.46	0.44	0.42	0.4	0.38	0.36	0.34
BW13	0.00	-0.01	-0.44	-0.71	-1.17	-1.3	-1.72	-2.08	-2.51	-2.8	-2.7	-3.19	-3.28	-3.53	-3.63	-3.77	-3.72	-3.92
BW14	0.00	-0.0005	-0.022	-0.0355	-0.0585	-0.065	-0.086	-0.104	-0.1255	-0.14	-0.135	-0.1595	-0.164	-0.1765	-0.1815	-0.1885	-0.186	-0.196
BW15	0.00	-0.0004	-0.0176	-0.0284	-0.0468	-0.052	-0.0688	-0.0832	-0.1004	-0.112	-0.108	-0.1276	-0.1312	-0.1412	-0.1452	-0.1508	-0.1488	-0.1568
NP11	0.00	0	0	0	0	0	0	0	0	-1.9	-0.8	-3.9	-4.6	-4.4	-5.9	-6.6	-8.3	-8.9
NP12	0.00	0	0	0	0	0	0	0	0	-0.38	-0.12	-0.78	-0.92	-0.88	-1.18	-1.32	-1.66	-1.78
NT11	0.00	-0.9	-1.9	-1.6	-3.3	-3.6	-4.3	-4.9	-6	-5.8	-6.5	-5.9	-4.9	-6.1	-6.1	-7	-6.7	-5.9
NT12	0.00	-0.18	-0.38	-0.32	-0.66	-0.72	-0.86	-0.98	-1.2	-1.16	-1.1	-1.18	-0.98	-1.22	-1.22	-1.4	-1.34	-1.18
PT11	0.00	-5.5	-7.5	-8.9	-9.2	-10.1	-11.4	-11.4	-11.5	-11.4	-11.5	-13	-10.5	-12.5	-12.7	-13.9	-11.4	-11.9
PT12	0.00	-1.1	-1.5	-1.78	-1.84	-2.02	-2.28	-2.28	-2.3	-2.28	-2.3	-2.6	-2.1	-2.5	-2.54	-2.78	-2.28	-2.38
PNS1	0.00	0	0	0	0	0	0	0	0	0	-0.2	-1	-1.4	-1.7	-1.8	-3.6	-3.3	-4.7
SPS1	0.00	-1.6	-5.6	-7	-9.5	-11.6	-14.1	-15.6	-17.6	-20.2	-20.3	-21	-21.9	-23	-24.4	-27.2	-28.1	-28.1
SPS2	0.00	-0.32	-1.1	-1.4	-1.9	-2.32	-2.82	-3.12	-3.52	-4.04	-4.06	-4.2	-4.38	-4.6	-4.88	-5.44	-5.62	-5.62
SPS3	0.00	-0.16	-0.55	-0.7	-0.95	-1.16	-1.41	-1.56	-1.76	-2.02	-2.03	-2.1	-2.19	-2.3	-2.44	-2.72	-2.81	-2.81
SPS4	0.00	-0.008	-0.0275	-0.035	-0.0475	-0.058	-0.0705	-0.078	-0.088	-0.101	-0.1015	-0.105	-0.1095	-0.115	-0.122	-0.136	-0.1405	-0.1405
SPS5	0.00	-0.0064	-0.022	-0.028	-0.038	-0.0464	-0.0564	-0.0624	-0.0704	-0.0808	-0.0812	-0.084	-0.0876	-0.092	-0.0976	-0.1088	-0.1124	-0.1124
SPN1	0.00	-5.1	-11.2	-15.3	-24.4	-32.3	-39.1	-44.4	-49.9	-51.7	-56	-58.8	-60.7	-63.1	-64.4	-66.9	-68.5	-69.7
SPN2	0.00	-1.02	-2.24	-3.06	-4.88	-6.46	-7.82	-8.88	-9.98	-10.34	-11.2	-11.76	-12.14	-12.62	-12.88	-13.38	-13.72	-13.94
SPN3	0.00	-0.51	-1.12	-1.53	-2.44	-3.23	-3.91	-4.44	-4.99	-5.17	-5.6	-5.88	-6.07	-6.31	-6.44	-6.69	-6.86	-6.97
SPN4	0.00	-0.0255	-0.056	-0.0765	-0.122	-0.1615	-0.1955	-0.222	-0.2495	-0.2585	-0.28	-0.294	-0.3035	-0.3155	-0.322	-0.3345	-0.343	-0.3485
SPN5	0.00	-0.0204	-0.0448	-0.0612	-0.0976	-0.1292	-0.1564	-0.1776	-0.1996	-0.2068	-0.224	-0.2352	-0.2428	-0.2524	-0.2576	-0.2676	-0.2744	-0.2788
SS1	0.00	0	0	0	0	0	0	0	0	-6.8	-8.1	-10.9	-13.5	-15.4	-18.9	-21.8	-24.4	-25.2
SS2	0.00	0	0	0	0	0	0	0	0	-1.36	-1.62	-2.18	-2.7	-3.08	-3.78	-4.36	-4.88	-5.04
SS3	0.00	0	0	0	0	0	0	0	0	-0.68	-0.81	-1.09	-1.35	-1.54	-1.89	-2.18	-2.44	-2.52
SS4	0.00	0	0	0	0	0	0	0	0	-0.034	-0.0405	-0.0545	-0.0675	-0.077	-0.0945	-0.109	-0.122	-0.126
SS5	0.00	0	0	0	0	0	0	0	0	-0.0272	-0.0324	-0.0436	-0.054	-0.0616	-0.0756	-0.0872	-0.0976	-0.1008

Fig. 8.20 Z displacement from 1990

	01/09/1990	01/09/1991	01/09/1992	01/09/1993	01/09/1994	01/09/1995	01/09/1996	01/09/1997	01/09/1998	01/09/1999	01/09/2000	01/09/2001	01/09/2002	01/09/2003	01/09/2004	01/09/2005	01/09/2006	01/09/2007	
BE11	80000.00	81200.00	82418.00	83654.27	84909.08	86182.72	87475.46	88787.59	90119.41	91471.20	92843.27	94235.91	95649.45	97084.20	98540.46	100018.57	101518.84	103041.63	104587.25
BE12	0.00	0.00	0.00	0.00	0.00	0.00	0.00	0.00	0.00	0.00	0.00	0.00	0.00	0.00	0.00	0.00	0.00	0.00	0.00
BE13	0.00	0.00	0.00	0.00	0.00	0.00	0.00	0.00	0.00	0.00	0.00	0.00	0.00	0.00	0.00	0.00	0.00	0.00	0.00
BE14	0.00	0.00	0.00	0.00	0.00	0.00	0.00	0.00	0.00	0.00	0.00	0.00	0.00	0.00	0.00	0.00	0.00	0.00	0.00
BE15	0.00	0.00	0.00	0.00	0.00	0.00	0.00	0.00	0.00	0.00	0.00	0.00	0.00	0.00	0.00	0.00	0.00	0.00	0.00
BW11	80000.00	81200.00	82418.00	83654.27	84909.08	86182.72	87475.46	88787.59	90119.41	91471.20	92843.27	94235.91	95649.45	97084.20	98540.46	100018.57	101518.84	103041.63	104587.25
BW12	0.00	0.00	0.00	0.00	0.00	0.00	0.00	0.00	0.00	0.00	0.00	0.00	0.00	0.00	0.00	0.00	0.00	0.00	0.00
BW13	0.00	0.00	0.00	0.00	0.00	0.00	0.00	0.00	0.00	0.00	0.00	0.00	0.00	0.00	0.00	0.00	0.00	0.00	0.00
BW14	0.00	0.00	0.00	0.00	0.00	0.00	0.00	0.00	0.00	0.00	0.00	0.00	0.00	0.00	0.00	0.00	0.00	0.00	0.00
BW15	0.00	0.00	0.00	0.00	0.00	0.00	0.00	0.00	0.00	0.00	0.00	0.00	0.00	0.00	0.00	0.00	0.00	0.00	0.00
NP11	1200.00	1224.00	1248.48	1273.45	1298.92	1324.90	1351.39	1378.42	1405.98	1434.11	1462.79	1492.05	1521.89	1552.33	1583.37	1615.04	1647.34	1680.29	1713.90
NP12	0.00	0.00	0.00	0.00	0.00	0.00	0.00	0.00	0.00	0.00	0.00	0.00	0.00	0.00	0.00	0.00	0.00	0.00	0.00
NT11	19200.00	19584.00	19975.68	20375.19	20782.70	21198.35	21622.32	22054.76	22495.86	22945.78	23404.69	23872.79	24350.24	24837.25	25333.99	25840.67	26357.49	26884.64	27422.33
NT12	0.00	0.00	0.00	0.00	0.00	0.00	0.00	0.00	0.00	0.00	0.00	0.00	0.00	0.00	0.00	0.00	0.00	0.00	0.00
PT11	5600.00	5712.00	5826.24	5942.76	6061.62	6182.85	6306.51	6432.64	6561.29	6692.52	6826.37	6962.90	7102.15	7244.20	7389.08	7536.86	7687.60	7841.35	7998.18
PT12	5600.00	5712.00	5826.24	5942.76	6061.62	6182.85	6306.51	6432.64	6561.29	6692.52	6826.37	6962.90	7102.15	7244.20	7389.08	7536.86	7687.60	7841.35	7998.18
PNS1	0.80	0.82	0.83	0.85	0.87	0.88	0.90	0.92	0.94	0.96	0.98	0.99	1.01	1.03	1.06	1.08	1.10	1.12	1.14
SPS1	19200.00	19584.00	19975.68	20375.19	20782.70	21198.35	21622.32	22054.76	22495.86	22945.78	23404.69	23872.79	24350.24	24837.25	25333.99	25840.67	26357.49	26884.64	27422.33
SPS2	0.00	0.00	0.00	0.00	0.00	0.00	0.00	0.00	0.00	0.00	0.00	0.00	0.00	0.00	0.00	0.00	0.00	0.00	0.00
SPS3	0.00	0.00	0.00	0.00	0.00	0.00	0.00	0.00	0.00	0.00	0.00	0.00	0.00	0.00	0.00	0.00	0.00	0.00	0.00
SPS4	0.00	0.00	0.00	0.00	0.00	0.00	0.00	0.00	0.00	0.00	0.00	0.00	0.00	0.00	0.00	0.00	0.00	0.00	0.00
SPS5	0.00	0.00	0.00	0.00	0.00	0.00	0.00	0.00	0.00	0.00	0.00	0.00	0.00	0.00	0.00	0.00	0.00	0.00	0.00
SPN1	19200.00	19584.00	19975.68	20375.19	20782.70	21198.35	21622.32	22054.76	22495.86	22945.78	23404.69	23872.79	24350.24	24837.25	25333.99	25840.67	26357.49	26884.64	27422.33
SPN2	0.00	0.00	0.00	0.00	0.00	0.00	0.00	0.00	0.00	0.00	0.00	0.00	0.00	0.00	0.00	0.00	0.00	0.00	0.00
SPN3	0.00	0.00	0.00	0.00	0.00	0.00	0.00	0.00	0.00	0.00	0.00	0.00	0.00	0.00	0.00	0.00	0.00	0.00	0.00
SPN4	0.00	0.00	0.00	0.00	0.00	0.00	0.00	0.00	0.00	0.00	0.00	0.00	0.00	0.00	0.00	0.00	0.00	0.00	0.00
SPN5	0.00	0.00	0.00	0.00	0.00	0.00	0.00	0.00	0.00	0.00	0.00	0.00	0.00	0.00	0.00	0.00	0.00	0.00	0.00
SS1	4800.00	4896.00	4993.92	5093.80	5195.67	5299.59	5405.58	5513.69	5623.97	5736.44	5851.17	5968.20	6087.56	6209.31	6333.50	6460.17	6589.37	6721.18	6855.58
SS2	0.00	0.00	0.00	0.00	0.00	0.00	0.00	0.00	0.00	0.00	0.00	0.00	0.00	0.00	0.00	0.00	0.00	0.00	0.00
SS3	0.00	0.00	0.00	0.00	0.00	0.00	0.00	0.00	0.00	0.00	0.00	0.00	0.00	0.00	0.00	0.00	0.00	0.00	0.00
SS4	0.00	0.00	0.00	0.00	0.00	0.00	0.00	0.00	0.00	0.00	0.00	0.00	0.00	0.00	0.00	0.00	0.00	0.00	0.00
SS5	0.00	0.00	0.00	0.00	0.00	0.00	0.00	0.00	0.00	0.00	0.00	0.00	0.00	0.00	0.00	0.00	0.00	0.00	0.00

Fig. 8.21 Loading of the buildings

5. A building file where the columns are the years: 1990 ...... 2007 (until the end of the existing data), and the lines are the different regions. For each region, we estimate the loading on the soil (Figs. 8.20, 8.21, and 8.22).

	01/09/1990	01/09/1991	01/09/1992	01/09/1993	01/09/1994	01/09/1995	01/09/1996	01/09/1997	01/09/1998	01/09/1999	01/09/2000	01/09/2001	01/09/2002	01/09/2003	01/09/2004	01/09/2005	01/09/2006	01/09/2007
BE11	90.40	117.76	102.86	112.08	121.76	84.72	98.16	106.64	93.92	140.96	140.16	114.88	93.28	103.60	107.76	109.28	107.76	109.28
BE12	20.91	27.23	23.79	25.92	28.16	19.59	22.24	24.68	21.72	32.60	32.41	26.57	21.57	23.96	24.92	25.27	24.92	25.27
BE13	1.70	2.21	1.83	2.10	2.28	1.59	1.80	2.00	1.78	2.64	2.63	2.15	1.75	1.94	2.02	2.05	2.02	2.05
BE14	5.68	8.41	6.31	6.48	7.26	5.32	5.18	6.44	5.02	8.15	7.30	7.61	6.50	6.44	6.88	8.51	6.88	8.51
BE15	90.82	134.53	100.99	103.74	116.10	85.18	82.88	102.98	80.28	130.43	116.74	121.79	104.06	102.98	110.14	136.13	110.14	136.13
BW11	22.70	33.63	25.26	25.94	29.02	21.30	20.72	25.74	20.06	32.61	29.18	30.45	26.02	25.74	27.54	34.03	27.54	34.03
BW12	24.97	37.00	27.77	28.53	31.93	23.43	22.79	28.32	22.07	35.87	32.10	33.49	28.62	28.32	30.29	37.44	30.29	37.44
BW13	0.00	0.00	0.00	0.00	0.00	0.00	0.00	0.00	0.00	0.00	0.00	0.00	0.00	0.00	0.00	0.00	0.00	0.00
BW14	9.10	9.90	6.50	6.50	5.80	6.50	8.50	7.80	7.60	9.90	8.20	5.40	6.30	7.60	7.30	9.00	7.30	9.00
BW15	14.58	16.86	13.38	13.76	15.50	15.34	17.24	18.88	16.32	19.31	20.84	20.40	15.94	18.17	17.57	17.41	17.57	17.41
NP11	3.64	4.22	3.35	3.44	3.88	3.84	4.31	4.72	4.08	4.83	5.21	5.10	3.98	4.54	4.39	4.35	4.39	4.35
NP12	6.86	7.94	6.30	6.48	7.30	7.22	8.12	8.88	7.68	9.09	9.80	9.60	7.50	8.55	8.27	8.19	8.27	8.19
NT11	1.72	1.98	1.57	1.62	1.82	1.80	2.03	2.22	1.92	2.27	2.45	2.40	1.88	2.14	2.07	2.05	2.07	2.05
NT12	0.96	0.48	0.64	0.64	1.52	1.04	1.12	1.68	0.96	1.92	1.52	1.12	1.20	1.60	1.36	1.36	1.36	1.36
PT11	0.24	0.12	0.16	0.16	0.38	0.26	0.28	0.42	0.24	0.48	0.38	0.28	0.30	0.40	0.34	0.34	0.34	0.34
PT12	0.00	0.00	0.00	0.00	0.00	0.00	0.00	0.00	0.00	0.00	0.00	0.00	0.00	0.00	0.00	0.00	0.00	0.00
PNS1	22.70	33.63	25.26	25.94	29.02	21.30	20.72	25.74	20.06	32.61	29.18	30.45	26.02	25.74	27.54	34.03	27.54	34.03
SPS1	24.97	37.00	27.77	28.53	31.93	23.43	22.79	28.32	22.07	35.87	32.10	33.49	28.62	28.32	30.29	37.44	30.29	37.44
SPS2	0.00	0.00	0.00	0.00	0.00	0.00	0.00	0.00	0.00	0.00	0.00	0.00	0.00	0.00	0.00	0.00	0.00	0.00
SPS3	9.10	9.90	6.50	6.50	5.80	6.50	8.50	7.80	7.60	9.90	8.20	5.40	6.30	7.60	7.30	9.00	7.30	9.00
SPS4	14.58	16.86	13.38	13.76	15.50	15.34	17.24	18.88	16.32	19.31	20.84	20.40	15.94	18.17	17.57	17.41	17.57	17.41
SPS5	3.64	4.22	3.35	3.44	3.88	3.84	4.31	4.72	4.08	4.83	5.21	5.10	3.98	4.54	4.39	4.35	4.39	4.35
SPN1	6.86	7.94	6.30	6.48	7.30	7.22	8.12	8.88	7.68	9.09	9.80	9.60	7.50	8.55	8.27	8.19	8.27	8.19
SPN2	1.72	1.98	1.57	1.62	1.82	1.80	2.03	2.22	1.92	2.27	2.45	2.40	1.88	2.14	2.07	2.05	2.07	2.05
SPN3	0.96	0.48	0.64	0.64	1.52	1.04	1.12	1.68	0.96	1.92	1.52	1.12	1.20	1.60	1.36	1.36	1.36	1.36
SPN4	0.24	0.12	0.16	0.16	0.38	0.26	0.28	0.42	0.24	0.48	0.38	0.28	0.30	0.40	0.34	0.34	0.34	0.34
SPN5	0.00	0.00	0.00	0.00	0.00	0.00	0.00	0.00	0.00	0.00	0.00	0.00	0.00	0.00	0.00	0.00	0.00	0.00
SS1	2	2	2	2	2	2	2	2	2	2	2	2	2	2	2	2	2	2
SS2	3	3	3	3	3	3	3	3	3	3	3	3	3	3	3	3	3	3
SS3	4	4	4	4	4	4	4	4	4	4	4	4	4	4	4	4	4	4
SS4	3	3	3	3	3	3	3	3	3	3	3	3	3	3	3	3	3	3
SS5	3	3	3	3	3	3	3	3	3	3	3	3	3	3	3	3	3	3

Fig. 8.22 A pumping table

8.3.5.2 The Calibration of the Model

The calibration of the parameters was performed for the hydro-geological model that assumes a full hydraulic connection between the cells. The software allows to open and/or to close a connection between the cells.

In the implementation of the multiple cell theory presented above to the Bangkok aquifers, the calibration process was separated into two steps, as follows:

1. Calibration of the hydrologic parameters for the flow model; and
2. Calibration of the consolidation parameters for the subsidence model.

8.3.5.3 The Parameter Calibration by the Head Estimation

The calibration of the parameters is done once the hydraulic connections between the cells have been defined. The calibrated parameters were as follows:

1. The storativity coefficients for each cell;
2. The transmissivity coefficients between connected cells; and
3. The transmissivity coefficients between the aquifer cells and the external cells.

Also, these coefficients permit the estimation of the vertical water transfer between cells of different depths in the aquifer. This estimation requires historic data on the piezometric heads in their respective adjacent cells. In previous cell models (Bachmat 1995), the terms of both the return flow and inter-aquifer water transfer could not be explicitly included in the parameter-estimation process. Trial estimates of the contribution of these terms were made by adding a constant to the flow equation of each cell or to the cells occupying the domain with an assumed inter-aquifer contact. Such an approach introduces additional parameters into the calibration process.

- An estimate of the long-term mean annual yield of the basin; and
- An estimate of the depth of the groundwater in any cell of the basin as a function of the entire set of annual pumping rates in all the cells under steady state conditions, to be supplemented with the local head loss in the cell.

The parameters described above were calibrated using the calibration algorithm described in Sect. 8.2. Different algorithms for the solution of the calibration problem presented in Chap. 7, were tested using the MATLAB algorithms library. The least-square algorithms tested were (see Zeitoun and Pinder 1993):

1. The singular value decomposition algorithm (SVD) without constraints on the parameters.
2. The conjugate gradient algorithm with non-negativity constraints (CGPC).
3. The conjugate gradient with bounded constraints (CGBC). The lower bounds are zero, and the upper bounds are fixed by the user.

8.3.6 An Analysis of the Uniformity of the Subsidence Data

1. For each computation cell, we are able to distinguish between the effect of groundwater pumping and the loading of the buildings.
2. The results of simulations and analyses such as groundwater levels for each aquifer and land subsidence due to each one of the three causes (natural consolidation, urbanization, and groundwater depletion) are presented separately.
3. The present section analyses the history of subsidence in the Bangkok region.
4. The implications regarding future compaction and subsidence at the Bangkok region is presented in the reference (Zeitoun and Wakshal 2009). Field measurements and simulations of aquifer-system compaction at the Bangkok region have been used to constrain estimates of the critical properties that govern compaction of the aquifers. Formulations of vertical compression and expansion of the

aquifer system to the applied stresses of groundwater level changes has been used to simulate a historical subsidence and to predict possible future subsidences for selected cases of future groundwater level changes in the Bangkok region.
5. The estimation of the natural consolidation is a difficult task due to the fact that this effect is very slow in the range of time of our measurements. We tried to estimate the size of the natural consolidation by using data collected in a zone where pumping is very low and the building is not significant (a virgin area).

In order to validate the Biot model using the compartmental approach, we first implemented the model on a large-scale model. The selected zones were chosen close to the various provinces with an analysis from 1990 to 2007.

In order to receive representative results from the model, we performed first an analysis of data uniformity for the subsidence data and the water-level data.

Given the subsidence data collected from the 36 stations (Fig. 8.24), we performed a statistical analysis for all the stations located in the same province. The statistical criterion was that the coefficient of variation of the rate of subsidence will be less than 10%.

The coefficient of variation is the ratio between the standard deviation and the average value.

In the following, we will present the results of the data analysis carried out on a selected area of uniform behavior. In the province where the criterion is not satisfied, we have split the province into two parts.

In this section, we will present the analysis of the subsidence data. After the data analysis, the seven provinces were separated as follows (see Fig. 8.23):

1. The city of Bangkok: It contains the zones BE (Bangkok East) and BW (Bangkok west);
2. The Nonthaburi Province (NT);
3. The Samutprakarn Province It contains the zones (SPS and SPN);
4. The Samut Sakhon Province (SSS);
5. The Pathum Thani Province (PT);
6. The Phra Nakhon Si Ayutthaya Province (PNS); and
7. The Nakhon Pathom Province (NP) (Table 8.7).

- The Bangkok area, separated into East and West with respect to the Chao Phraya River. Thus, the various cells are: BE11, BE12, BE13, BE14, and BE15 (for the first five aquifers), and for the western part: BW11, BW12, BW13, BW14, and BW15.
- The Samu Phrakarn area, separated into its northern and southern parts.
- The Samu Sakmonit region, separated into East and West.

The different stations are described in the following Fig. 8.24.

The following figures show the results (Fig. 8.25).

All of the subsidence data analysis was performed on the first layer. We also used the subsidence data measured in depth (Figs. 8.3 and 8.26).

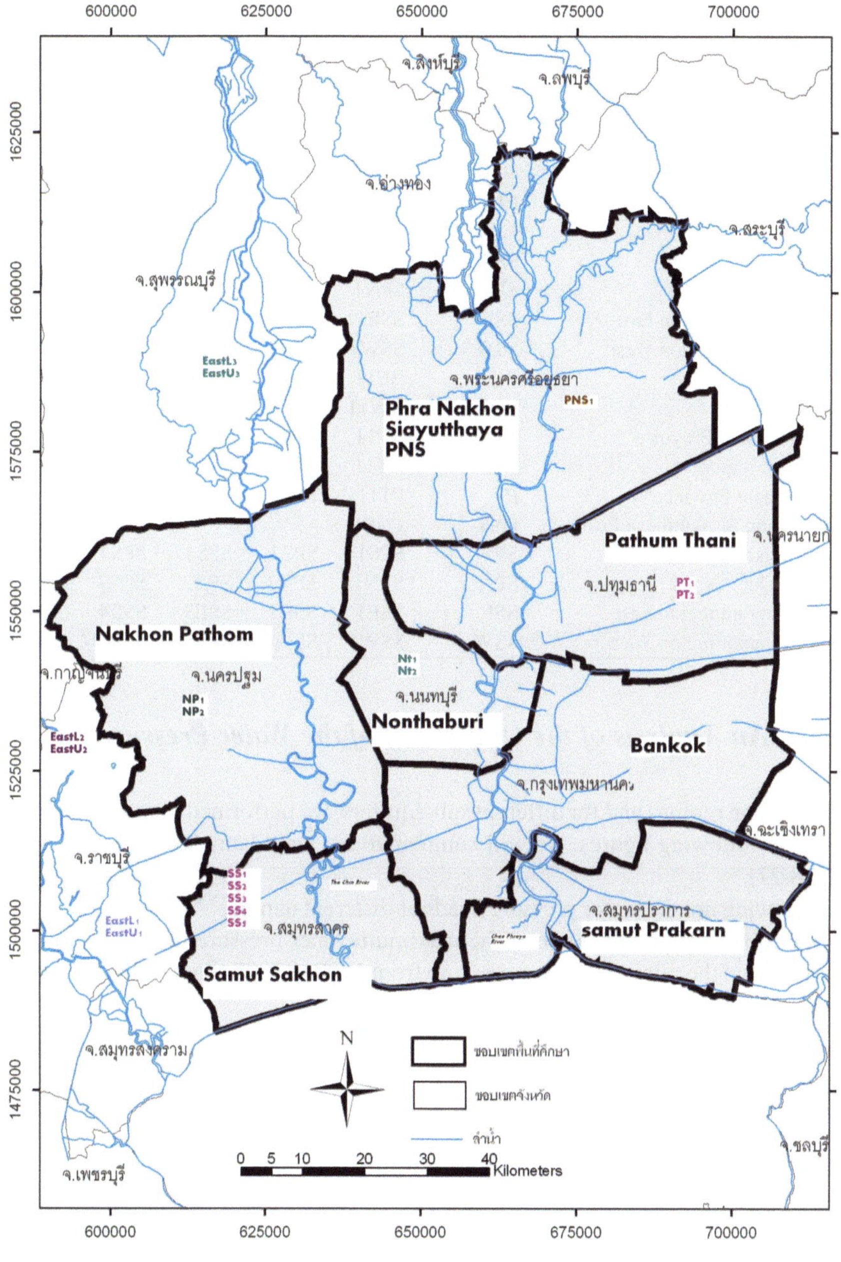

Fig. 8.23 Compartmental cells in the Bangkok region

Table 8.7 A list of regions in the cell model

Province	Cell name	Layer1	Layer2	Layer3	Layer4	Layer5
Bangkok East	BE	BE11	BE12	BE13	BE14	BE15
Bangkok West	BW	BW11	BW12	BW13	BW14	BW15
Nakhon Pathom Province	NP	NP11	NP12			
Nonthaburi region	NT	NT11	NT12			
Pathum Thani Province	PT	PT11	PT12			
Phra Nakhon Si Ayutthaya Province	PNS	PNS1				
Samu Phrakarn South	SPS	SPS1	SPS2	SPS3	SPS4	SPS5
Samu Phrakarn North	SPN	SPN1	SPN2	SPN3	SPN4	SPN5
Samu Sakmonit region East	SSE	SSE1	SSE2	SSE3	SSE4	SSE5
Samu Sakmonit region West	SSW	SSW1	SSW2	SSW3	SSW4	SSW5
Bangkok East	BE	BE11	BE12	BE13	BE14	BE15
Bangkok West	BW	BW11	BW12	BW13	BW14	BW15
Nakhon Pathom Province	NP	NP11	NP12			
Nonthaburi region	NT	NT11	NT12			
Pathum Thani Province	PT	PT11	PT12			
Phra Nakhon Si Ayutthaya Province	PNS	PNS1				
Samu Phrakarn South	SPS	SPS1	SPS2	SPS3	SPS4	SPS5
Samu Phrakarn North	SPN	SPN1	SPN2	SPN3	SPN4	SPN5
Samu Sakmonit region East	SSE	SSE1	SSE2	SSE3	SSE4	SSE5
Samu Sakmonit region West	SSW	SSW1	SSW2	SSW3	SSW4	SSW5

8.3.7 An Analysis of the Uniformity of the Water Pressure

For the same regions and for different sub-aquifers, we performed a similar analysis.

In the following figures, we will summarize the well data used in this analysis (Fig. 8.27).

Following are the water pressure heads at different depths.

The following graphs describe the maximum water pressure, the average water pressure, and the minimum water pressure from 1990 to 2007 (Figs. 8.28, 8.29, 8.30, and 8.31).

The following table summarizes the uniformity analysis performed for both the settlement data in the upper layer and the water pressure in the various sub-aquifers. The coefficient of variation (CV) is the ratio between the standard deviation and the average value (Table 8.8).

The results of this statistical analysis show that the behavior of the average value of both the settlement and the water pressure inside the various selected regions is similar. Therefore, a compartmental model that takes into consideration the average water pressure and the settlement inside each zone will represent a valid model for the Bangkok Basin.

The choice of the cells in the compartmental model was based on this analysis.

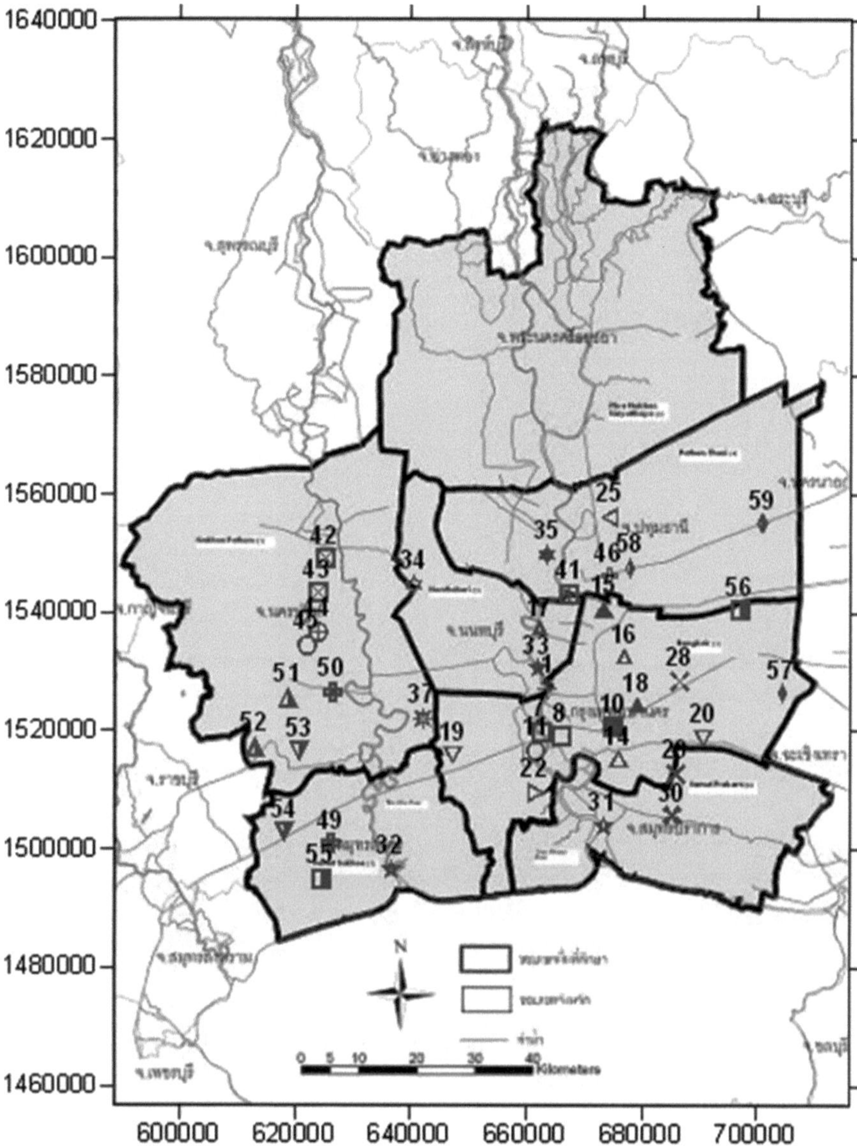

Fig. 8.24 A subsidence station map

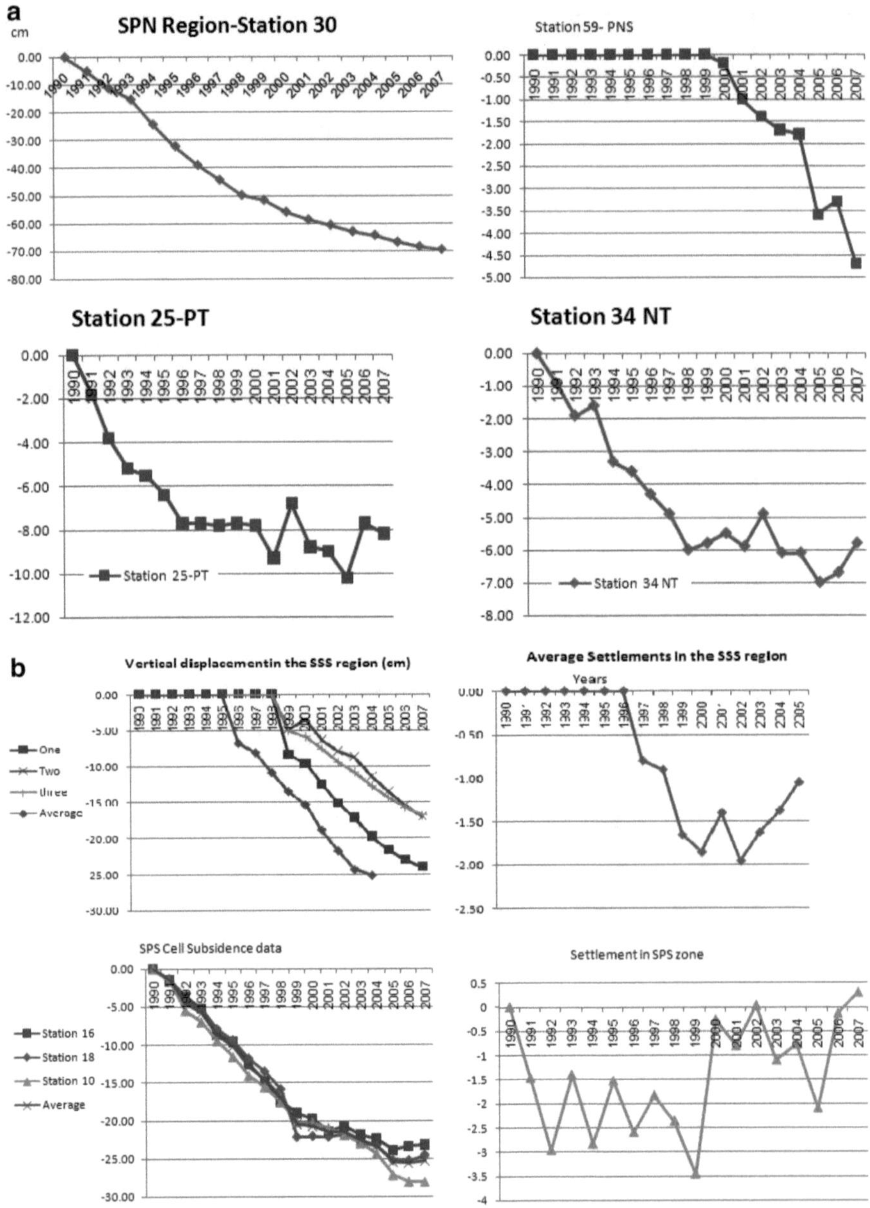

Fig. 8.25 (**a**) Subsidence data in various regions (**b**) Settlements and subsidence in various regions (**c**) Subsidence in the NP region (**d**) Subsidence in the Bangkok region

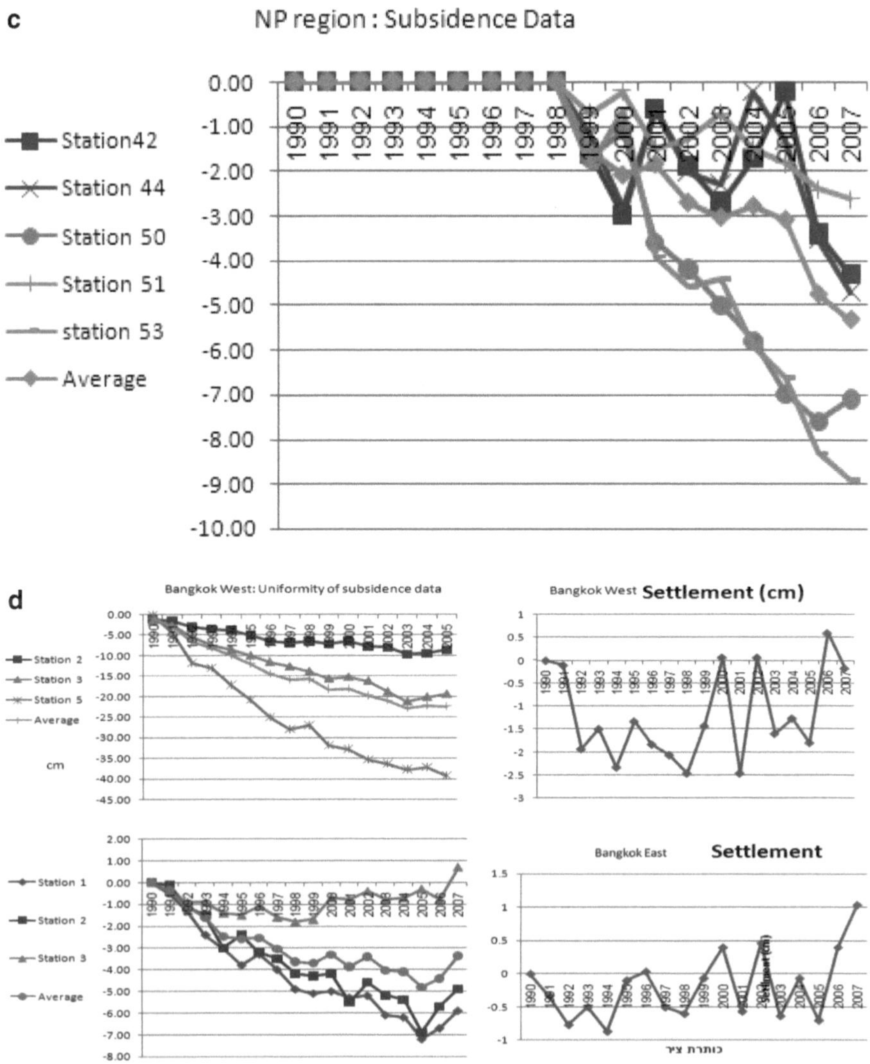

Fig. 8.25 (continued)

8.3.8 *The Results of the Model*

8.3.8.1 The Calibration Results

(a) The Groundwater Flow Calibration

The following figures describe the calibration results for each compartmental cell (Figs. 8.32, 8.33, 8.34, and 8.35).

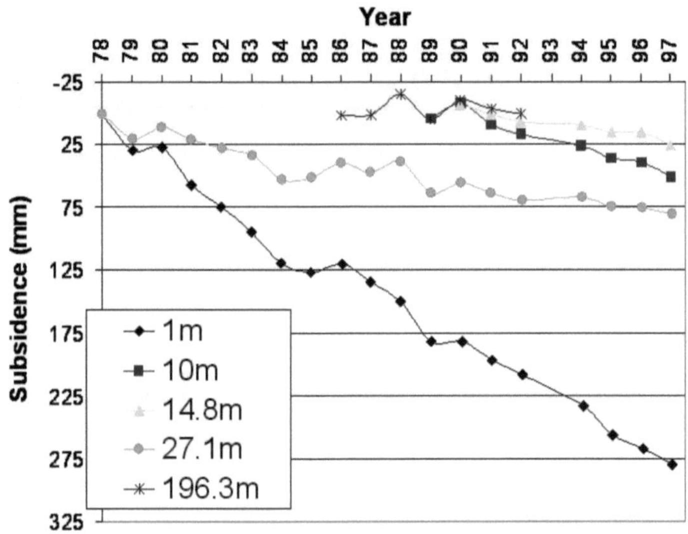

Fig. 8.26 Subsidence in a deeper layer in the Bangkok region

8.4 Summary

1. The calibration of the water pressure was performed on the average value of the compartmental cell.
2. The standard deviation of the difference between computed water pressure and measured water pressure for each year during the calibration period was less than 10% of the average value in most of the cells.
3. The behavior of the computed value follows the behavior of the measured data during the year in most of the cells.
4. Layer 5 in most of the regions gives bad results.
5. The SSS region and in the north, the SPN region, must be improved.
6. The global water mass balance and the cell mass balance are also available.

(b) Subsidence Calibration

1. The calibration of the subsidence data for each cell was performed on the average value of each compartmental cell.
2. The standard deviation of the difference between the computed subsidence and the measured subsidence for each year during the calibration period was less than 10% of the average value in most of the cells.
3. The behavior of the computed value follows the behavior of the measured data during the year in most of the cells.
4. The SSS region and in the north the SPN region must be improved.

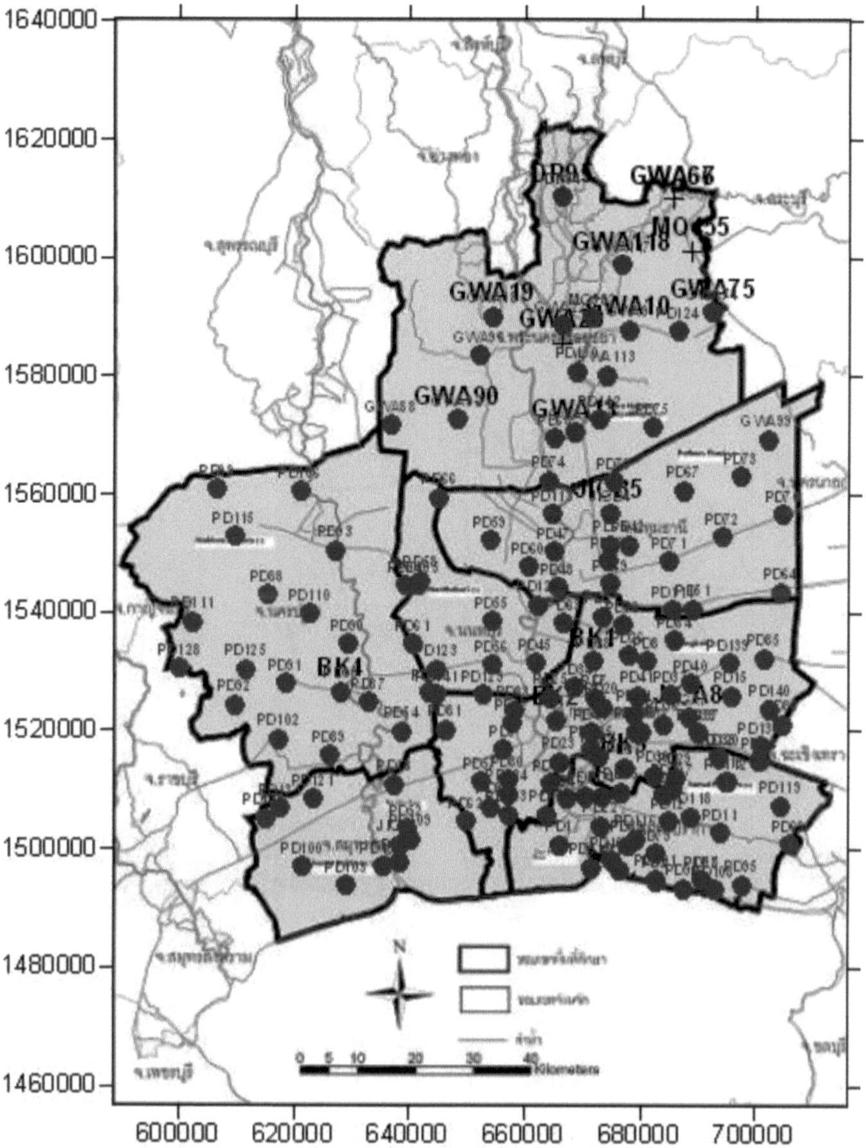

Fig. 8.27 (**a**) A well map: well locations in Layers 1 and 2 (**b**) A well map: well locations in Layers 3 and 4 (**c**) A well map: well locations in Layers 5, 6, 7, and 8

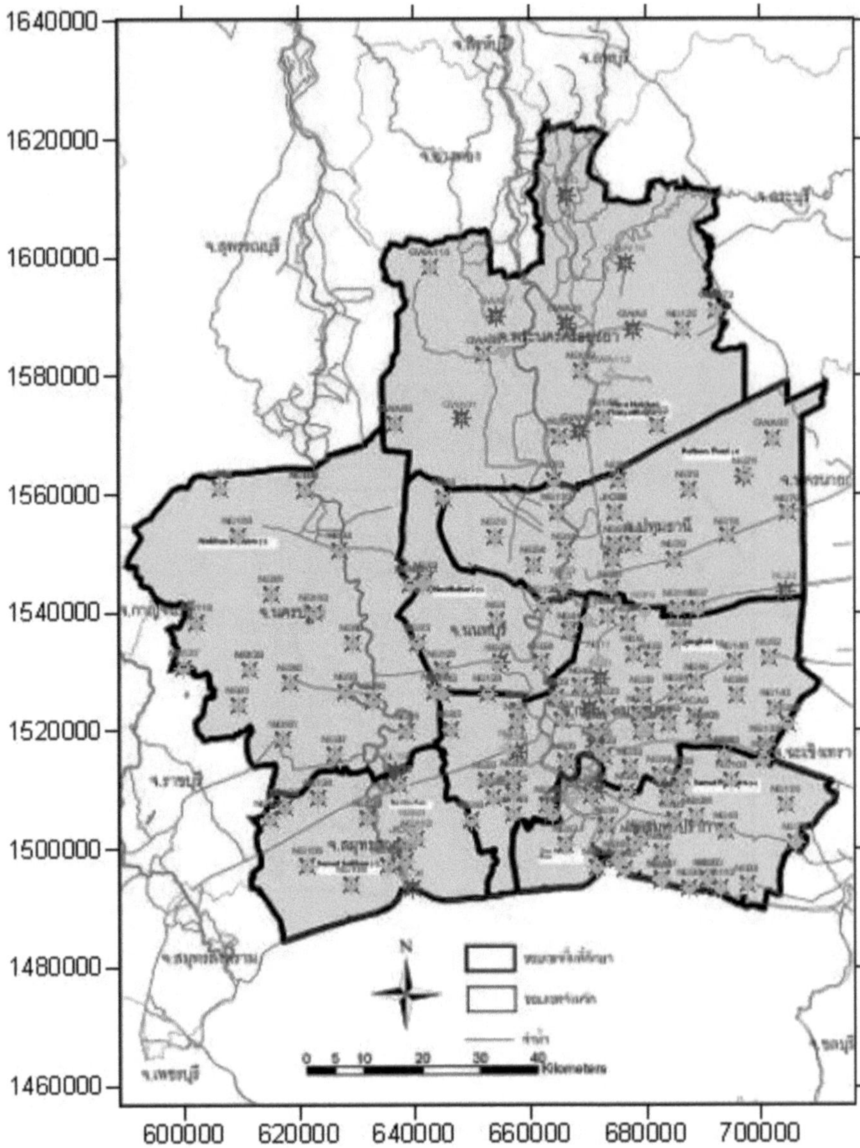

Fig. 8.27 (continued)

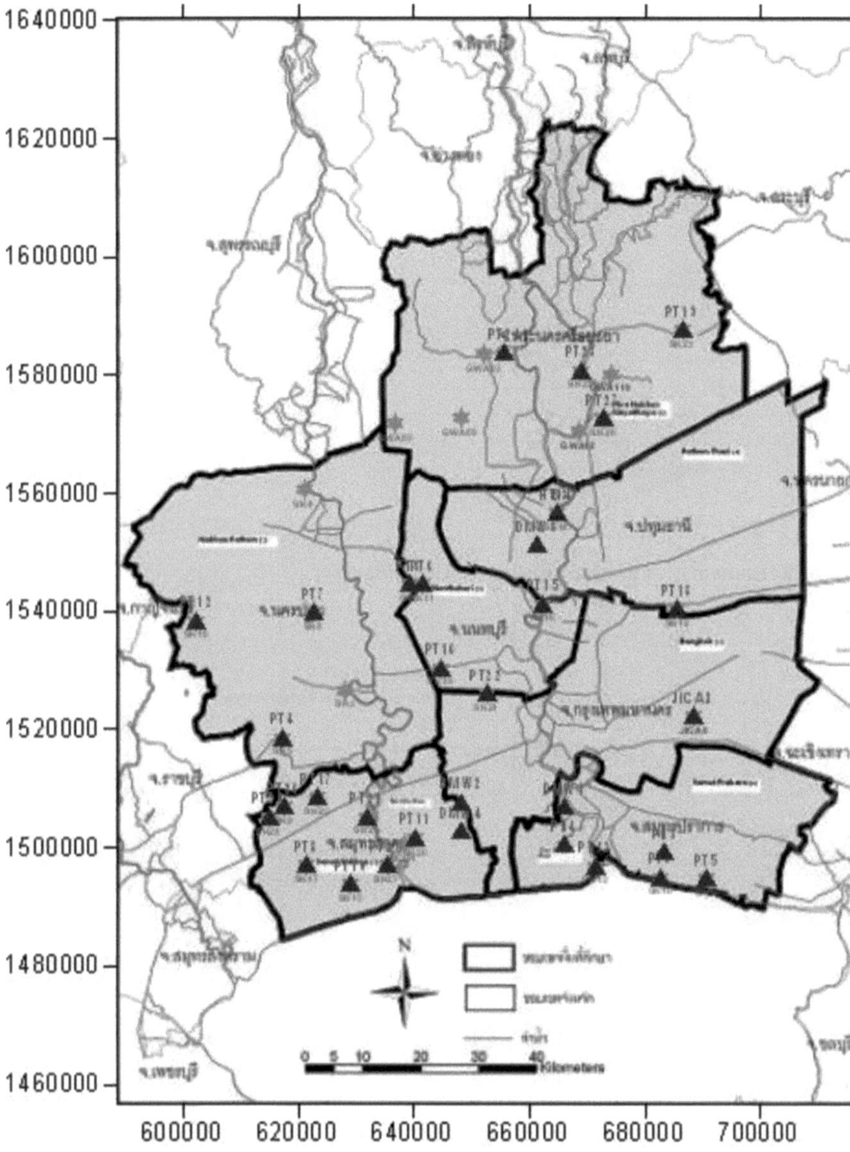

Fig. 8.27 (continued)

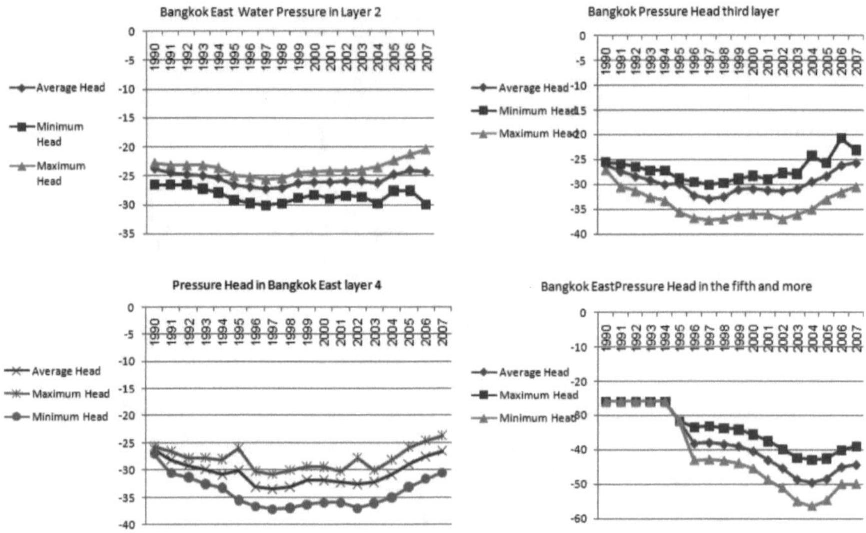

Fig. 8.28 Uniformity of the water head in the city of Bangkok

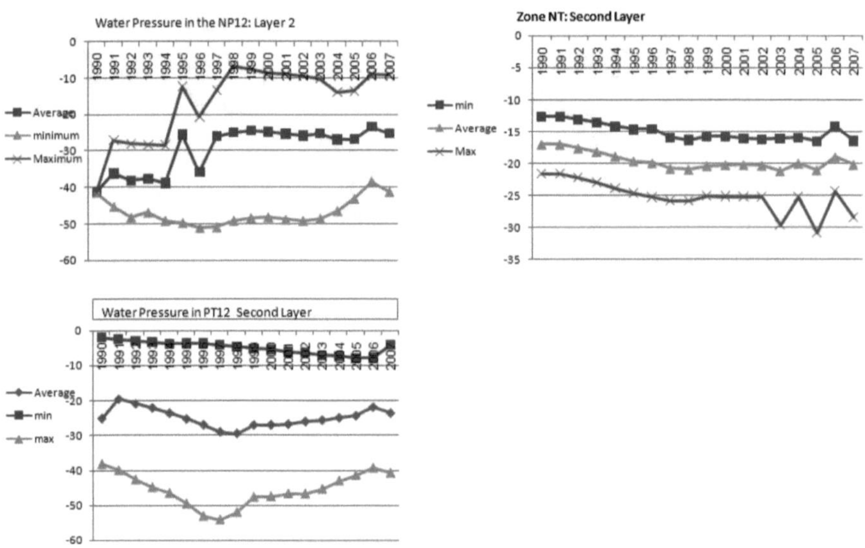

Fig. 8.29 Uniformity of the water head in the NT region

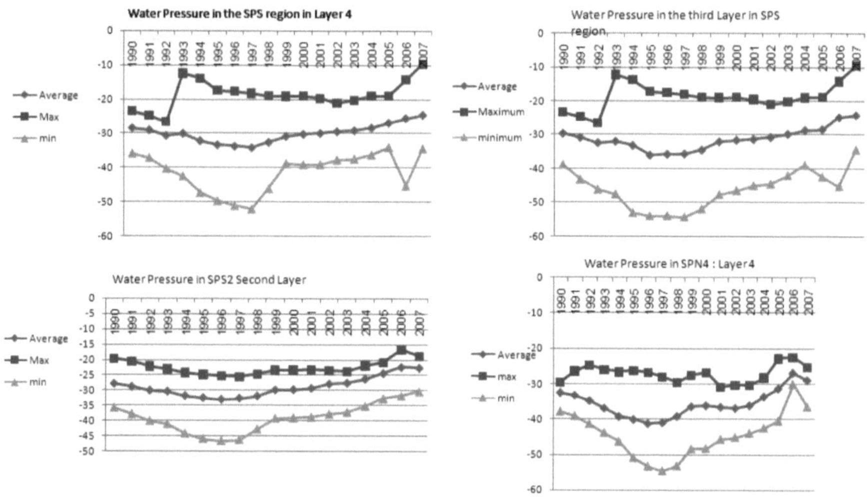

Fig. 8.30 Uniformity of the water head in the SPS region

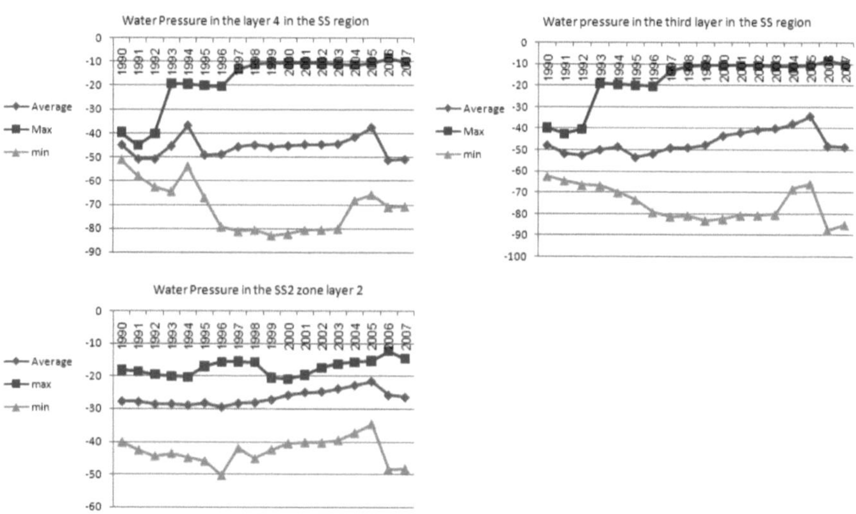

Fig. 8.31 Uniformity of the water head in the SS region

Table 8.8 A uniformity test for the subsidence and the water pressure data

Pressure head	Settlement (CV)	Area
5	3	BE
6	5	BW
5	2	PNS
3	2	NT
2	3	NP
4	6	SSS
2	4	PT
4	3	SPS
3	2	SPN

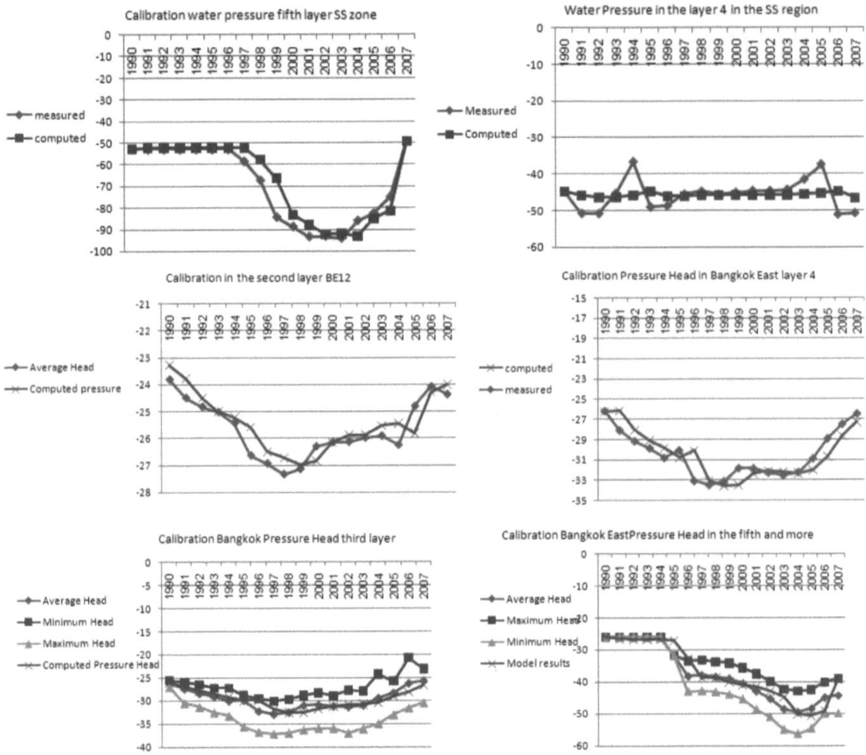

Fig. 8.32 The calibration of the water pressure using the compartmental model (SS region)

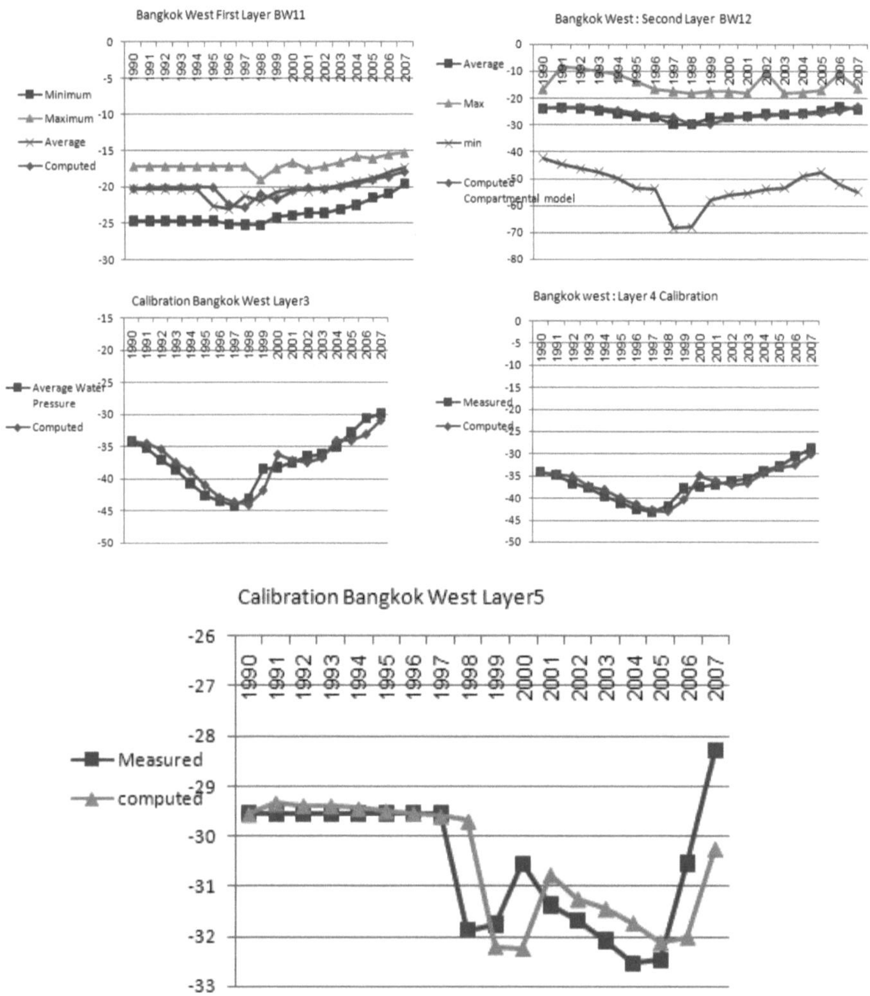

Fig. 8.33 The calibration of the water pressure using the compartmental model (BW region)

The software computes also the relative proportion due to building loadings and water pressure. Typical results of the software are presented in Fig. 8.36a, b, c.

8.4.1 The Results of the Large-Scale Model

In the following table, we have summarized the first results of our estimation.

For each cell, the table shows the total displacement and the contribution of each one of the causes in 2008 (Table 8.9).

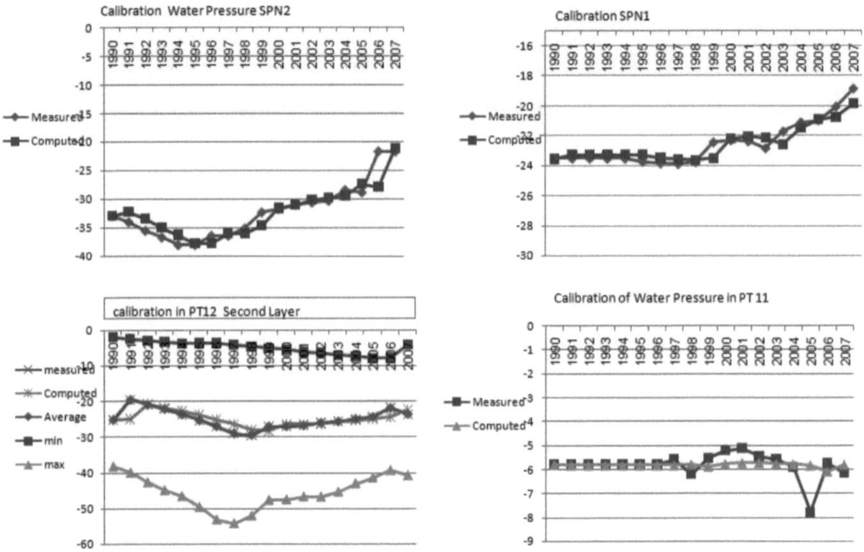

Fig. 8.34 The calibration of the water pressure using the compartmental model (the northern part)

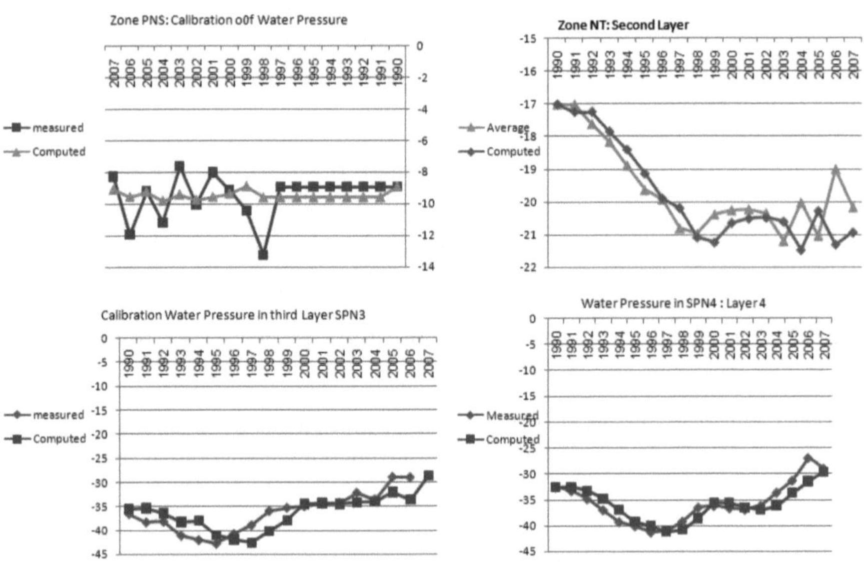

Fig. 8.35 The calibration of the water pressure using the compartmental model (the southern part)

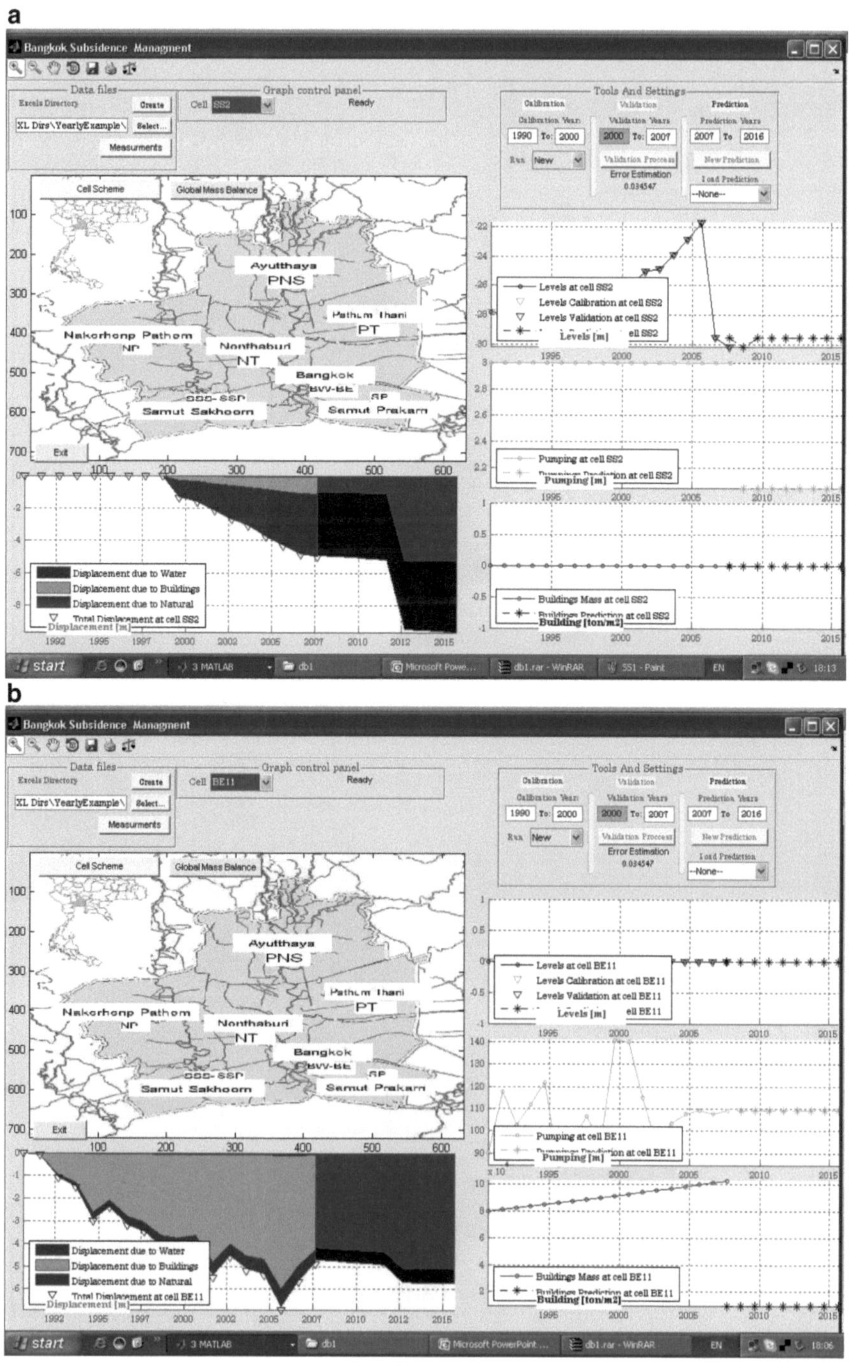

Fig. 8.36 (**a**) Results of the simulation in the second layer in the SS region (**b**) The results of the simulation in the Bangkok east region (Layer 1) (**c**) The results of the simulation in the Bangkok East region (Layer 2) (**d**) The results of the simulation in the SS region (Layer 1)

c

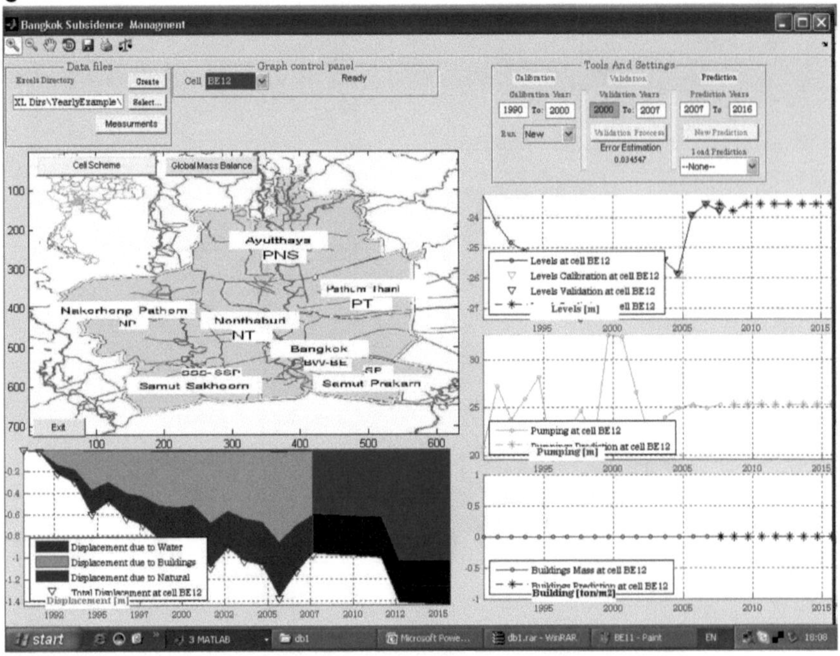

d

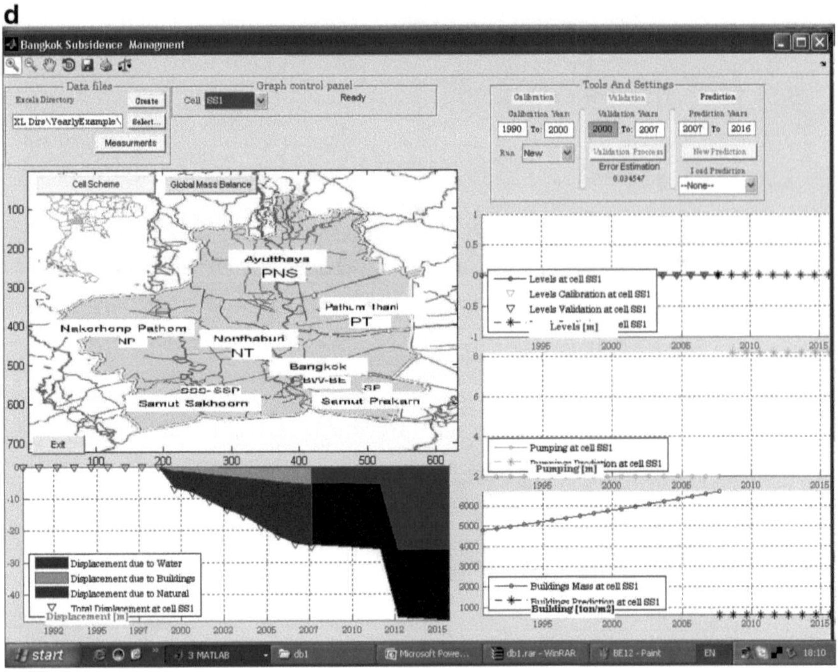

Fig. 8.36 (continued)

Table 8.9 The simulation results in 2008

Area	Settlement	Water (%)	Building (%)	Natural
BE	−1.0	50	50	Negligible
BW	−1.1	55	45	Negligible
PNS	−0.001	27	73	Negligible
NT	−1.2	22	88	Negligible
NP	−1.5	22	88	Negligible
SSS	−4.0	30	70	Questionable
PT	−0.1	10	90	Negligible
SPS	−2.1	30	70	Negligible
SPN	−1.7	35	65	Negligible

References

Asian Institute of Technology (1982) Investigation of land subsidence caused by deep well pumping in the Bangkok area: comprehensive report 1978–1981. NEB. Pub. 1982–002

Bachmat Y (1995) A compartmental model for the Yarkon Taninim aquifer. Report of the Hydrological Service of Israel

Bear J (1972) Dynamics of fluids in porous media. Dover, New York

Japan International Cooperation Agency (JICA) (1995) The study on management of groundwater and land subsidence in the Bangkok metropolitan area and its vicinity. Final report submitted to the Department of Mineral Resources, Bangkok, Thailand

Magnan J-P (2001) Cours de mécanique des sols et des roches. Ecole Nationale des Ponts et Chaussées Vol. 2

Mongsrijun S, Srisuksawad K, Lorsirirat K, Nantawisarakul T (2008) Radioisotope determination in sediments in Lam Phra Phloeng Dam, Northeastern Thailand, using gamma ray spectrometry. In: Proceedings of the international conference on applied geophysics, Chiang Mai, Thailand, 12–13 Nov 2008

Piancharoen C, Chuamthaisong C (1976) Groundwater of Bangkok metropolis. In: Proceedings of international hydro-geological conference, Budapest, Thailand, pp 510–526

Piancharoen C, Chuamthaisong C (1983) Groundwater of Bangkok metropolis. In: Proceedings of international hydro-geological conference, Budapest, Thailand, pp 510–526

Ramnarong V (1976) Pumping tests for Nakhon Luang and Bangkok aquifers. Groundwater Open File Report no. 90, Department of Mineral Resources, Bangkok, Thailand

Ramnarong V (1999) Evaluation of groundwater management in Bangkok: positive and negative. In: Chilton J (ed) Groundwater in the urban environment. Balkema A.A, Lisse

Ramnarong V, Buapeng S (1991) Mitigation of groundwater crisis and land subsidence in Bangkok. J Thai Geosci 2:125–137

Ramnarong V, Buapeng S (1992) Groundwater resources of Bangkok and its vicinity: impact and management. In: Piancharoen C (ed) Proceedings of a national conference on geologic resources of Thailand: potential for future development, Bangkok, Thailand

Zeitoun DG, Pinder GF (1993) An optimal control least squares method for solving coupled flow transport systems. Water Resour Res 29:217–227

Zeitoun DG, Wakshal E (2009) The analysis of subsidence in the Bangkok area and its vicinity. Report DGR, Bangkok, Thailand

Chapter 9
Conclusions

The maximum permissible land subsidence (or consolidation) is a constraint in various management problems such as: groundwater management (Ramnarong and Buapeng 1991; Brozovic et al. 2006), and the planning of town and the laws on building construction (Corwin et al. 1991). In order to develop a legal framework to litigation, it is essential that direct and indirect causes of land subsidence effects can be quantified with sufficient accuracy from a technical aspect.

It is, therefore, necessary to quantify the contribution of each cause to the soil subsidence of the ground surface in the urban areas in cities.

Most existing methods and software applications treat the subsidence problem by analyzing one of the causes. For example, over-pumping creates large-scale subsidence, while building loading creates only local subsidence/consolidation.

In this book, we have presented a unified approach to the problem of subsidence, which is particularly important in urban areas where several factors are causing the soil subsidence.

In this approach, we used the coupled system of the Biot equations to describe the change in water pressure and the displacement field from initial states of stress existing in the soil. We derived various superposition principles that allowed us to analyze the various types of loading separately.

Closed-form solutions of the coupled system of equations are discussed. Then, the coupled system of the Biot equations is approximated using a volume cell approach. In this advanced cell model (ACM), the cells are built close to the hydrogeology of the aquifer, and the stated variables are both the water level, the pumping rate in the cells, and the vertical displacement in the cell. A cell model is simple, easy to use, and allows the easy checking of various management paradigms.

The methodology includes the following the three main models:

- **A groundwater flow model** that includes:
  - a replenishment model; and
  - a hydro-geological model

D.G. Zeitoun and E. Wakshal, *Land Subsidence Analysis in Urban Areas*,
Springer Environmental Science and Engineering, DOI 10.1007/978-94-007-5506-2_9,

   1. Global mass balance
   2. Cell mass balance

- **A soil mechanics model** that includes:
  - the initial stress in the soil;
  - Terzaghi's principle of effective stress; and
  - a Biot model (full three dimensional)
- **A land-use model**
  - an estimation of the building load from the initial year; and
  - the rate of population growth under various assumptions

The present work describes the design of a computer model based on a multi-cell approach. In its functionality, the model implemented in the software is similar to a finite element and/or a finite difference groundwater models such as in MODFLOW (Harbaugh et al. 2000; Harbaugh 2005).

The main difference between these classical numerical models and the present approach is the scale of the basic cell.

In classical numerical models, the basic elements or grid discretizations are areas inside a given geological layer; while in the present approach, the cell represents the hydrological behavior of a whole region. Of course, the inherent difficulty of our approach is the definition of the area that is covered by a given cell. The main criterion for this choice is the similar hydrological behavior of the cell area, which requires a previous analysis of the hydrological data.

9.1 Use of the ACM Method

This book presents the theoretical basis and its mathematical implementation of the ACM method. It also briefly presents the development of the user interface connected with the scientific software MATLAB, emphasizing the generality of the mathematical approach and the software. This approach may be used for the flow estimation of any type of groundwater aquifer.

Basically, the present cell model may be used for two purposes:

1. as an independent tool for regional analysis; and
   The output of the model permits the following:
   - hydrological results presented on the administrative cells;
   - planning scenarios;
   - an existing flow inter-cell; and
   - river or springs fluxes

2. as a pre-model that determines a sub-model.
   The ACM model may be used as a first model for the determination of boundary conditions of a sub-regional model where accurate models such as MODFLOW are used. The use of this type of techniques exists in:

   - the multi-domain method; and
   - the down-scale method.

9.2 Hydrological Aspects of the Approach

The hydrological model implemented at this stage considers only the annual change of water levels, and therefore, the seasonal change of water levels is not analyzed at this stage. This is not a restriction of the model; the time step of the data is an input parameter of the current software. However, a shorter time step requires more detailed data for the water level and the pumping rates.

The way the software was designed is that the user first lists the cells and defines the hydrological connections (a hydro-geological model) and then, the user calibrates the model parameters. The calibration of the model parameters, such as cell storativity, cell transmissivity, rain coefficients, etc., is done once the hydro-geological model has been defined. The advantage of the present approach is the ability to analyze various possible hydrological connections between cells and to analyze the best calibration. Afterwards, the present approach may be used as a tool for testing the hydro-geological model of the aquifer.

In classical numerical modeling, the difficult issue in the modeling is the zoning part where the hydrologist (the user) must define zones of material properties and/or zones of replenishment coefficients and hydrological connections between the parts of the aquifer.

The present simplified cell model may be utilized for an initial analysis before implementing a detailed finite element modeling of the same aquifer.

In the classical hydro-geological modeling of groundwater aquifers, the modeling approach takes place after the building of the geo-hydrological models has been performed by geologists.

The computer model is basically a quantitative translation of the geological model.

Unfortunately, the geological approach does not always permit a full understanding of the hydrological connection. Therefore, a hydrological analysis is generally needed in order to understand the hydrological connections between each of the parts of the aquifer; and this must be done before using precise computer modeling tools, such as MODFLOW or other finite element models (Babu et al. 1993, 1997). The multiple-cell model presented in this approach may be used for the hydrological analysis of the connections between each part of the aquifer.

9.3 The ACM Method Versus High-Resolution Numerical Models

As mentioned above, there is a series of well-known difficulties with the use of numerical models for the hydrological behavior of a whole basin:

9.3.1 Uncertainty Due to Boundary Conditions

The ACM approach allows introducing into the model recharge input and boundary conditions on a large scale, which enables receiving better estimates of the mass balance in the aquifer.

The groundwater pumping component, difficult to estimate from indirect methods, becomes the default closure term in many water balances. However, a rigorous evaluation of this approach is lacking.

Groundwater recharge and pumping are spatially distributed basin-wide processes. Data availability on the various water balance components needed for closure at sufficiently detailed spatio-temporal scales is generally limited.

Most input components must be estimated, extrapolated, or aggregated from sparsely distributed point-wise measurements of several spatio-temporal varying processes. The choice of process complexity in the conceptual model underlying a basin water balance computation is limited, even though highly complex advanced models are readily available to estimate individual flux components in these basins. Simple 'tipping bucket' or storage capacity-based distributed hydrologic process models of the water balance have been successfully applied to compute a basin water yield (Milly 1994; Jothityangkoon et al. 2001; Atkinson et al. 2002; Labadie 2004).

Therefore, a combined use of the ACM model and a high-resolution model may be of interest. In this case, the ACM model will be used as a pre-model for large-scale modeling, and the high resolution model may be used for a specific region, using the boundary conditions created by the ACM model.

The ACM model enables the creation of accurate boundary conditions for a sub-region analysis, such as with the multi-domain method.

9.3.2 The Difficulty of the Replenishment Estimation

Another interesting feature of the ACM method is the possibility of inserting a replenishment model on a regional scale. In a large aquifer, rain stations are not available at each place. Therefore, rain data are scarce, and we need to associate a given station with a large region.

In the ACM model, we can calibrate the replenishment model at a regional scale independent of the global flow model. This calibration may also be used as an input for a more accurate numerical model.

References

Atkinson SE, Woods RA, Sivapalan M (2002) Climate and landscape controls on water balance model complexity over changing timescales. Water Resour Res 38:1314. doi:10.1029/2002WR001487

Babu DK, Pinder GF, Niemi A, Ahlfeld DP, Stothoff SA, Zeitoun DG (1993) Chemical transport by three-dimensional density-dependent groundwater flows. Princeton University report 01/93/243

Babu DK, Pinder GF, Niemi A, Ahlfeld DP, Stothoff SA (1997) Chemical transport by three-dimensional groundwater flows. Report 84-WR-3. Princeton University, Princeton, NJ

Brozovic N, Sunding DL, Zilberman D (2006) Optimal management of groundwater over space and time. In: Berga D, Goetz R (eds) Frontiers in water resource economics, natural resource management and policy series, vol 29. Springer, New York

Corwin EJ, Alhadeff SC, Oggel SP, Shlemon RJ (1991) Earth fissures, urbanization and litigation. In: Proceedings of the fourth international symposium on land subsidence, May 1991. IAHS Publication – 200

Harbaugh AW (2005) MODFLOW-2005, the US geological survey modular ground-water model – the ground-water flow process techniques and methods 6-A16. US Geological Survey, Reston

Harbaugh AW, Banta ER, Hill MC, McDonald MG (2000) MODFLOW-2000, the US geological survey modular ground-water model – user guide to modularization concepts and the ground-water flow process. US geological survey open-file-report 00–92

Jothityangkoon C, Sivapalan M, Farmer DL (2001) Process controls of water balance variability in a large semi-arid catchment: downward approach to hydrological model development. J Hydrol 254:174–198

Labadie JW (2004) Optimal operation of multi-reservoir systems: state-of-the-art review. J Water Res Pl-ASCE 130:93–111

Milly PCD (1994) Climate, soil water storage, and the average annual water balance. Water Resour Res 30:2143–2156

Ramnarong V, Buapeng S (1991) Mitigation of groundwater crisis and land subsidence in Bangkok. J Thai Geosci 2:125–137

Index

D.G. Zeitoun and E. Wakshal, *Land Subsidence Analysis in Urban Areas*,
Springer Environmental Science and Engineering, DOI 10.1007/978-94-007-5506-2,

J

Q

U

MIX
Papier aus verantwortungsvollen Quellen
Paper from responsible sources
FSC® C105338

If you have any concerns about our products,
you can contact us on
ProductSafety@springernature.com

In case Publisher is established outside the EU,
the EU authorized representative is:
Springer Nature Customer Service Center GmbH
Europaplatz 3, 69115 Heidelberg, Germany

Printed by Libri Plureos GmbH
in Hamburg, Germany